Framework 7

MATHS S

David Capewell Westfield School, Sheffield

Marguerite Comyns Queen Mary's High School, Walsall

Gillian Flinton All Saints Catholic High School, Sheffield

Geoff Fowler Maths Strategy Manager, Birmingham

Kam Grewal-Joy Mathematics Consultant

Derek Huby Mathematics Consultant

Peter Johnson Wellfield High School, Leyland, Lancashire

Penny Jones Mathematics Consultant, Birmingham

Jayne Kranat Langley Park School for Girls, Bromley

Ian Molyneux St. Bedes RC High School, Ormskirk

Peter Mullarkey School Improvement Officer, Manchester

Nina Patel Ifield Community College, West Sussex

OXFORD
UNIVERSITY PRESS

Great Clarendon Street, Oxford OX2 6DP

Oxford University Press is a department of the University of Oxford.
It furthers the University's objective of excellence in research, scholarship,
and education by publishing worldwide in

Oxford New York

Auckland Cape Town Dar es Salaam Hong Kong Karachi
Kuala Lumpur Madrid Melbourne Mexico City Nairobi
New Delhi Shanghai Taipei Toronto

With offices in

Argentina Austria Brazil Chile Czech Republic France Greece
Guatemala Hungary Italy Japan Poland Portugal Singapore
South Korea Switzerland Thailand Turkey Ukraine Vietnam

Oxford is a registered trade mark of Oxford University Press
in the UK and in certain other countries

First published 2002

British Library Cataloguing in Publication Data

Data available

ISBN 978-0-19-914848-6

10 9 8 7 6

Typeset by Mathematical Composition Setters Ltd.

Printed at Cayfosa Quebecor, Spain.

Acknowledgements
The photograph on the cover is reproduced courtesy of Pictor International (UK).

The publishers and authors would like to thank the following for permission to use photographs and other copyright material: Corbis UK, pages 1, 12, 165 and 235, Robert Harding, pages 12, 18 and 230, Empics, page 29, Anthony Blake Photo Library, pages 39 and 68, Pictor International UK, page 157.

Figurative artwork by Jeff Anderson.

The authors would like to thank
Sarah Caton, Karen Greenway, Lyn Lynam, David Shiers and Karl Warsi for their help in compiling this book.

About this book

This book has been written specifically for students who have gained Level 2 or 3 at the end of KS2. It is designed to help students raise their achievement to Level 4. The content is based on the Year 5 and 6 teaching objectives from the Primary Framework and each unit provides access to Year 7 objectives.

The authors are experienced teachers and maths consultants, who have been incorporating the approaches of the Framework into their teaching for many years and so are well qualified to help you successfully introduce the objectives in your classroom.

The book is made up of units which follow the Support tier of the medium term plans that complement the Framework document, following the pitch, pace and progression.

The units are:

A1	Sequences and functions	1–14
N1	Number calculations	15–28
S1	Perimeter and area	29–38
N2	Fractions, decimals and percentages	39–52
D1	Statistics and probability	53–66
A2	Expressions and formulae	67–78
S2	Angles and shapes	79–86
D2	Handling data	87–98
N3	Multiplication and division	99–116
A3	Functions and graphs	117–130
S3	Triangles and quadrilaterals	131–142
N4	Percentages, ratio and proportion	143–154
A4	Linear equations	155–164
S4	Transformations	165–178
N5	More number calculations	179–196
D3	Analysing statistics	197–208
D4	Probability experiments	209–216
A5	Equations and graphs	217–234
S5	Polygons	235–248

Each unit comprises double page spreads which should take a lesson to teach. These are shown on the full contents list.

Problem solving is integrated throughout the material as suggested in the Framework.

How to use this book

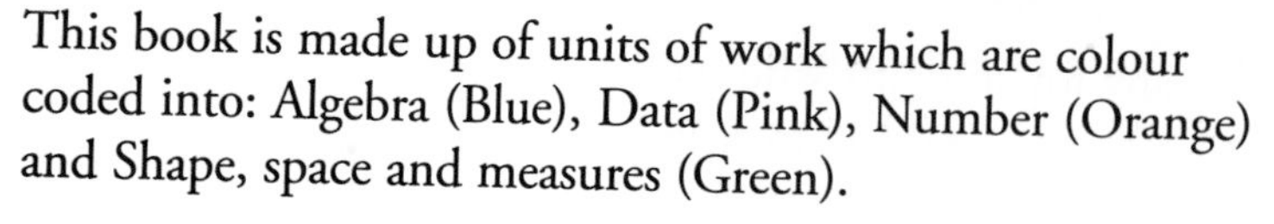

This book is made up of units of work which are colour coded into: Algebra (Blue), Data (Pink), Number (Orange) and Shape, space and measures (Green).

Each unit of work starts with an overview of the content of the unit, as specified in the Primary Framework document, so that you know exactly what you are expecting to learn.

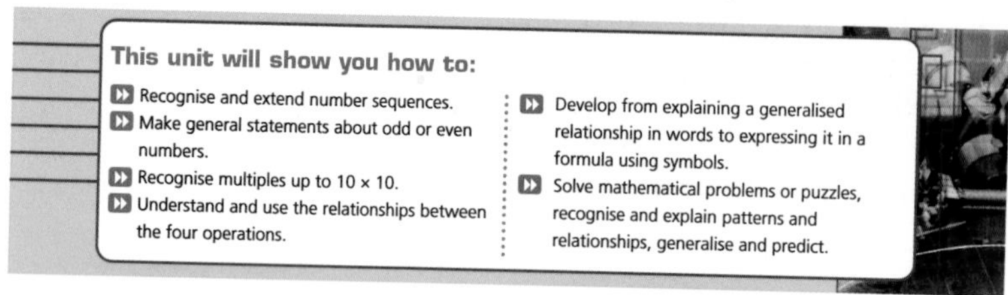

The first page of a unit also highlights the skills and facts you should already know and provides Check in questions to help you revise before you start so that you are ready to apply the knowledge later in the unit:

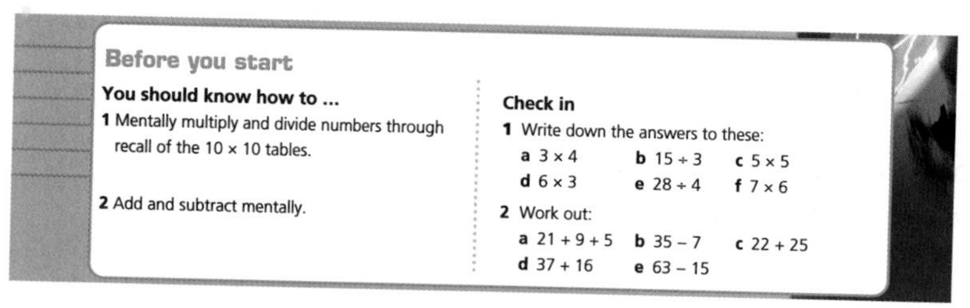

Inside each unit, the content develops in double page spreads which all follow the same structure.

The spreads start with a list of the learning outcomes and a summary of the keywords:

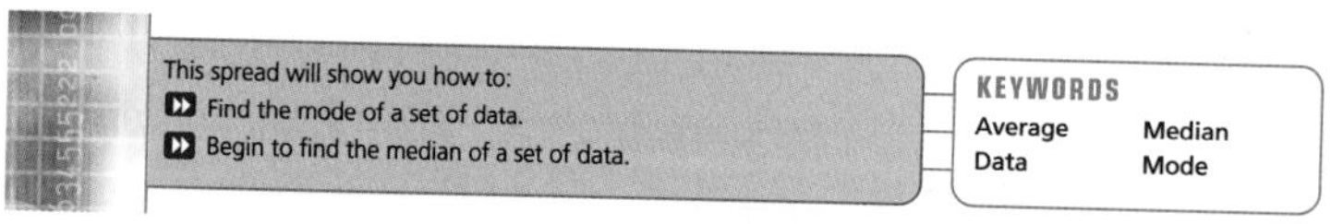

The keywords are summarised and defined in a Glossary at the end of the book so you can always check what they mean.

Key information is highlighted in the text so you can see the facts you need to learn.

▸ Division is the opposite of multiplication: $5 \times 3 = 15$ so $15 \div 5 = 3$.

Examples showing the key skills and techniques you need to develop are shown in boxes. Also hint boxes show tips and reminders you may find useful:

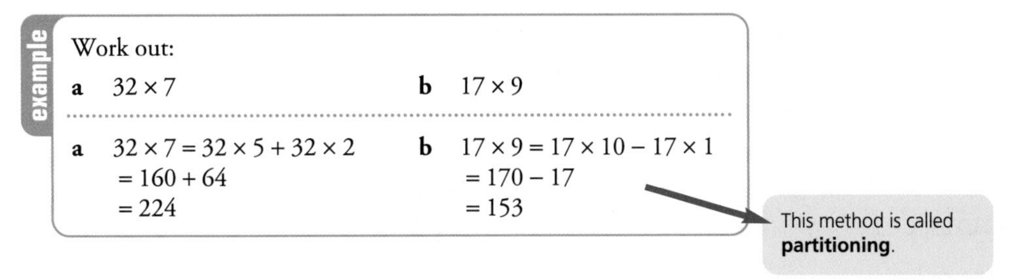

example

Work out:

a 32×7 **b** 17×9

a $32 \times 7 = 32 \times 5 + 32 \times 2$
$= 160 + 64$
$= 224$

b $17 \times 9 = 17 \times 10 - 17 \times 1$
$= 170 - 17$
$= 153$

This method is called **partitioning**.

Each exercise is carefully graded, set at three levels of difficulty:

1. The first few questions provide lead-in questions, revising previous learning.
2. The questions in the middle of the exercise provide the main focus of the material.
3. The last few questions are challenging questions that provide a link to the Year 7 learning objectives.

At the end of each unit is a summary page so that you can revise the learning of the unit before moving on.

Check out questions are provided to help you check your understanding of the key concepts covered and your ability to apply the key techniques.

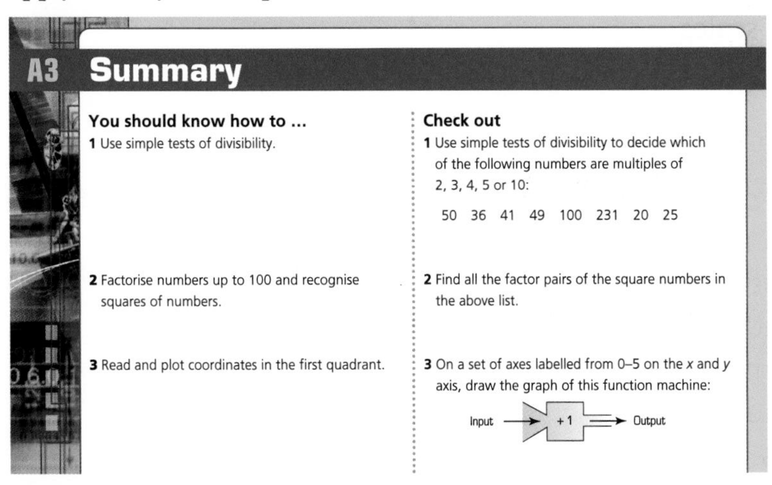

A3 Summary

You should know how to ...	Check out
1 Use simple tests of divisibility.	**1** Use simple tests of divisibility to decide which of the following numbers are multiples of 2, 3, 4, 5 or 10: 50 36 41 49 100 231 20 25
2 Factorise numbers up to 100 and recognise squares of numbers.	**2** Find all the factor pairs of the square numbers in the above list.
3 Read and plot coordinates in the first quadrant.	**3** On a set of axes labelled from 0–5 on the x and y axis, draw the graph of this function machine:

The answers to the Check in and Check out questions are produced at the end of the book so that you can check your own progress and identify any areas that need work.

Contents

A1 Sequences and functions 1–14

A1.1 Introducing sequences 2
A1.2 Sequences and rules 4
A1.3 Sequences in diagrams 6
A1.4 Function machines 8
A1.5 Finding the function 10
A1.6 Functions and algebra 12
Summary 14

N1 Number calculations 15–28

N1.1 Place value and ordering 16
N1.2 Negative numbers 18
N1.3 Negative numbers – addition and subtraction 20
N1.4 Mental strategies 22
N1.5 Adding and subtracting decimals 24
N1.6 Using a calculator 26
Summary 28

S1 Perimeter and area 29–38

S1.1 Perimeter and area 30
S1.2 More perimeter and area 32
S1.3 Measurement and scales 34
S1.4 Three-dimensional shapes 36
Summary 38

N2 Fractions, decimals and percentages 39–52

N2.1 Understanding fractions 40
N2.2 Equivalent fractions 42
N2.3 Adding and subtracting fractions 44
N2.4 Fractions and decimals 46
N2.5 Fractions of an amount 48
N2.6 Fractions, decimals and percentages 50
Summary 52

D1 Statistics and probability 53–66
D1.1 Finding the average 54
D1.2 The mean 56
D1.3 Interpreting diagrams 58
D1.4 Introducing probability 60
D1.5 Calculating probabilities 62
D1.6 Experimental probability 64
Summary 66

A2 Expressions and formulae 67–78
A2.1 Using letter symbols 68
A2.2 Rules of algebra 70
A2.3 Simplifying expressions 72
A2.4 Substitution 74
A2.5 Using formulae 76
Summary 78

S2 Angles and shapes 79–86
S2.1 Time for a change 80
S2.2 Angles and lines 82
S2.3 Coordinates and shapes 84
Summary 86

D2 Handling data 87–98
D2.1 Discussing statistical methods 88
D2.2 Collecting data 90
D2.3 Organising the data 92
D2.4 Displaying your results 94
D2.5 Interpreting your diagrams 96
Summary 98

N3 Multiplication and division 99–116
- N3.1 Number and measures 100
- N3.2 Order of operations 102
- N3.3 Mental methods 104
- N3.4 Multiplying by partitioning 106
- N3.5 Multiplying on paper 108
- N3.6 Dividing on paper 110
- N3.7 Dividing with remainders 112
- N3.8 Calculator methods 114
- Summary 116

A3 Functions and graphs 117–130
- A3.1 Factors and primes 118
- A3.2 Patterns in numbers 120
- A3.3 Squares and triangles 122
- A3.4 Functions and multiples 124
- A3.5 Graphs of functions 126
- A3.6 Using a table of values 128
- Summary 130

S3 Triangles and quadrilaterals 131–142
- S3.1 Measuring angles 132
- S3.2 Finding angles 134
- S3.3 Drawing angles 136
- S3.4 Angles in triangles 138
- S3.5 2-D drawings of 3-D shapes 140
- Summary 142

N4 Percentages, ratio and proportion 143–154
- N4.1 Fraction, decimal and percentage equivalents 144
- N4.2 Finding simple percentages 146
- N4.3 Finding harder percentages 148
- N4.4 Proportion 150
- N4.5 Introducing ratio 152
- Summary 154

A4 Linear equations 155–164
A4.1 Using algebraic expressions 156
A4.2 Algebraic operations 158
A4.3 Using brackets 160
A4.4 Solving equations 162
Summary 164

S4 Transformations 165–178
S4.1 Reflection symmetry 166
S4.2 Reflecting shapes 168
S4.3 Reflecting in all four quadrants 170
S4.4 Translating shapes 172
S4.5 Rotation 174
S4.6 Transformations 176
Summary 178

N5 More number calculations 179–196
N5.1 Rounding 180
N5.2 Factors, multiples and primes 182
N5.3 Multiplying and dividing mentally 184
N5.4 Standard written calculations 186
N5.5 Standard written division 188
N5.6 Using equivalent fractions 190
N5.7 Converting fractions, decimals and percentages 192
N5.8 Calculating parts of quantities 194
Summary 196

D3 Analysing statistics 197–208
D3.1 Planning the data collection 198
D3.2 Constructing statistical diagrams 200
D3.3 Comparing data using diagrams 202
D3.4 Describing data using statistics 204
D3.5 Communicating results 206
Summary 208

D4 Probability experiments 209–216

D4.1	Theoretical probability	210
D4.2	Experimental probability	212
D4.3	Comparing experiment with theory	214
	Summary	216

A5 Equations and graphs 217–234

A5.1	Solving equations	218
A5.2	Using formulae	220
A5.3	Formulae using letters	222
A5.4	Generating sequences	224
A5.5	Spot the function	226
A5.6	Drawing graphs	228
A5.7	Graphs of formulae	230
A5.8	All four quadrants	232
	Summary	234

S5 Polygons 235–248

S5.1	More angle facts	236
S5.2	Constructing triangles	238
S5.3	Constructing squares	240
S5.4	Reflection symmetry	242
S5.5	Rotational symmetry	244
S5.6	Tessellating shapes	246
	Summary	248

Glossary 249
Answers 263
Index 273

A1 Sequences and functions

This unit will show you how to:

- Recognise and extend number sequences.
- Make general statements about odd or even numbers.
- Recognise multiples up to 10 × 10.
- Understand and use the relationships between the four operations.
- Develop from explaining a generalised relationship in words to expressing it in a formula using symbols.
- Solve mathematical problems or puzzles, recognise and explain patterns and relationships, generalise and predict.

Many natural patterns grow according to a sequence.

Before you start

You should know how to …

1 Count on or back in steps of any size.

2 Use × and ÷ facts for 2, 5, 10 tables.

3 Use a 10 × 10 multiplication grid.

Check in

1 **a** Count on from 2 in steps of 3:
2, _, _, _, _

b Count back from 20 in steps of 4:
20, _, _, _, _

2 Write down the answer to each question:

a 2 × 10 = __ **b** 18 ÷ 2 = _
c 4 × 5 = __ **d** 35 ÷ 5 = __
e 40 ÷ 4 = __ **f** 60 ÷ 10 = _
g 3 × 10 = __ **h** 40 ÷ 5 = _

3 Use a multiplication grid to work out the answer to each question:

a 24 ÷ 6 = _ **b** 35 ÷ 7 = __
c 5 × __ = 35 **d** _ × 6 = 36
e __ ÷ 4 = 36 **f** 72 ÷ 9 = _
g 8 × __ = 48 **h** 7 × __ = 56

A1.1 Introducing sequences

This spread will show you how to:

- Recognise and extend number sequences.

KEYWORDS

Generate | Sequence
Rule | Term

When numbers follow a pattern or rule they are in a **sequence**.

Amy starts with 10.
She generates sequences by following different rules:

add 4	divide by 2	take away 3
$10 + 4 = 14$	$10 \div 2 = 5$	$10 - 3 = 7$
$14 + 4 = 18$	$5 \div 2 = 2.5$	$7 - 3 = 4$
$18 + 4 = 22$	$2.5 \div 2 = 1.25$	$4 - 3 = 1$

The numbers she produces are:

10, 14, 18, 22   10, 5, 2.5, 1.25   10, 7, 4, 1

These lists of numbers are sequences.

- **Each number in a sequence is called a term.**

In the sequence 10, 14, 18, 22 ...

The first term is 10 and the third term is 18.

You can generate the numbers in a sequence using a **rule**.

example

Generate the first three terms of these sequences:

a start at 3 and add 5   **b** start at 40 and divide by 2

a the first term is 3
the second term is $3 + 5 = 8$
the third term is $8 + 5 = 13$

b the first term is 40
the second term is 20
the third term is 10

You can show your results in a table:

Term	1	2	3
Number	3	8	13

Term	1	2	3
Number	40	20	10

Exercise A1.1

1 Describe these sequences in words:
Copy and complete the sentences.

a 2, 4, 6, 8, ... Start at ______ and ______ ______.
b 3, 5, 7, 9, ... Start at ______ and ______ ______.
c 5, 10, 15, 20, ... Start at ______ and ______ ______.
d 20, 18, 16, 14, ... Start at ______ and ______ ______.
e 30, 27, 24, 21, ... Start at ______ and ______ ______.

2 Write down the first five terms of these sequences:

a The first term is 2. The rule is +7.
b The first term is 100. The rule is −10.
c The first term is 12. The rule is double.
d The first term is 25. The rule is +4.
e The first term is 80. The rule is ÷2.
f The first term is 4. The rule is ×3.
g The first term is 200. The rule is ÷2.
h The first term is 1. The rule is ×3.
i The first term is 2. The rule is ×5.
j The first term is 5. The rule is ×10.

3 Amy starts with 16 and generates sequences using the following rules:

a add 3 **b** multiply by 2 **c** divide by 2

For each of these sequences write down the first five terms. Show your results in a table like this:

Term	1	2	3	4	5
a	16				
b	16				
c	16				

4 **Investigation**
The house numbers on Joe's street are arranged like this:

2	4	6	8	10	12
11	9	7	5	3	1

Joe lives at number 7.
a What number house is opposite Joe's?
b Describe the sequence of house numbers on the side opposite Joe's house.

The street is extended and the houses are renumbered like this:

2	4	...	...	18	20
19	17	...	...	3	1

c What number does Joe live opposite now?

d What number does Jack live at if he lives opposite number 11?
Investigate further.

A1.2 Sequences and rules

This spread will show you how to:

- Recognise and extend number sequences.

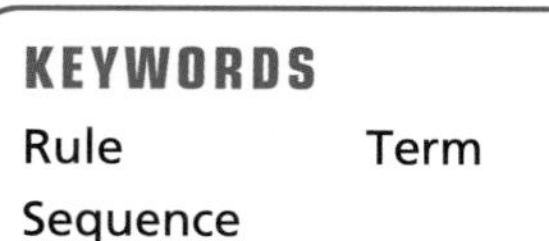

KEYWORDS

Rule
Term
Sequence

Lauren is trying to describe a number sequence to Sorcha.

Lauren needs to tell Sorcha the first term.

▶ To describe a sequence you need a start number and a **rule**.

example

Describe this sequence using a rule: 4, 9, 14, 19, ...

The dots ... show the sequence goes on forever.

The sequence starts at 4 and goes up in 5s.
The rule is you start at 4 then add 5.

▶ You can use the rule of a sequence to find more terms.

example

Find the next three terms of these sequences:

a 3, 7, 11, 15

b 24, 21, 18, 15

a The first term is 3.

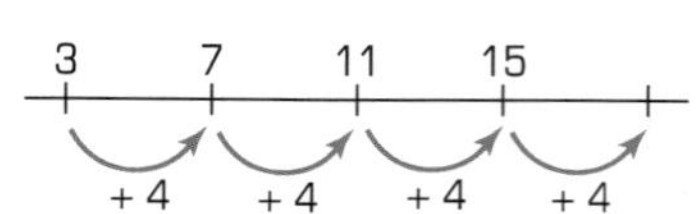

The sequence goes up in 4s

The next terms are: $15 + 4 = 19$
$19 + 4 = 23$
$23 + 4 = 27$

b The first term is 24.

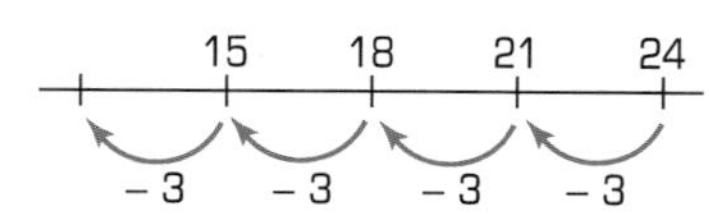

The sequence goes down in 3s

The next terms are: $15 - 3 = 12$
$12 - 3 = 9$
$9 - 3 = 6$

Exercise A1.2

1 Copy and complete these statements for each of these sequences:

- The first term is _____
- The rule is _____ _____
- The next three terms are _____, _____, _____

a 2, 5, 8, 11
b 2, 3, 4, 5
c 20, 18, 16, 14, 12
d 5, 7, 9, 11, 13
e 2, 10, 18, 26
f 3, 6, 9, 12, 15
g 30, 26, 22, 18, 14
h 20, 40, 60, 80
i 7, 21, 63, 189
j 0, 6, 12, 18, 24

You may need a calculator for part i.

2 Describe these sequences in words.
Use your rule to find the next two terms.
a 25, 21, 17, 13 ...
b 3, 10, 17, 24 ...
c 16, 19, 22, 25 ...
d 5, 10, 20, 40 ...

3 Write down the first five terms of these sequences:
a start at 4, each term is 2 bigger than the one before it
b the first term is 7, and each term is 4 larger
c the first term is 19, and each term is 3 smaller
d the first term is 6, and each term is 1.5 larger
e the second term is 13, and each term is 7 larger

4 a Make up a sequence where the second term is 5. Describe it in words.
Here is an example: 3, 5, 7, 9, 11.
b Make up a sequence where the second term is 3. Describe it in words.
c Make up a sequence where the second term is 2. Describe it in words.
d Make up a sequence where the third term is 5. Describe it in words.

5 A sequence starts: 1, 2,

The sequence could be: 1, 2, 3, 4, 5, ...
The first term is 1 and the rule is + 1.

The sequence could also be: 1, 2, 4, 8, 16, ...
The first term is 1 and the rule is × 2.

Find two different ways to continue each of these sequences.
Describe each of your rules.
a 2, 4, __, __, __ or 2, 4, __, __, __
b 1, 3, __, __, __ or 1, 3, __, __, __
c 5, 10, __, __, __ or 5, 10, __, __, __

A1.3 Sequences in diagrams

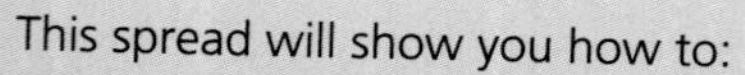

This spread will show you how to:

- Make general statements about odd or even numbers.
- Recognise multiples up to 10 × 10.

KEYWORDS
Rule
Sequence
Term

Natasha and Connie are building a bench.
As they build it they notice something ...

1, 3, 5 ... are odd numbers.

2, 4, 6 ... are even numbers.

The pictures show why odd numbers are called odd numbers!

You can illustrate other number sequences using picture patterns:

example

Use multilink cubes to describe how this number sequence grows:
3, 5, 7, 9, ...

Start with 3 cubes ...

... add 2 cubes ...

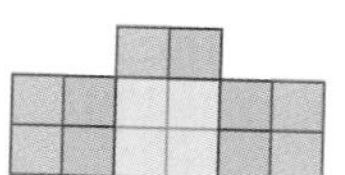

... add another 2 ...

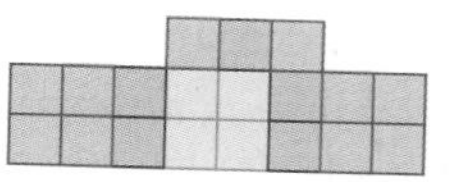

... and another 2

The first term is 3. The rule is add 2.

▶ You can describe picture patterns using number sequences

example

Describe this sequence in words.

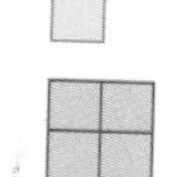
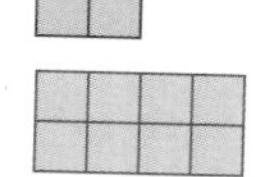

The first term is 1.
The rule is double or ×2

Exercise A1.3

1 Describe these sequences in words.
Use the rules to find the next two numbers in each sequence.

a

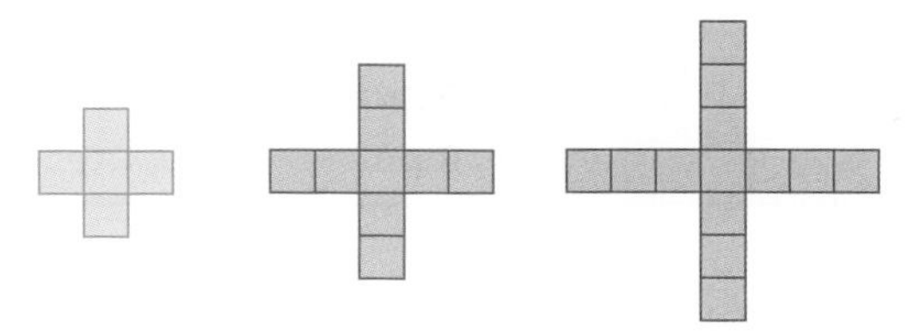

b

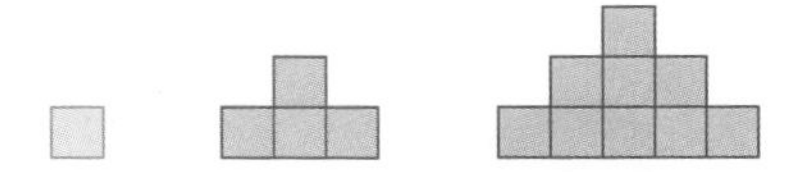

2 **a** How many matches are in the third term of this sequence?

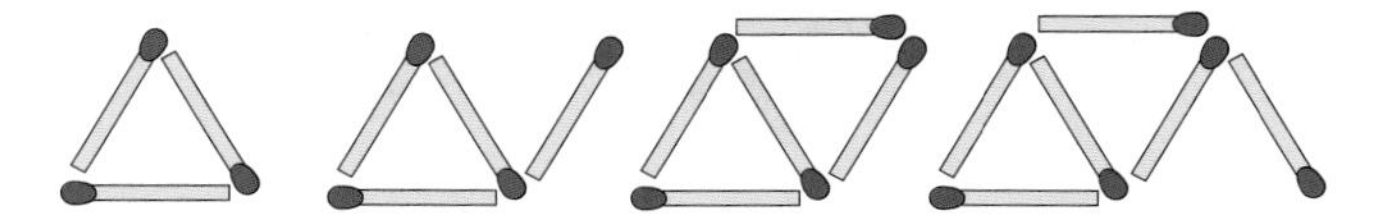

b Draw the next two patterns in the sequence.
c How many matches will be in the 8th pattern?

3 Look at the pattern in this grid.

7	9	11	
10			
	15	17	
16			

Fill in the empty squares.

4 You need a set of multilink cubes.
This pattern describes the sequence of even numbers:

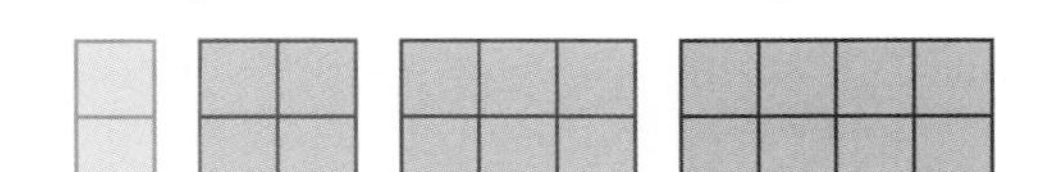

Design a pattern to describe each of these sequences:

a odd numbers
b three times table
c five times table

For each sequence, sketch the first four patterns.

A1.4 Function machines

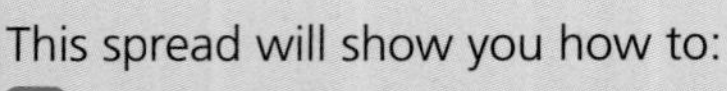

This spread will show you how to:

- Recognise multiples up to 10 × 10.
- Know multiplication facts up to 10 × 10 and derive corresponding division facts.

KEYWORDS
Function machine
Function
Input
Output

Harry the housekeeper can't remember his 7 times table – but he needs it to do the housekeeping!

He uses his calculator to work out the total cost of his milk, bread and newspaper for the week:

He inputs the price, multiplies by 7, then the calculator outputs the answer.

You can show Harry's calculations using a function machine:

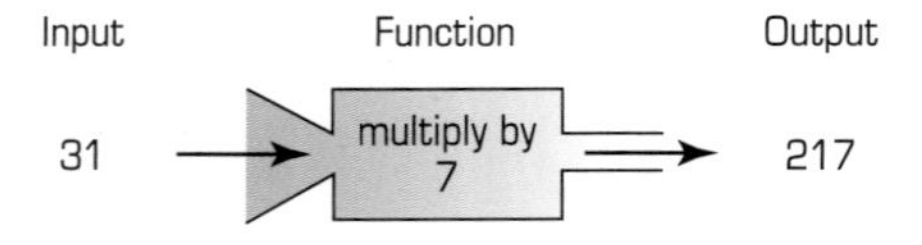

- In a function machine:
- The **input** is the value you put in to the machine.
- The machine performs the **function**.
- The **output** is the result that the machine puts out.

This function machine multiplies everything by 2:

You put in a number The machine multiplies it by 2. It puts out the result.

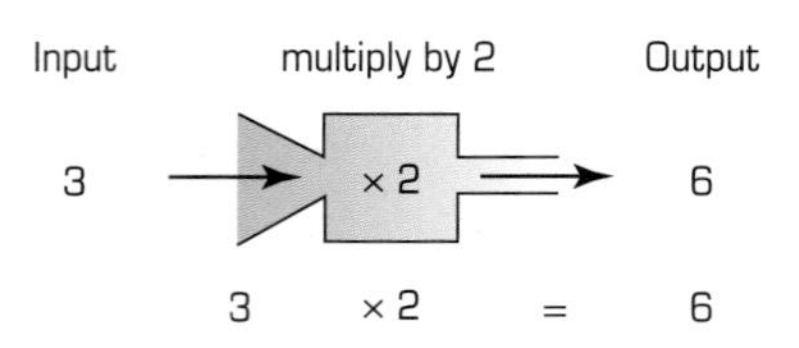

Exercise A1.4

1 For these function machines, find the outputs for each of the inputs.

a

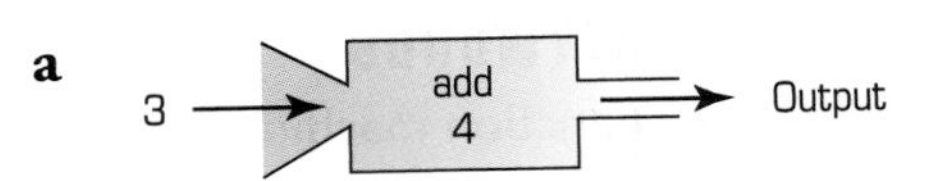

b

5 → multiply by 3 → Output

c

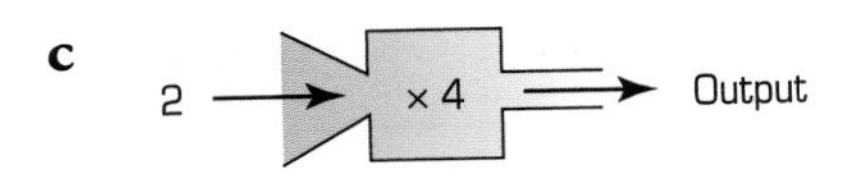

d

4 → × 12 → Output

2 For these function machines, find the outputs for each of the inputs 1, 2 and 3.

Hint: Input 1, then 2, then 3.

a

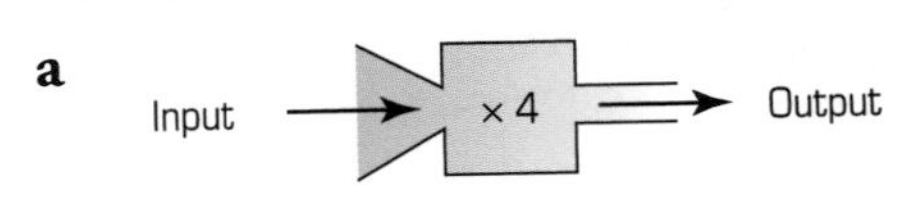

b

Input → + 3 → Output

c

Input → ÷ 2 → Output

d

Input → − 1 → Output

3 These function machines contain two operations.
Follow the example to find the outputs for the given inputs.

Example

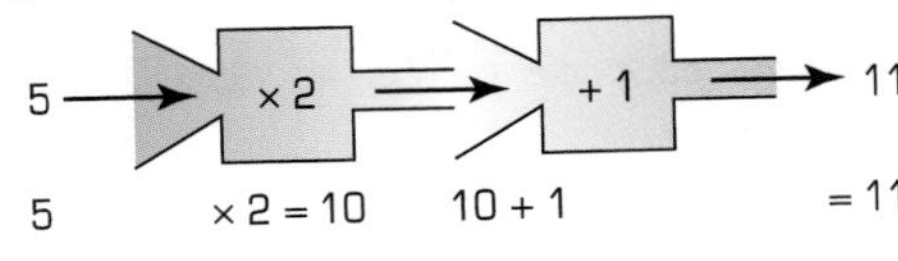

a

2 → × 3 → + 2 → Output

b

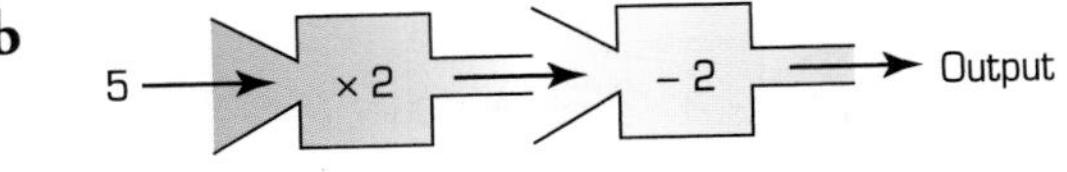

c

6 → ÷ 3 → + 4 → Output

d

8 → + 4 → ÷ 3 → Output

4 **Investigation**

a Input the values 1, 2 and 3 and work out the outputs for this function machine.

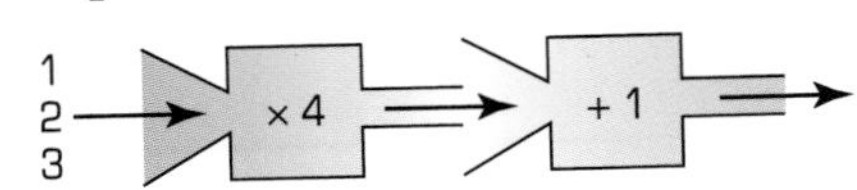

b Reverse the order of the functions (add 1 first).
Work out the outputs.
Are they the same as in (a)?

c *If you change the order of the operations you get a different output.*
Investigate this statement for the functions in question 3.

A1.5 Finding the function

This spread will show you how to:

- Understand and use the relationships between the four operations.

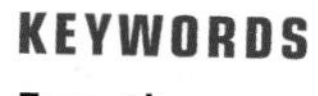

KEYWORDS

Function machine

Function

Input

Output

Harry's spilt coffee all over his sheet and can't read some of the numbers:

He knows that the function machine for any of the items looks like this:

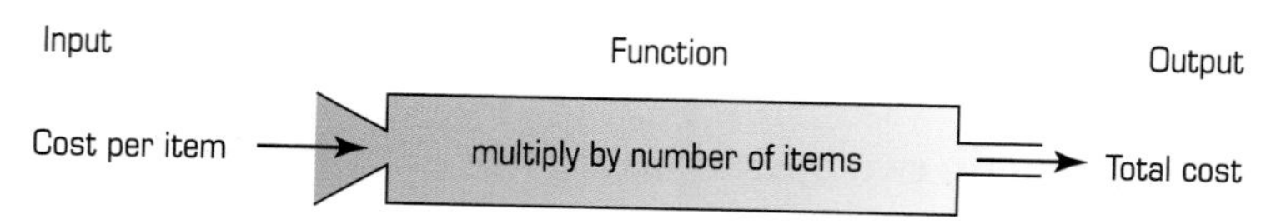

He uses the function machine to work out the values he can't read:

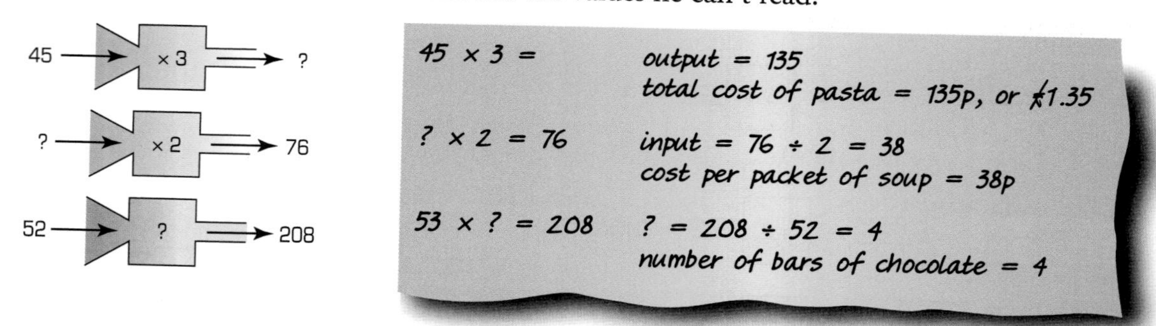

example

What input gives an output of 10 for this machine?

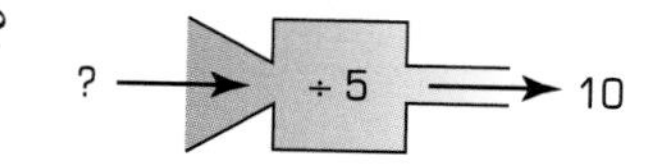

$10 \times 5 = 50$ so $50 \div 5 = 10$ The input is 50.

There may be more than one possible function in a function machine.

example

What is the function in this machine?

$2 \times 3 = 6$ and $2 + 4 = 6$. It could be $\times 3$, or $+4$.

Exercise A1.5

1 For each of these function machines, work out the inputs for the given output.

a

b

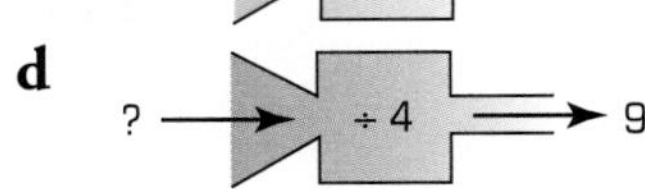

c ? → − 17 → 25

d ? → ÷ 4 → 9

2 In this function machine, the input has been multiplied by 3:

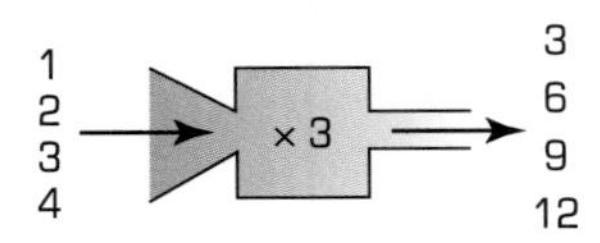

Find the function for each of these machines.

a 1, 2, 3, 4 → 12, 24, 36, 48

b

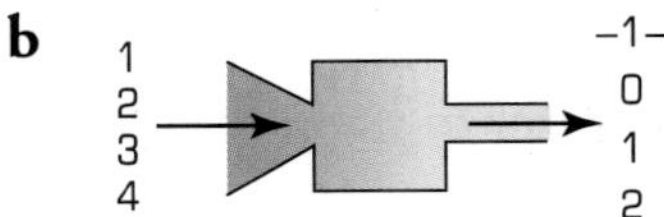

c

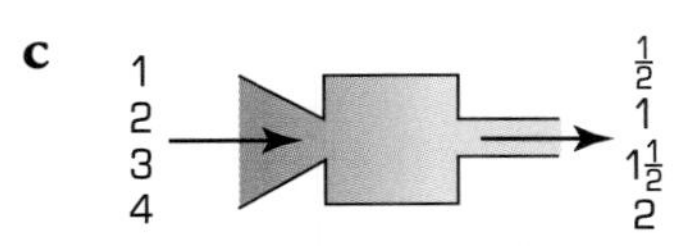

d

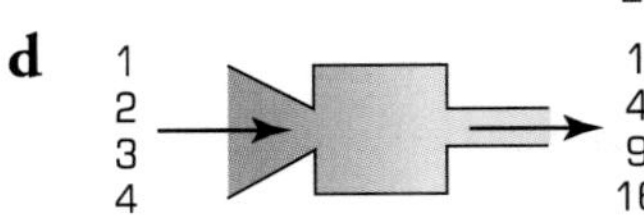

3 **Function Guessing Game**

- Player 1 thinks of a function.
- Player 2 writes down an input number, for example: 1
- Player 1 works out the output, and writes it down like this: $1 \rightarrow 5$
- Player 2 thinks of another number, for example: 3. Player 1 writes the output.
- Continue like this:

  ... until Player 2 guesses the function correctly. In this case, the function is 'add 4'.

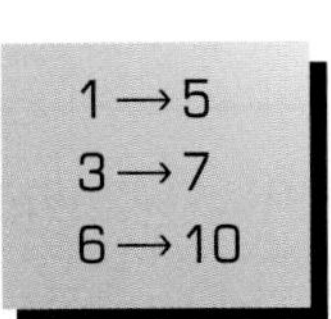

- Swap roles. Player 2 now thinks of a function.
- The winner is the player who has guessed the functions the quickest.

4 **Investigation**

a Write down the outputs for this function machine.

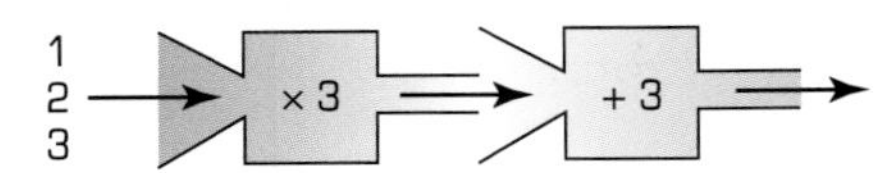

b Write down the outputs for this function machine:

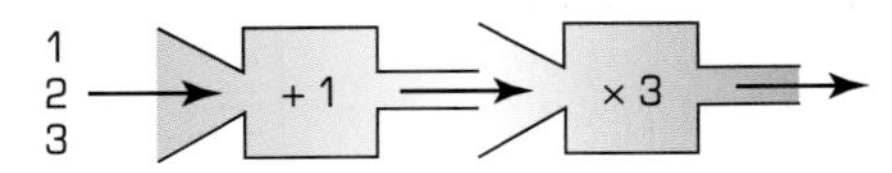

c What do you notice?

Can you find another pair of functions that give the same outputs?

A1.6 Functions and algebra

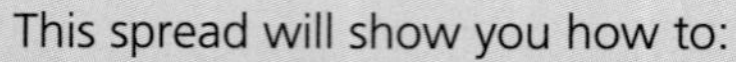

This spread will show you how to:

- Develop from explaining a generalised relationship in words to expressing it in a formula using letters as symbols.

KEYWORDS
Algebra Variable
Value

Harry's housekeeping list changes slightly each week.
He uses letters to show that the number of items can vary.
In Harry's list, the number of items is a **variable**.

- **A variable is a value that can change.**
- **You can use a letter to stand for a variable.**

For example, you could say that there are:

s grains of sugar in a bag.

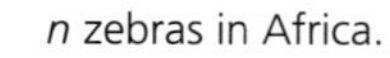

n zebras in Africa.

c days until Christmas.

When you use letters to stand for numbers, you are using **algebra**.

example

There are n zebras in Africa. Two more are born. How many are there now?

There are n zebras plus two zebras: you write $n + 2$ zebras.

- In algebra, you write $3 \times x$ as $3x$.

example

There are s grains of sugar in a bag. How many grains of sugar are there in 5 bags?

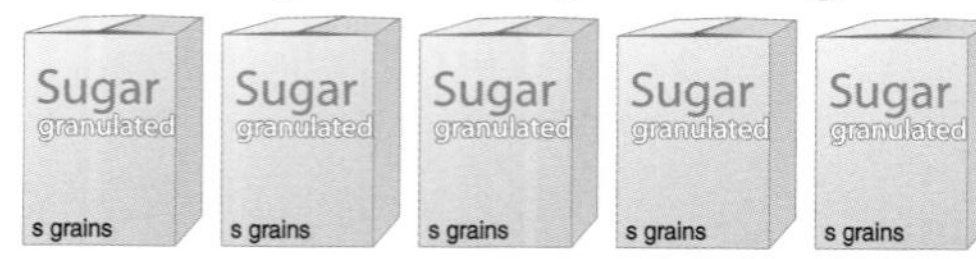

There are 5 lots of s grains, or $5 \times s$.
In algebra you write $5s$.

Exercise A1.6

1 Write a sentence using a suitable letter to represent each of these variables.

a The number of hairs on a cat.
b The number of words in a newspaper.
c The number of skateboards in America.
d The number of trees in a forest.
e The number of mobile phones in the UK.

> For example:
> The number of cars in the world.
> There are c cars in the world.

2 Look at Harry's housekeeping list in the example on page 12.
You could write the total cost of pasta as $45x$.

a Write the weekly cost of chocolate without multiplication signs.
b Write the total weekly cost of all three items using algebra.

3 Lake Smalldrop contains p fish, q boats, and r ducks.
In Lake Bigwater, there is five times the amount of everything.
How many:

a fish **b** boats **c** ducks
are there in Lake Bigwater?

Lake Median contains 20 less ducks, 100 less fish and 5 less boats than Lake Bigwater.

Using algebra, write how many
d fish, **e** boats **f** ducks
there are on Lake Median.

4 Write the outputs for these function machines using algebra.

For example

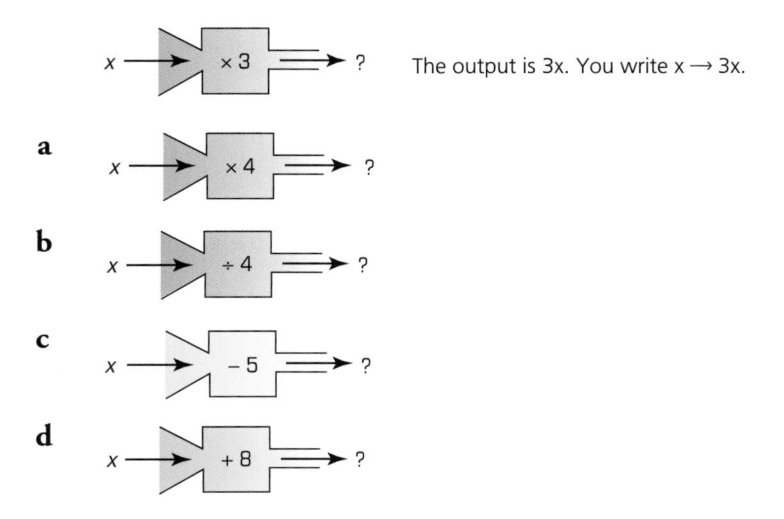

Summary

You should know how to …

1 Recognise and extend number sequences.

Check out

1 Write the first five terms of these sequences:

a The first term is 0.
The rule is +4

b The first term is 1.
The rule is ×2

c The first term is 10.
The rule is −3

d The second term is 15.
The rule is +6

Describe each of the sequences in words.
Find the next three terms using your rule.

e 3, 5, 7,9

f 20, 17, 14, __, __, __

g 2, 4, 8, __, __, __

h 0, −1, −2, −3, __, __, __

2 Develop from explaining a generalised relationship in words to expressing it in a formula using symbols.

2 Write a sentence using a suitable letter to represent each of these variables:

a The number of dogs in a park.

b The amount of chocolate eaten in each week in the UK.

c The number of people who own a skateboard.

d The number of fleas on a cat.

e The number of fish in the sea.

3 Solve mathematical problems or puzzles, recognise and explain patterns and relationships, generalise and predict.

3 Here is a sequence of square patterns:

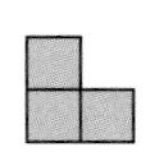
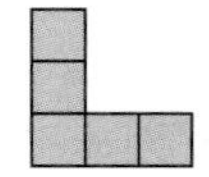

a How many squares are there in the 3rd pattern?

b Draw the next two patterns in the sequence.

c How many squares are there in the 6th pattern?

d How many squares will there be in the 20th pattern?

1 Number calculations

This unit will show you how to:

- Read and write whole numbers in figures and words, and know what each digit represents.
- Use decimal notation for tenths and hundredths.
- Know what each digit represents in a number with up to two decimal places.
- Order a set of numbers.
- Calculate a temperature rise and fall across 0°C.
- Use informal paper and pencil methods to support, record or explain additions and subtractions.
- Consolidate mental methods of calculation.
- Use standard column procedures to add and subtract whole numbers.
- Develop calculator skills and use a calculator effectively.
- Estimate by approximating, then check result.
- Choose and use appropriate number operations to solve problems, and appropriate ways of calculating.

Quick calculations can save you money!

Before you start

You should know how to ...

1 Write numbers up to 100 in digits or words.

Check in

1 **a** Write 43 in words
b Write eighty seven in digits.
c Write fourteen in digits
d Write 60 in words

2 Order positive numbers on a number line.

2 Order these numbers on a number line marked from 0 to 20:
11 5 7 3 17 14 6
Which number is in the middle?

3 Know complements to 100.

3 Write down the complements to 100 of these numbers:
a 60 **b** 75 **c** 81 **d** 37

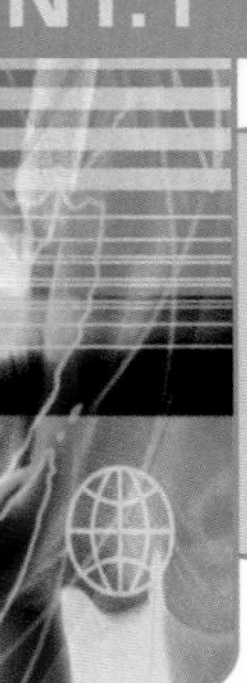

N1.1 Place value and ordering

This spread will show you how to:

- Read and write whole numbers in figures and words.
- Know what each digit represents in decimal numbers.
- Use decimal notation for tenths and hundredths.
- Order a set of mixed numbers.

KEYWORDS

>, < | Hundredth
Digit | Place value
Order | Tenth
Decimal number

▶ **Each digit in a number has a value that depends on its place in the number. This is its place value.**

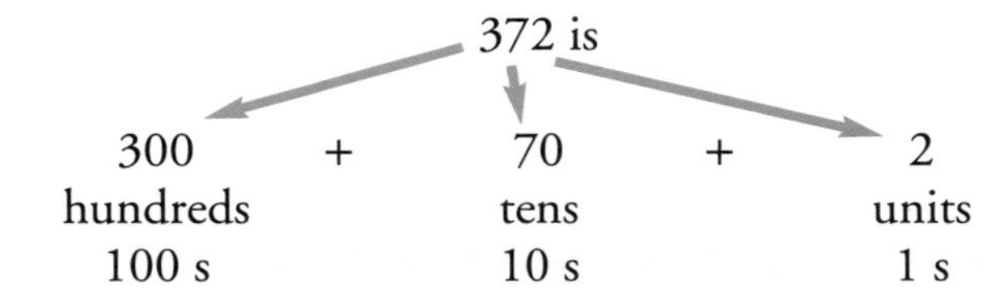

You say three hundred and seventy two

▶ **A decimal number has a whole number part and a fractional part.**

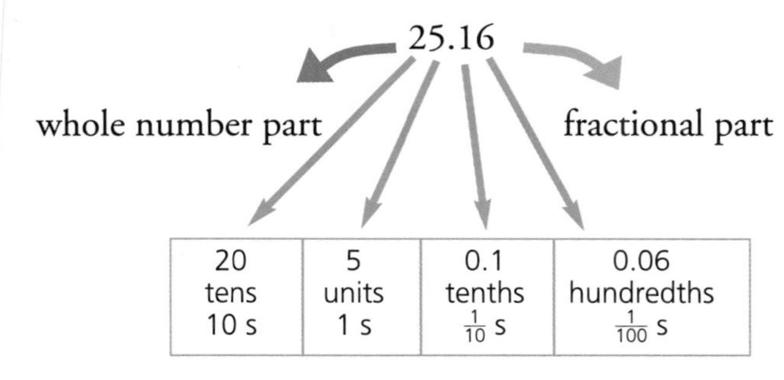

20	5	0.1	0.06
tens	units	tenths	hundredths
10 s	1 s	$\frac{1}{10}$ s	$\frac{1}{100}$ s

25.16 is more than 25 but less than 26.

You say twenty five point one six

▶ **Every number can be represented as a position on a number line.**

example

Which is bigger: 5.47 or 5.8?

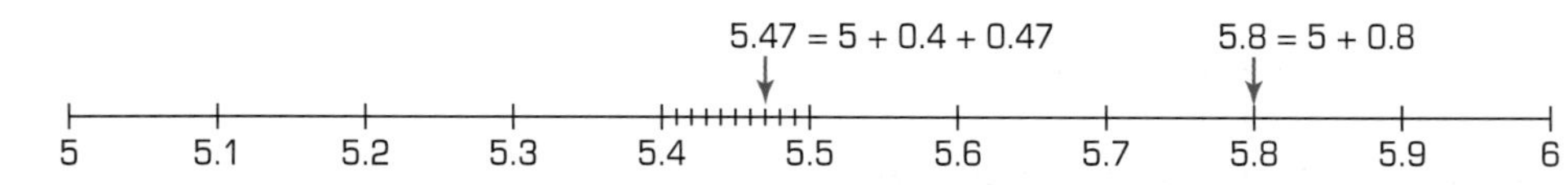

5.8 is bigger than 5.47.

Exercise N1.1

1 Write these numbers using digits:

a eighty five
b six hundred and thirty nine
c nine hundred and five point three
d nought point eight seven
e five point seven eight
f thirty nine point nought one

2 Use words to write these numbers as you would say them:

a 73 **b** 1092 **c** 34.01
d 325 **e** 0.57 **f** 109.63

3 For each set of numbers A, B and C:

A	35	33	36	38	39
B	3.2	3.4	3.8	3.5	3.7
C	3.12	3.24	3.42	3.06	3.7

a Write down the biggest number.
b Write down the smallest number.
c Write the numbers in size order starting with the smallest.

4 Copy and complete these pairs of numbers, using one of the signs:

> means greater than, < means less than

to show which number is bigger.
The first two are done for you.

a	4.2	>	2.1
b	3.6	<	5.2
c	5.8		5.7
d	12.23		12.56
e	8.32		8.09
f	0.3		0.25
g	0.57		0.06
h	9.25		9.6
i	13.06		12.91

5 Using the digits 4 0 5 9 make:

a the largest number
b the smallest number
c the number nearest to 5000
d the number nearest to 10 000

6 What number lies halfway between:

a 13 and 15 **b** 2 and 3 **c** 0.4 and 0.42?

7 These are the results of the 100 metres race at Superspeed School:

Amy	13.3 s
Belinda	14.1 s
Claudia	13.4 s
Danni	13.1 s
Elle	14.0 s

Put the runners in order from fastest to slowest.

8 Put these amounts in order from smallest to largest:

a £4.27, £4.31, 428p, 440p
b 1.82 m, 1.9 m, 185 cm, 1 m 88 cm

Negative numbers

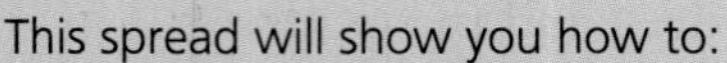

This spread will show you how to:

- Calculate a temperature rise and fall across 0°C.
- Order a set of positive and negative integers.

KEYWORDS

Decrease, Negative, Difference, Positive, Increase

A thermometer shows how hot or cold a temperature is:

Numbers above 0 are **positive** numbers.

Numbers below 0 are **negative** numbers.

The temperature scale is a number line.

▶ **Negative numbers can be represented on a number line.**

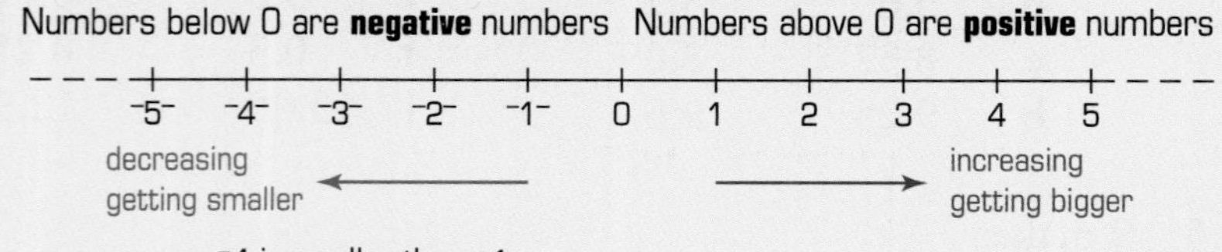

$^-4$ is smaller than $^-1$.

You can use a number line to help you solve problems involving negative numbers.

example

One night in London, the temperature is $^-4$°C.
The next day it rises to 8°C.
How many degrees does it rise?

From the number line you can see that:
The temperature rises by 12°C.

You can count in 1s from $^-4$ to 8.

$1 + 1 + 1 + 1 + 1 + 1 + 1 + 1 + 1 + 1 + 1 + 1 = 12$

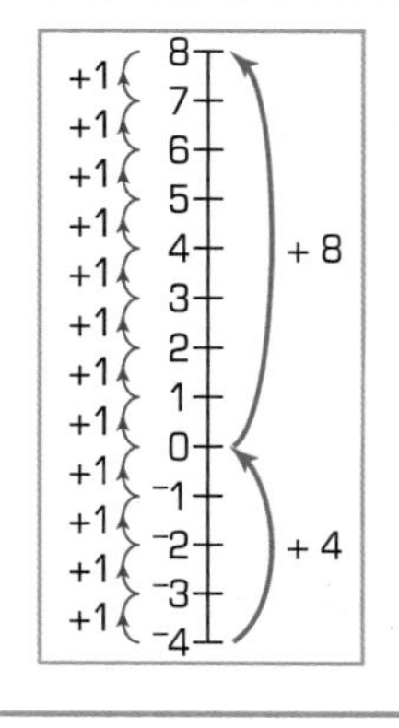

Or count in stages: first the negative part then the positive part.

$4 + 8 = 12$

Exercise N1.2

1 Here are six thermometers each showing different temperatures:

A B C D E F

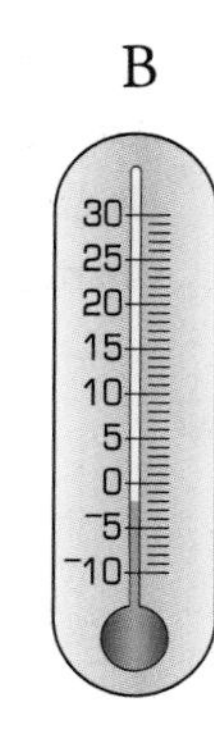

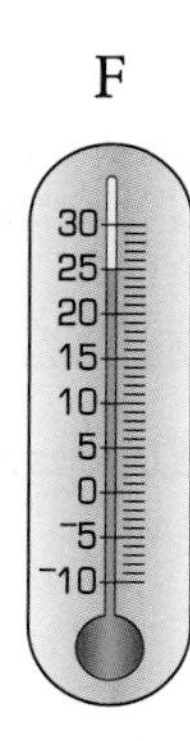

a Which thermometer shows the hottest temperature?
b Which thermometer shows the coldest temperature?
c What is the temperature on thermometer B?
d Which thermometers show less than 0 °C?

2 Write these numbers in words:
a 3603 **b** 21 335 **c** 10 070 **d** $^{-}12.5$

3 Write these numbers in figures:
a seventy three
b minus two hundred and thirty
c three thousand, six hundred and two
d minus eleven point three

4 a Write this list of numbers in order of size, smallest first:
$^{-}3$, 4, 2, $^{-}1$, 1
b Draw a number line from $^{-}4$ to 4 and label the numbers on it.

5 From this list of numbers
9 $^{-}3$ $^{-}5$ 0 $^{-}1$ 7
write down:
a all the positive numbers
b all the negative numbers
c the numbers in order of size, starting with the smallest

5 Copy and complete these tables.

a

Start	Increase	Finish
3 °C	Rises by 6 °C	9 °C
8 °C	Rises by 5 °C	
5 °C	Rises by 12 °C	
$^{-}8$ °C	Rises by 7 °C	
$^{-}3$ °C	Rises by 6 °C	

b

Start	Decrease	Finish
18 °C	Falls by 7 °C	
13 °C	Falls by 6 °C	
$^{-}1$ °C	Falls by 5 °C	
2 °C	Falls by 9 °C	$^{-}7$ °C
$^{-}6$ °C	Falls by 6 °C	

N1.3 Negative numbers – addition and subtraction

This spread will show you how to:

- Find the difference between a positive and negative integer, or two negative integers, in context.

KEYWORDS
Add
Subtract
Negative
Positive

You use negative numbers in many situations in real life ...

... describing the temperature

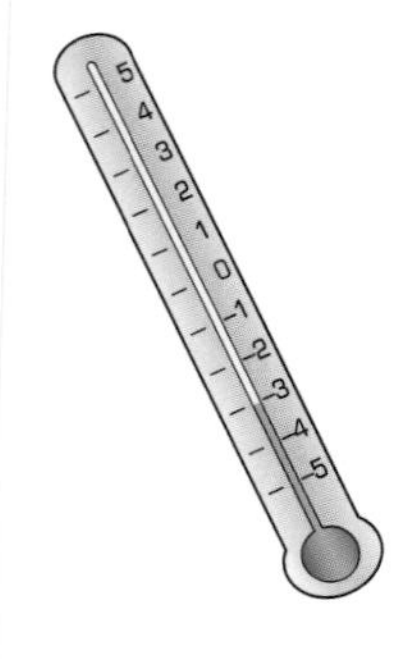

$^{-}3$°C is a temperature of 3°C below zero

... storing frozen food

These burgers must be stored below $^{-}20$°C to stay frozen

... describing depths

$^{-}2$ is two floors below ground level

You can use a number line to help you add and subtract using negative numbers.

- If you add it gets bigger.
- If you subtract it gets smaller.

example

Work out:

a $^{-}1° - 8°$ **b** $2° - 7°$ **c** $^{-}4° + 9°$

Using a number line:

a Start at $^{-}1$ and take away 8:

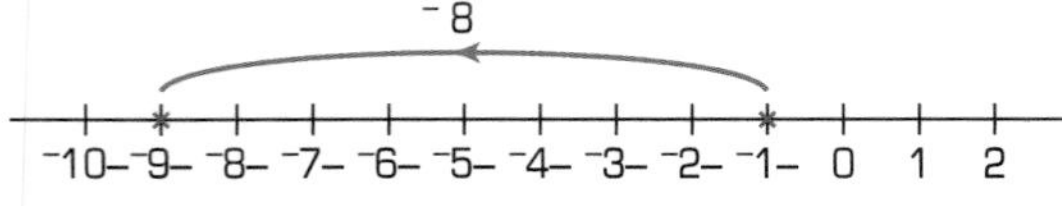

$^{-}1° - 8° = {}^{-}9°$

b Start at 2 and take away 7:

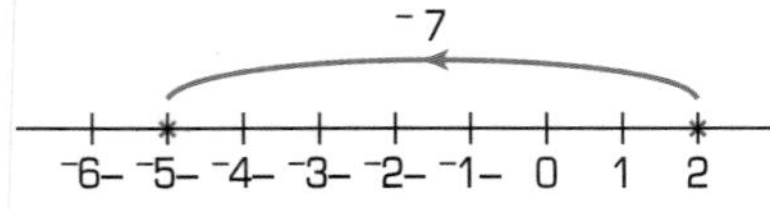

$2° - 7° = {}^{-}5°$

c Start at $^{-}4$ and add on 9:

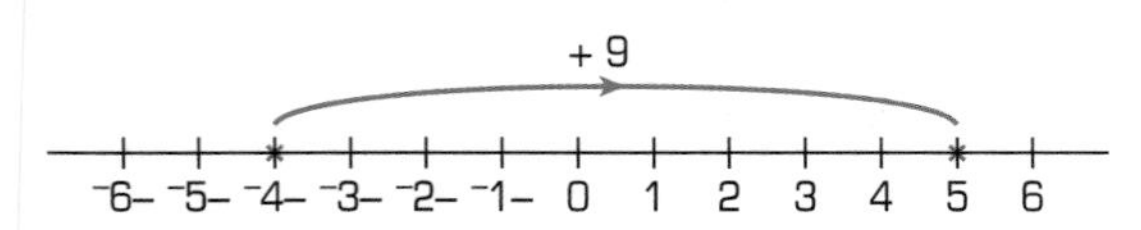

$^{-}4° + 9° = 5°$

Exercise N1.3

1 Write an addition or a subtraction and solve each of these problems:

a The temperature is $^{-}3$ degrees. It rises by 5 degrees. What is the new temperature?
b The temperature is 4 degrees. It falls by 6 degrees. What is the new temperature?
c The temperature is $^{-}7$ degrees. It rises by 8 degrees. What is the new temperature?
d The temperature is 12 degrees. It falls by 15 degrees. What is the new temperature?

2 Calculate the following. You should be able to do some of these questions mentally.

a $7 - 11$ **b** $30 + 46$ **c** $^{-}20 + 30$
d $^{-}18 - 20$ **e** $11 - 20$ **f** $^{-}38 - 70$
g $^{-}160 + 29$ **h** $^{-}30 - 75$ **i** $^{-}185 + 260$

3 Copy and complete this table:

1st term	Rule	2nd term	3rd term	4th term
15	Subtract 5	10	5	0
6	Subtract 4			
$^{-}9$	Add 3			
$^{-}3$	Subtract 5			

4 The table shows the temperatures in eight cities one night.

City	Temp °C	City	Temp °C
London	5	Paris	7
Madrid	2	Rome	2

City	Temp °C	City	Temp °C
Moscow	$^{-}8$	Sydney	15
New York	12	Warsaw	$^{-}12$

How many degrees ...

a warmer than London was Sydney?
b colder than Paris was Rome?
c difference was there between New York and Moscow?
d difference was there between Madrid and Warsaw?

5 Extend this pattern:

$$2 + 2 = 4$$
$$2 + 1 = 3$$
$$2 + 0 = 2$$
$$2 + {}^{-}1 = 1$$
$$2 + {}^{-}2 = 0$$
$$2 + {}^{-}3 = {}^{-}1$$

6 Write down the next three numbers in each of these sequences:

a 2, 7, 12, 17, 22, _, _, _
b 25, 21, 17, 13, 9, _, _, _
c $^{-}15$, $^{-}13$, $^{-}11$, $^{-}9$, $^{-}7$, _, _, _
d 6, 3, 0, $^{-}3$, $^{-}6$, _, _, _
e $^{-}30$, $^{-}22$, $^{-}14$, $^{-}6$, _, _, _
f 23, 12, 1, $^{-}10$, $^{-}21$, _, _, _

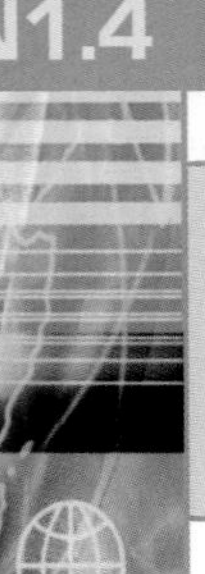

N1.4 Mental strategies

This spread will show you how to:
- Consolidate mental strategies for addition and subtraction.
- Use informal pencil and paper methods to support, record or explain additions and subtractions.

KEYWORDS

Difference, Add, Estimate, Subtract, Partition

There are some easy ways to add and subtract in your head. Here are two of them:

Partitioning

You break down a number into place value parts:

For example:	253 + 325	742 – 329
▶ Estimate the answer first:	300 + 300 = 600	700 – 300 = 400
▶ Write the larger number first in addition calculations:	325 + 253	742 – 329
▶ Break the smaller number into place value parts:	325 + 200 + 50 + 3	742 – 300 – 20 – 9
▶ Add/subtract the hundreds:	525 + 50 + 3	442 – 20 – 9
▶ Add/subtract the tens:	575 + 3	422 – 9
▶ Add/subtract the units:	578	413
So	253 + 325 = 578	742 – 329 = 413

Check your estimate!

Picture the number line:

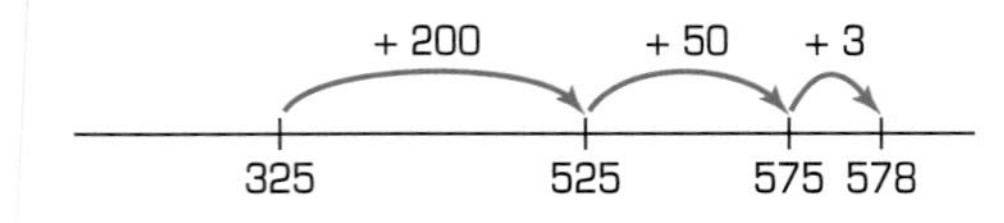

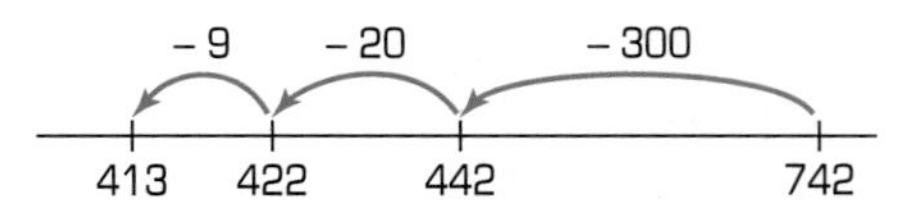

Compensation

You round a number and then later compensate for the rounding.

For example:	276 + 398	651 – 394
▶ Estimate first:	300 + 400 = 700	700 – 400 = 300
▶ Round one number and write the compensation:	276 + 400 – 2	651 – 400 + 6
▶ Use the rounding:	676 – 2	251 + 6
▶ Then compensate:	674	257
So:	276 + 398 = 674	651 – 394 = 257

Picture the number line:

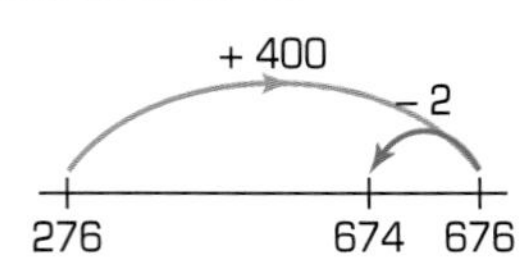

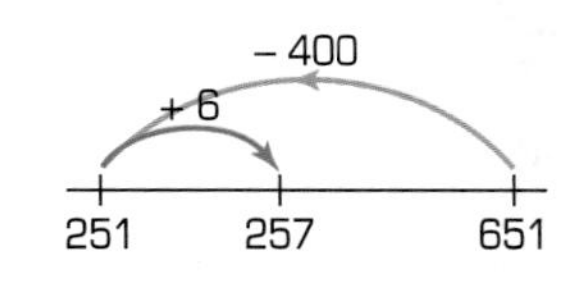

Exercise N1.4

Use **partitioning** to answer questions 1–3.

1 Add together these money amounts in your head.

a 23p and 5p **b** 53p and 4p **c** 32p and 13p **d** 43p and 23p
e 36p and 32p **f** 25p and 63p **g** £41 and £44 **h** £71 and £28

2 Find the difference between these money amounts:

a 26p and 5p **b** 39p and 7p **c** 36p and 13p **d** 46p and 24p
e 65p and 35p **f** 59p and 43p **g** £87 and £24 **h** £67 and £36

3 Work out these – they get harder as you go through!

a $123 + 52$ **b** $323 + 46$ **c** $158 - 32$ **d** $356 - 23$
e $26 + 321$ **f** $454 - 323$ **g** $125 + 132$ **h** $304 + 234$
i $875 - 832$ **j** $5.2 + 3.4$ **k** $7.2 + 2.5$ **l** $3.8 - 2.4$
m $8.8 - 3.5$ **n** $12.2 + 5.3$ **o** $18.6 - 5.4$ **p** $2.26 + 0.33$

Use **compensation** to answer questions 4–6

4 Add these distances together:

a 34 cm and 19 cm **b** 45 cm and 17 cm
c 68 m and 28 m **d** 46 m and 39 m
e 35 m and 57 m **f** 67 m and 27 m

5 Find the difference between these money amounts:

a 53p and 19p **b** 74p and 18p
c 56p and 28p **d** 78p and 29p
e £45 and £27 **f** £63 and £37

6 Work out these – they get harder as you go through!

a $35 + 19$ **b** $123 + 29$ **c** $235 + 98$ **d** $81 - 28$
e $252 - 97$ **f** $528 + 195$ **g** $267 + 297$ **h** $453 - 297$
i $238 + 187$ **j** $466 + 285$ **k** $483 - 286$ **l** $732 - 589$

7 In these questions you must choose which mental method to use.
The best method uses fewer steps so you can do the calculation quicker!

a $45 + 93$ **b** $65 - 48$ **c** $163 - 88$ **d** $235 + 207$
e $187 + 235$ **f** $354 - 123$ **g** $4.8 - 2.9$ **h** $3.4 + 5.5$
i $8.3 - 6.8$ **j** $5.3 + 3.8$ **k** $12.4 + 5.8$ **l** $13.6 + 12.8$
m $7.23 + 2.34$ **n** $2.65 + 4.27$ **o** $8.82 - 3.98$ **p** $22.3 - 15.9$

Adding and subtracting decimals

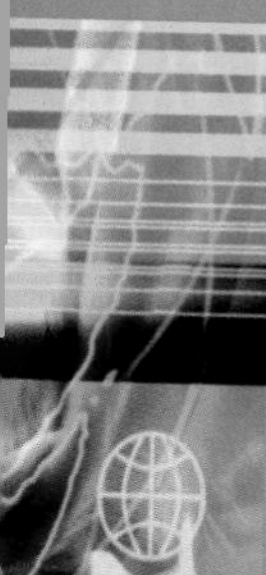

This spread will show you how to:

- Extend written methods to column addition and subtraction of numbers involving decimals.

KEYWORDS

Amount, Total, Calculate, Estimate, Sum, Decimal

Some numbers are too difficult to add or subtract in your head...

You can work out more difficult calculations on paper...

Or you can estimate the answer if you don't have any paper to hand...

- **When you calculate with numbers on paper, first line up the units. This holds the digits in their proper places.**

example

Jem has done some calculations on paper but he has got them wrong. Show him how to get them right.

a $352 + 34 = \begin{array}{r} 352 \\ +34 \\ \hline 692 \end{array}$

b $473 + 36 = \begin{array}{r} 473 \\ +36 \\ \hline 4109 \end{array}$

c $635 - 243 = \begin{array}{r} 635 \\ -243 \\ \hline 412 \end{array}$

a He should line up the units first:

$352 + 34 = \begin{array}{r} 352 \\ +34 \\ \hline 386 \end{array}$

b He should carry the '1' from the 10s column: $70 + 30 = 100$

$473 + 36 = \begin{array}{r} 473 \\ +36 \\ \hline 509 \\ {\scriptstyle 1} \end{array}$

c He should 'borrow' 10 from the 100s column: $130 - 40 = 90$

$635 - 243 = \begin{array}{r} {}^{5}\not{6}{}^{1}35 \\ -243 \\ \hline 392 \end{array}$

- **When you add and subtract decimals, line up the units first.**

example

Work out:

a $527.38 + 13.2$

b $159.6 - 41.13$

a Estimate first: $500 + 10 = 510$

Line up the units:

$\begin{array}{r} 527.38 \\ +13.2 \end{array}$

Fill in the blanks with 0s:

$\begin{array}{r} 527.38 \\ +13.20 \end{array}$

Now add:

$\begin{array}{r} 540.58 \end{array}$

b Estimate first: $160 - 40 = 120$

Line up the units:

$\begin{array}{r} 159.6 \\ -41.13 \end{array}$

Fill in the blanks with 0s:

$\begin{array}{r} 159.60 \\ -41.13 \end{array}$

Now subtract:

$\begin{array}{r} 118.47 \end{array}$

Exercise N1.5

Show all your working out in this exercise.

1 Calculate

a $364 + 34$	**b** $127 + 52$	**c** $831 + 63$	**d** $243 + 335$
e $137 + 532$	**f** $253 + 146$	**g** $387 - 32$	**h** $265 - 54$
i $558 - 41$	**j** $385 - 214$	**k** $957 - 725$	**l** $837 - 216$

2 Jerry has £243 and Jenny has £354.
How much do they have altogether?

3 George is 195 cm tall. His son is 132 cm tall.
How much taller is George than his son?

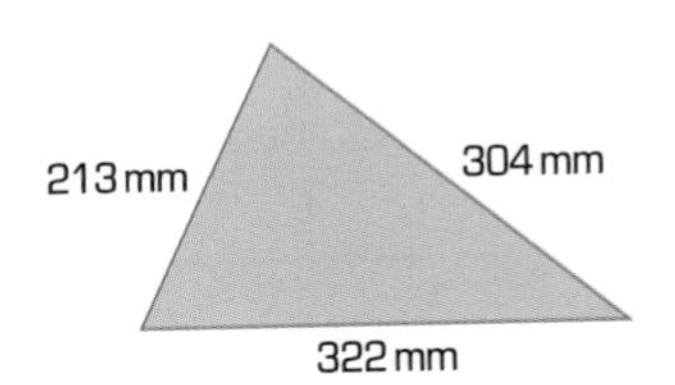

4 What do the three sides of this triangle add up to?

5 Total each of these money amounts:

a £173 and £216	**b** £203 and £562	**c** £185 and £253
d £32.35 and £61.42	**e** £53.07 and £15.21	**f** £25.52 and £14.37
g £18.53 and £18.32	**h** £25.63 and £22.83	

6 Work out these money problems:

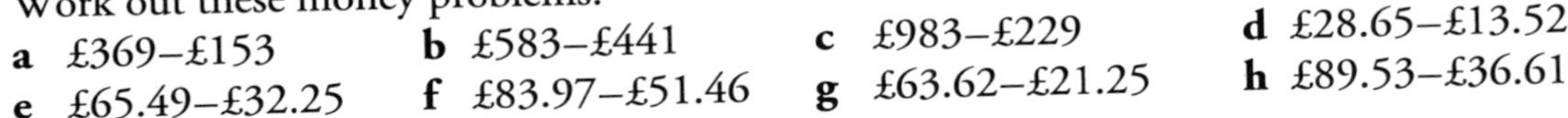

a £369–£153	**b** £583–£441	**c** £983–£229	**d** £28.65–£13.52
e £65.49–£32.25	**f** £83.97–£51.46	**g** £63.62–£21.25	**h** £89.53–£36.61

7 Will buys a saucepan costing £23.49 and a book costing £12.75.
How much does he spend altogether?

8 Tracey went shopping with £35.53 in her purse.
She spends £18.26 on clothes.
How much does she have left?

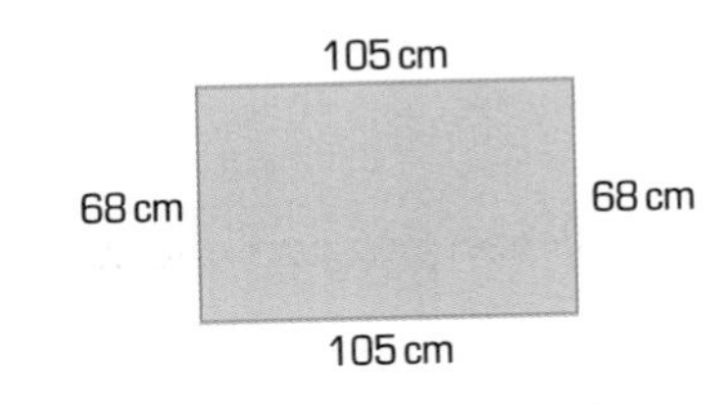

9 What is the sum of the four sides of this rectangle?

10 Calculate:

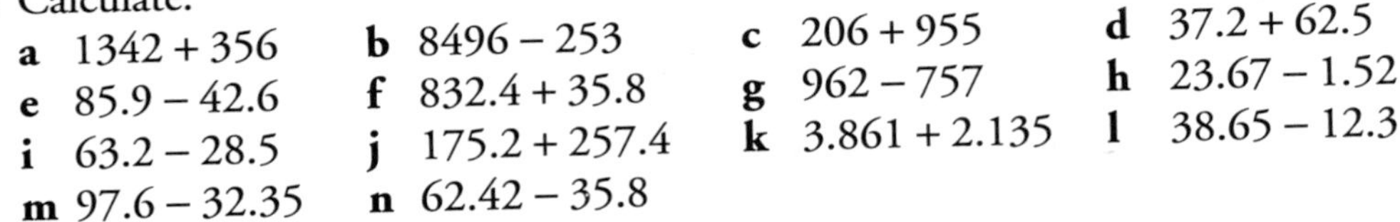

a $1342 + 356$	**b** $8496 - 253$	**c** $206 + 955$	**d** $37.2 + 62.5$
e $85.9 - 42.6$	**f** $832.4 + 35.8$	**g** $962 - 757$	**h** $23.67 - 1.52$
i $63.2 - 28.5$	**j** $175.2 + 257.4$	**k** $3.861 + 2.135$	**l** $38.65 - 12.3$
m $97.6 - 32.35$	**n** $62.42 - 35.8$		

11 Make a number as close to 2000 as possible by adding two of these numbers:

832 1257 952 1038

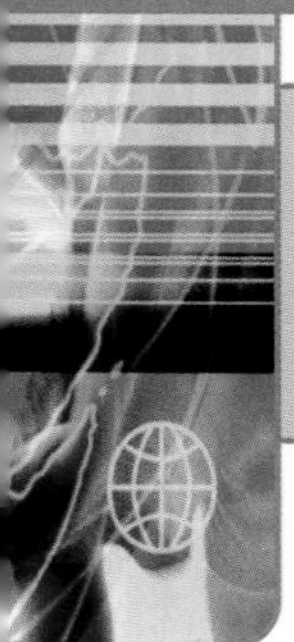

N1.6 Using a calculator

This spread will show you how to:
- Estimate by rounding then check the result.
- Develop calculator skills and use a calculator effectively.

KEYWORDS
Approximate, Nearest, Between, Round, Calculator

Before you use a calculator it is useful to approximate so that you can check whether your answer is right.

▶ **To approximate, round each number to the nearest whole number, nearest 10 or nearest 100.**

example

Round 367.4 to the nearest:

a 100 **b** 10 **c** unit

a 367.4 has 3 hundreds.
It is between 300 and 400.

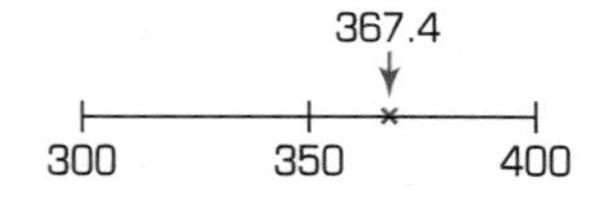

367.4 is closer to 400.

367.4 is 400 to the nearest 100

b 367.4 has 6 tens.
It is between 360 and 370.

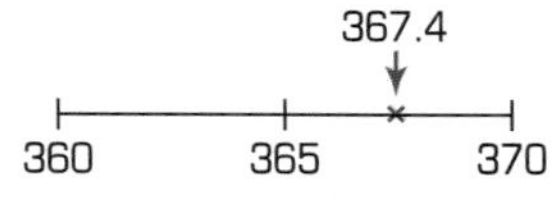

367.4 is closer to 370.

367.4 is 370 to the nearest 10

c 367.4 has 7 units.
It is between 367 and 368.

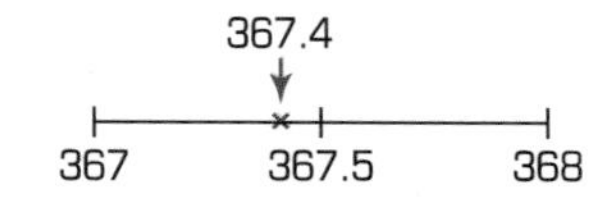

367.4 is closer to 367.

367.4 is 367 to the nearest unit

There are many times when a calculator is useful.

▶ **You can check your answer to a calculation.**

For example:	Check that	365 + 92 = 457
	Input	3 6 5 + 9 2 =
	Display	4 5 7

▶ **You can do calculations that are too difficult to do mentally.**

For example:	Work out	4325 + 3653
	Input	4 3 2 5 + 3 6 5 3 =
	Display	7 9 7 8

Check you know how your calculator works by following the examples to get the display shown.

▶ **You can do calculations that take too long to do on paper.**

For example:	Work out	32.65 + 2.98 + 33.67 + 123.7
	Input	3 2 . 6 5 + 2 . 9 8 + 3 3 . 6 7 + 1 2 3 . 7 =
	Display	1 9 3

▶ **You can do lots of calculations really quickly.**

Exercise N1.6

Remember to approximate before calculating.

1 Work out the answers to these calculations on paper.
Check your answers using a calculator.

a 325 + 124 **b** 857 − 235 **c** 258 + 735
d 1354 + 3235 **e** 9268 − 3154 **f** 10 012 − 2573

2 Work out the answers to these money problems on paper.
Check your answers using a calculator.

a £12.32 + £13.55 **b** £16.25 + £23.20 **c** £29.75 − £15.26
d £43.52 − £32.38 **e** £52.63 + £9.21 **f** £41.07 + £15.93

3 Use a calculator to work out:

a 3463 + 2761 + 1926 **b** 5732 + 3581 − 926 **c** 632 + 35.7 + 92.8

4 From this list of numbers

57 93 64 82

use your calculator to work out:

a Which two numbers add to 150? **b** Which two numbers add to 157?
c Which three numbers add to 214? **d** Which two numbers have a difference of 18?

5 Solve these problems using a mental or written method, using a calculator where appropriate.

a A shirt costs £15, a tie costs £13 and a book costs £6.50.
Is £34.80 enough to pay for all these items? Explain your answer.

b These are the weights of ingredients in a recipe:

flour	200 g	1 egg	24.5 g
sugar	100 g	1 litre milk	230 g
vanilla	0.6 g		

What is the total weight of the ingredients?
How many grams less than 1 kilogram is this?

1 kg = 1000 g

6 **Make me zero**

- The first player enters a number into the calculator.
- Players take it in turns to subtract a number from this, but only one digit can be changed at a time.
For example, 3.62 − 0.**6** = 3.**0**2 is allowed because only the 6 has changed.
- The display must never show a negative number.
- The winner is the player who makes the display on the calculator zero.

Example:

Player	Button	Display
1		83.62
2	−80	3.62
1	−0.6	3.02
2	−3	0.02
1	−0.01	0.01
2	−0.01	0

Player 2 wins.

Summary

You should know how to ...

1 Use decimal notation for tenths and hundredths.

2 Calculate a temperature rise and fall across 0°C.

3 Consolidate mental methods of calculation.

4 Extend written methods to column addition and subtraction of numbers involving decimals.

5 Choose and use appropriate number operations to solve problems, and appropriate ways of calculating.

6 Order a set of measurements.

Check out

1 For the number 3.56:

a write it in words
b what does the 3 represent?
c what does the 6 represent?
d what is the nearest whole number?

2 Work out:

a 2°C + 7°C b ⁻2°C + 7°C
c ⁻3°C + 12°C d ⁻3°C – 2°C

3 Work these out in your head:

a 19 + 35 b 52 – 17
c £5.20 – £1.70 d £1.90 + £3.50
e 205 + 134 f 5.7 – 2.8

4 Work these out on paper:

a 257 + 193 b 517 + 289
c 580 + 69 + 10.7 d 627 – 189

5 Choose the most appropriate methods to use to solve each of these problems.

a 19 + 67
b 37 – 18
c 125 + 210
d 920 + 430
e 982 – 795
f 127 + 976
g 1758 + 1921
h £5.27 – £1.53
i 86 + 197
j 737 – 298
k 5.2 – 1.8
l 693 – 287

6 Place these lengths in order from smallest to longest:

1.35 m 1.53 m 2.35 m
1.33 m 1.55 m 1.5 m

1 Perimeter and area

This unit will show you how to:

- Use, read and write standard metric units and their abbreviations, and relationships between them.
- Suggest suitable units and measuring equipment to estimate or measure length.
- Measure and draw lines to the nearest mm.
- Record estimates and readings from scales to a suitable degree of accuracy.
- Understand area is measured in square centimetres (cm^2).
- Understand and use the formula in words 'length × width' for the area of a rectangle.
- Measure and calculate perimeters of rectangles and regular polygons.
- Calculate perimeters of simple compound shapes.
- Visualise 3-D shapes from 2-D drawings and identify different nets for an open cube.
- Solve mathematical problems or puzzles, recognise and explain patterns and relationships, generalise and predict.

A white line marks the perimeter of this pitch.

Before you start

You should know how to …

1 Calculate the area of shapes by counting methods.

2 Measure and draw lines to the nearest centimetre.

3 Read from a scale.

Check in

1 Find the area of the red shape.
Copy and complete:
Area = ________ squares.

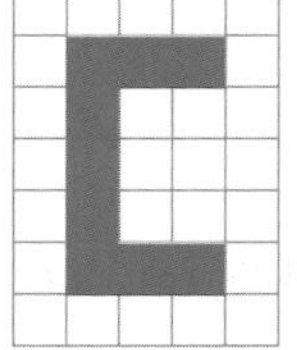

2 Measure the length of this nail.

3 What reading does the pointer give?

0 ├─┼─┼─┼─┼─┼─┼─┼─┼─┼─┤ 10 (pointer at the 7th division)

S1.1 Perimeter and area

This spread will show you how to:
- Calculate the perimeter and area of simple shapes.
- Understand that area is measured in square centimetres.

Keywords
Area
Length
Perimeter
Unit
Width

The **perimeter** of the pitch is the distance around the edge.

The longer the perimeter, the more paint you need to mark it!

The **area** of the pitch is the space it covers.

The bigger the area, the bigger the pitch and the more grass you have to mow!

- **The perimeter of a shape is the distance around the edge. Perimeter is measured in mm, cm, m or km.**
- **The area of a shape is the amount of space it covers. Area is measured in squares: mm^2, cm^2, m^2 or km^2.**

example

Find the perimeter and area of each rectangle.

a

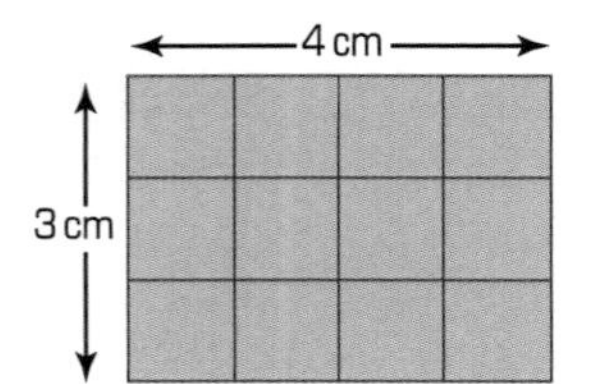

For the perimeter add the lengths of the edges:
3 cm + 4 cm + 3 cm + 4 cm
Perimeter = 14 cm

For the area count the squares:
there are 12 squares so the area is 12 cm^2.

You can also multiply to find the area:
3 cm × 4 cm = 12 cm^2

b

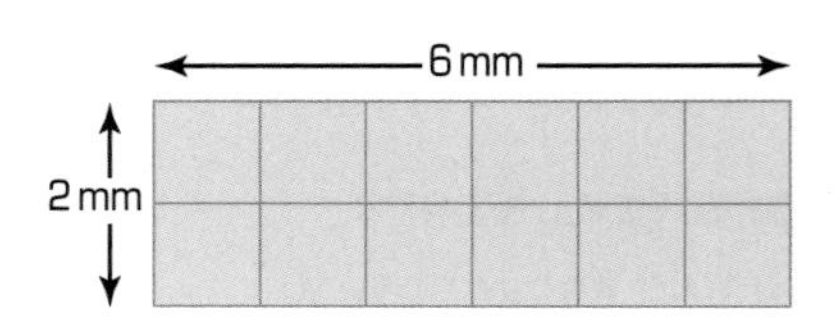

For the perimeter add the lengths of the edges:
2 mm + 6 mm + 2 mm + 6 mm
Perimeter = 16 mm

For the area count the squares:
there are 12 squares so the area is 12 mm^2.

You can also multiply to find the area:
2 mm × 6 mm = 12 mm^2.

Exercise S1.1

1 **a** Copy these letters onto centimetre-squared paper.

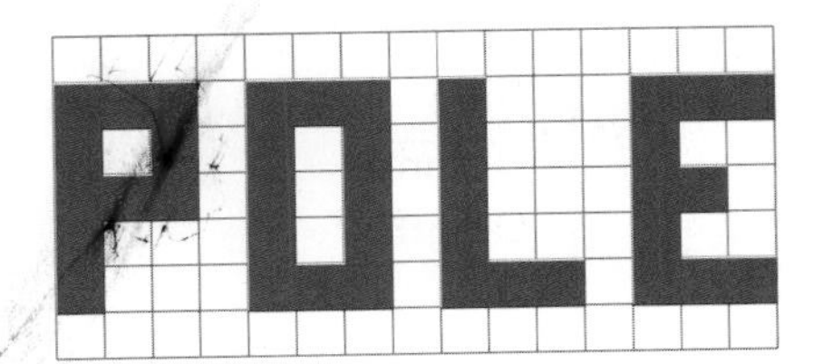

b The perimeter of the P is 16 cm. Find the perimeter of the other letters.

c The area of the P is 10 cm^2. Find the area of the other letters.

2 **a** Find the perimeter and area of these shapes.

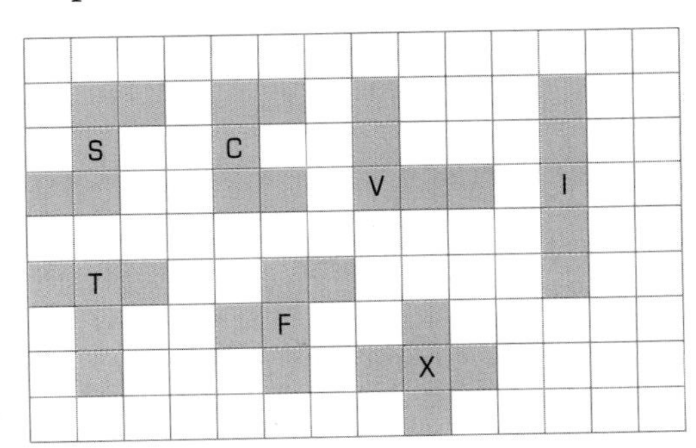

b Find the perimeter and area of these rectangles.

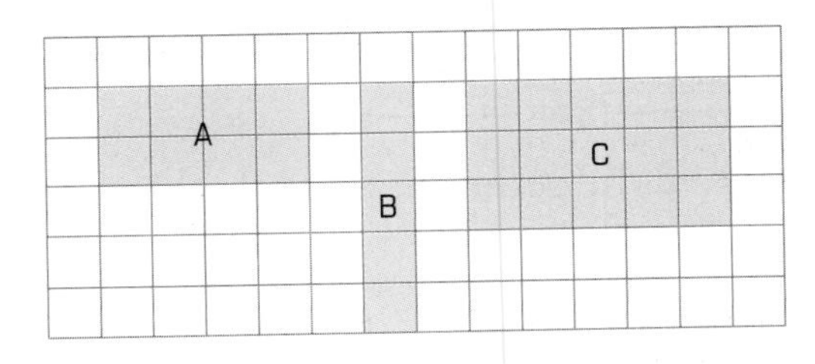

3 Find the perimeter and area of these shapes.

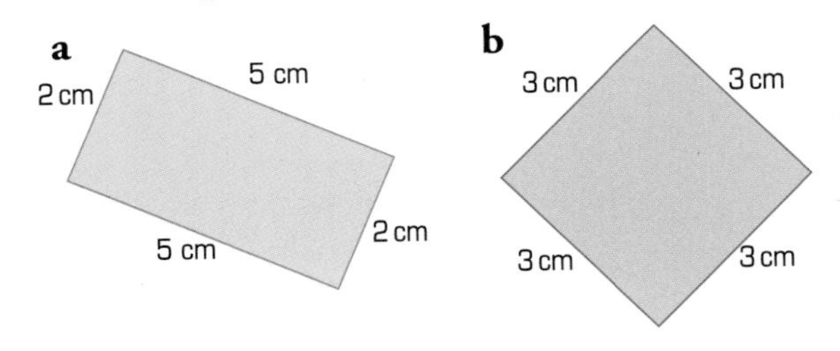

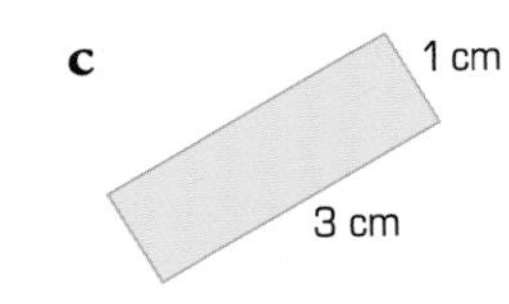

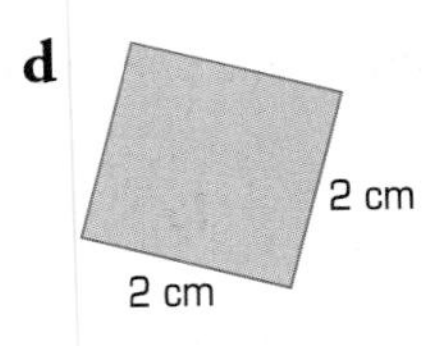

4 This rectangle has a perimeter of 10 cm:

4 cm

1 cm | 1 cm

4 cm

a Copy and complete this table of results for rectangles with a perimeter of 10 cm.

b Calculate the area of each of the rectangles.

length	width	perimeter
1 cm	4 cm	10 cm
2 cm		10 cm
3 cm		10 cm

length	width	perimeter
4 cm		10 cm
3.5 cm		10 cm
4.5 cm		10 cm

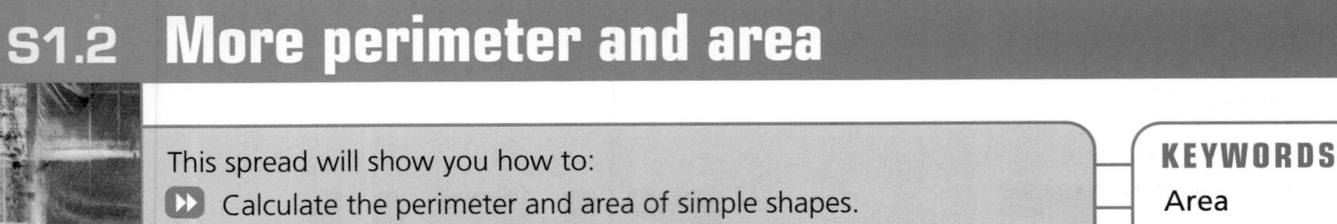

S1.2 More perimeter and area

This spread will show you how to:

- Calculate the perimeter and area of simple shapes.
- Understand and use the formula in words for the area of a rectangle.

KEYWORDS

Area, Perimeter, Distance, Width, Length

▶ **The perimeter of a shape is the distance around the edge.**

This L shape is made from two rectangles:

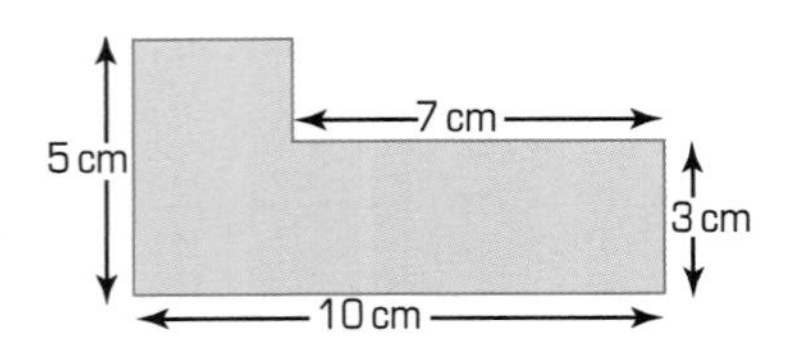

1
2

You find the perimeter by adding the lengths of the edges. First find the missing lengths:

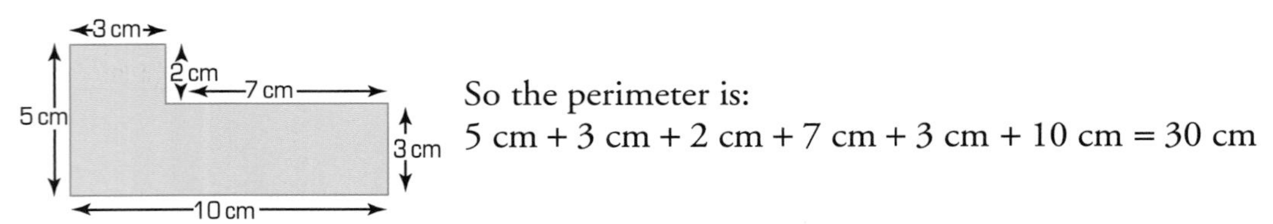

So the perimeter is:
5 cm + 3 cm + 2 cm + 7 cm + 3 cm + 10 cm = 30 cm

To find the area of the L shape add the areas of the two rectangles.
Imagine the rectangles are covered by a cm square grid:

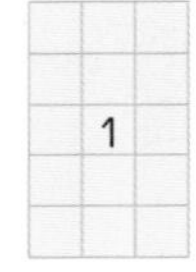

Area = 5 cm × 3 cm = 15 cm^2

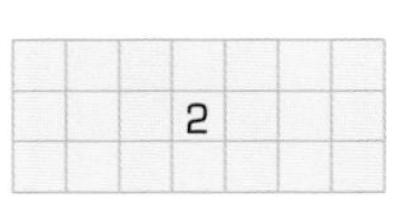

Area = 7 cm × 3 cm = 21 cm^2

The 2 shows there are 2 dimensions: length and width

Area of the L shape is 15 cm^2 + 21 cm^2 = 36 cm^2

▶ **Area of a rectangle = length × width**

example

A rectangle has an area of 18 cm^2. The length is 6 cm, what is the width?

Area = 18 cm^2
6 cm

Length × width = 18 cm^2
6 × 3 = 18 so the width must be 3 cm.

Exercise S1.2

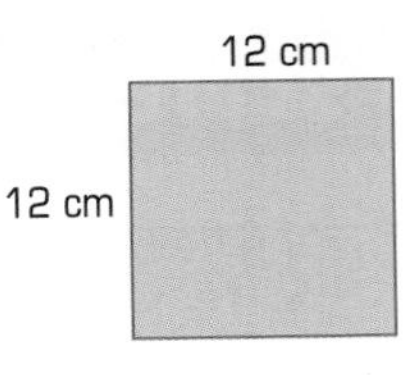

1 The side of a square is 12 cm.
  a Calculate the perimeter of the square.
  b Calculate the area of the square.

2 A playing card measures 9 cm by 6 cm.
What is the area of the playing card?

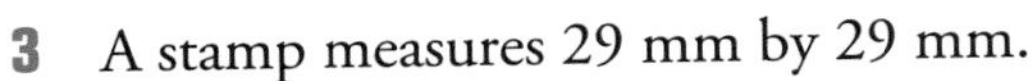

3 A stamp measures 29 mm by 29 mm.
  a What is the area of one stamp?
  b What is the area of a sheet of 12 stamps?

4 Find the missing lengths in each diagram.

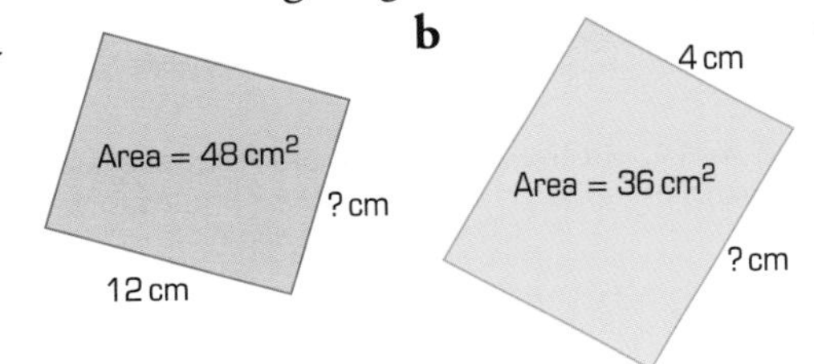

c
? cm
4 cm
Perimeter = 20 cm
4 cm
? cm

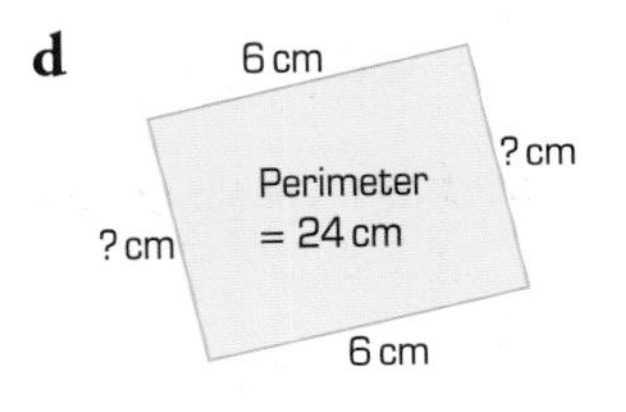

5 The width of a rectangle is 6 cm.
The perimeter of the rectangle is 20 cm.
Calculate the area of the rectangle.

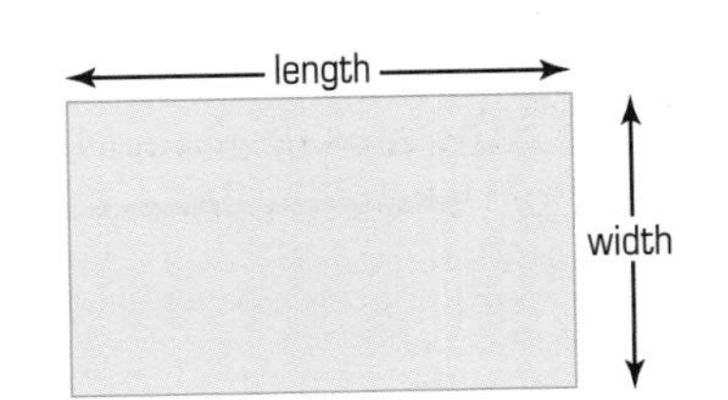

6 **a** In words, find the perimeter of the rectangle.
  b In words, find the area of the rectangle.

7 **a** Copy and complete this table of results for rectangles.

length	width	area
1 cm	18 cm	18 cm^2
2 cm		18 cm^2
3 cm		18 cm^2

length	width	area
6 cm		18 cm^2
9 cm		18 cm^2
18 cm		18 cm^2

You can draw rectangles on squared paper to help you.
  b Calculate the perimeter of each of the rectangles.

8 Find the perimeter and area of these shapes.

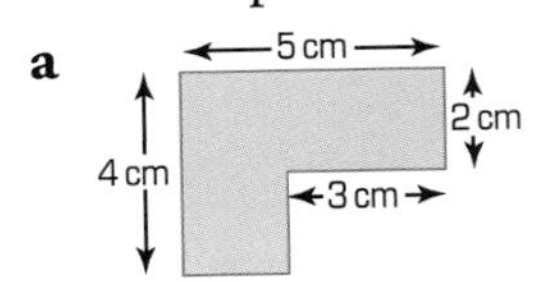

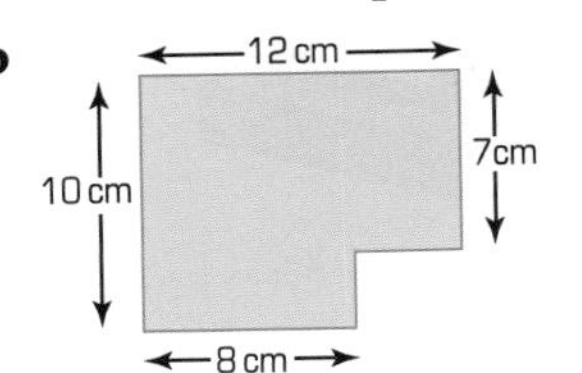

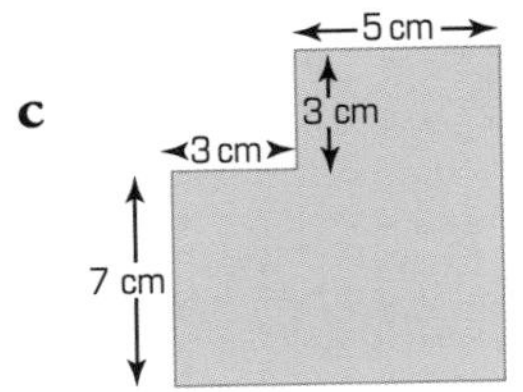

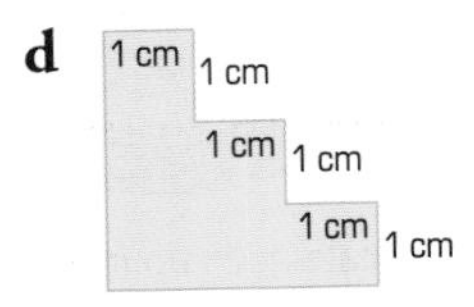

Measurement and scales

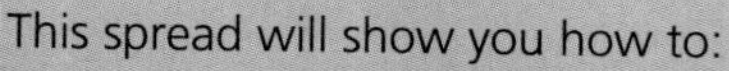

This spread will show you how to:

- Use standard metric units.
- Suggest suitable units to estimate or measure length.
- Measure and draw lines to the nearest millimetre.
- Record estimates and readings from scales.

KEYWORDS

Distance	Ruler
Length	Width
Measure	Scale

▶ **You can measure lengths and distances using metric units:**

Millimetres (mm)

1 mm is about the width of a grain of sand

Centimetres (cm)

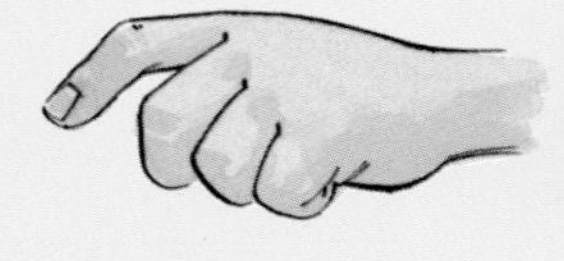

1 cm is about the width of your little finger nail

Metres (m)

1 m is about the width of a door

Kilometres (km)

1 km takes about 15 minutes to walk

You can measure and draw short lengths accurately using a ruler.

example

a Measure this line to the nearest mm.

b Draw a line of length 2 cm 3 mm.

a Make sure you measure from the 0 mark:

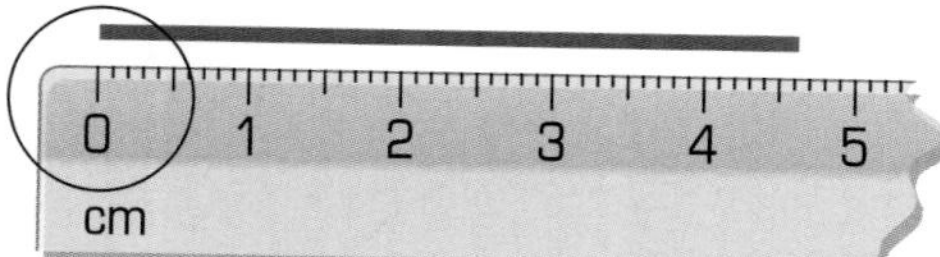

The line is 4 cm and 6 mm long.
That is the same as 4.6 cm or 46 mm.

b Start to draw from 0:

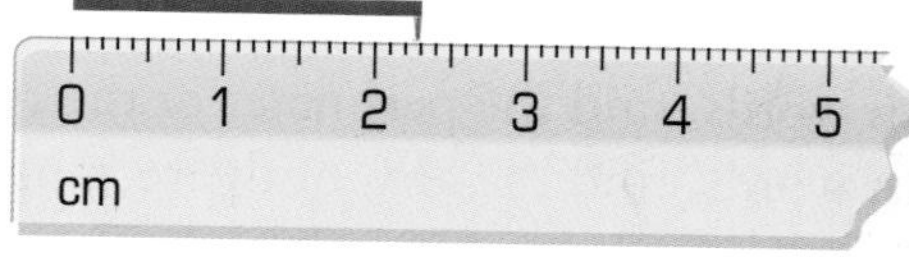

2 cm 3 mm is the same as 2.3 cm or 23 mm.

You should also be able to read other scales:

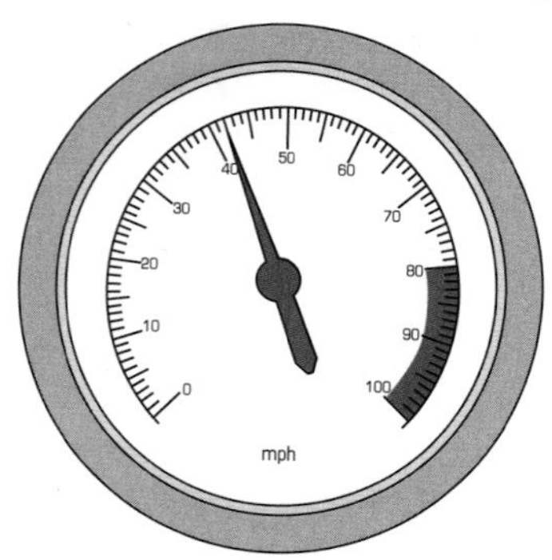

This scale goes up in 10s
Each mark is 1 mph
The reading shows 42 mph

This scale goes up in 100s
Each mark is 25 g
The reading shows 225 g

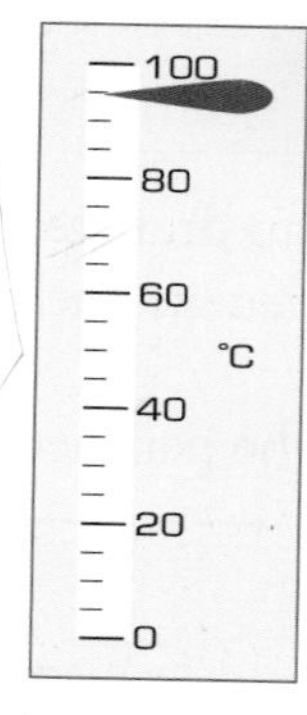

This scale goes up in 20s
Each mark is 5 °C
The reading shows 95 °C

Exercise S1.3

1 Write down the values shown on each of these scales:

a

0 2 4 6 8 9 10 m

b

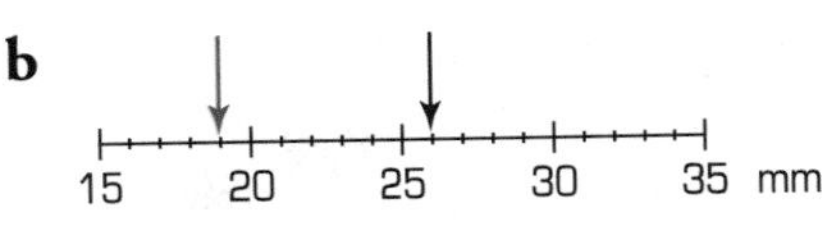

c

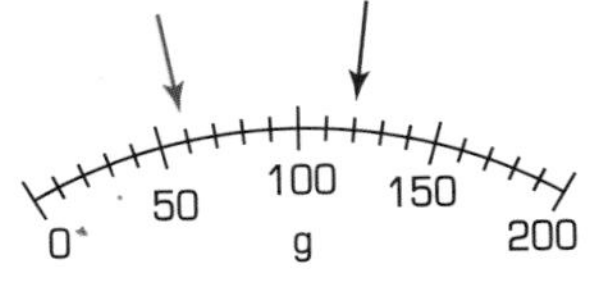

d

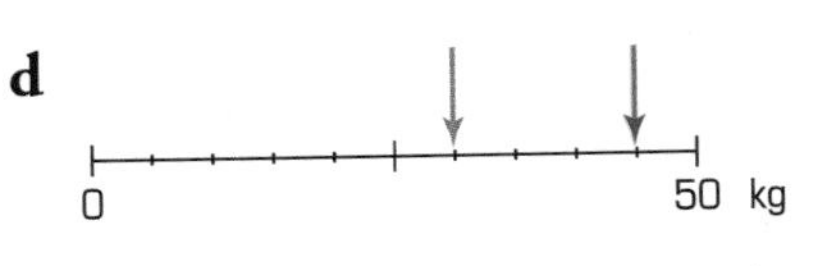

e

0 10 20 30 40 50 mm

f

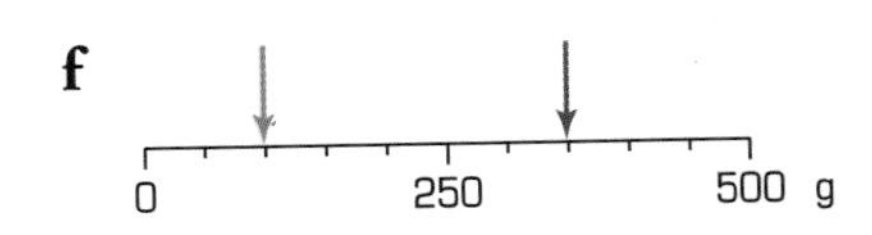

2 **a** Measure these lines to the nearest mm.

A

B

b Draw lines of length 2.8 cm and 4 cm 3 mm.

3 Which line is longer: A or B?
Guess first, and then measure them to see if you guessed correctly.

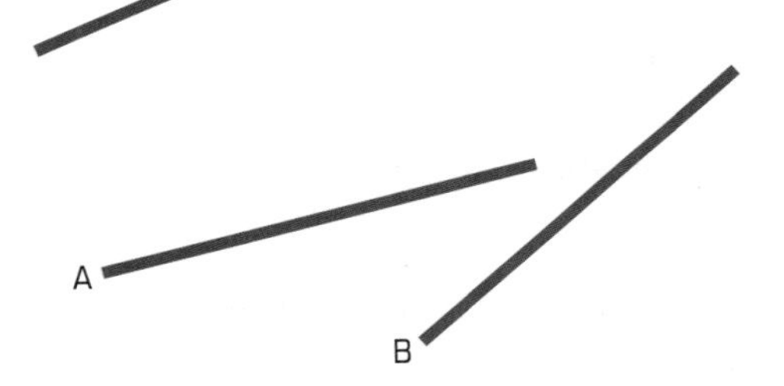

4 Which units would you use to measure each of these lengths?
Choose from mm, cm, m, km.

a The perimeter of this page
b The perimeter of a field
c The perimeter of a sports arena
d The perimeter of a stamp
e The perimeter of a room
f The perimeter of England

5 Which units would you use to measure each of these areas?
Choose from mm^2, cm^2, m^2, km^2.

a The area of this page
b The area of a field
c The area of a carpet
d The area of a pinhead
e The area of a table top
f The area of England

6 Copy and complete:
The length of my little finger is about ______ times the width of my fingernail so the length is approximately ______ cm.
The height of the classroom door is about ______ times the width of the door so the height is approximately ______ m.

7 Measure the length and width of:
a your text book **b** your calculator
In each case find the area and perimeter.

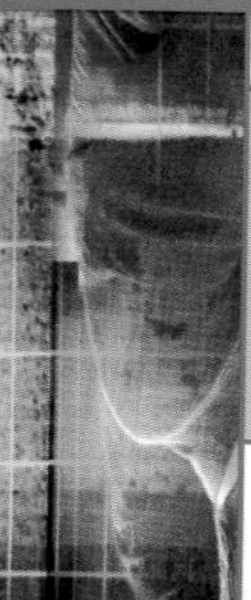

S1.4 Three dimensional shapes

This spread will show you how to:
- Identify different nets for an open cube.
- Visualise 3-D shapes from 2-D drawings.

KEYWORDS
3-D
Net
Surface area

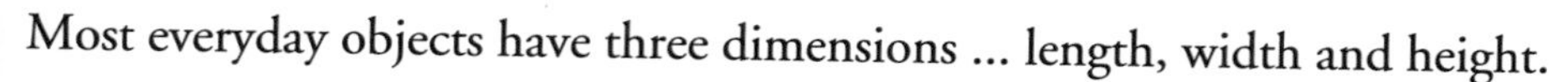

Most everyday objects have three dimensions ... length, width and height.

Some common 3-D shapes are:

Cubes

All faces are the same size

Cuboids

Opposite faces are the same

Most 3-D packages start out as flat shapes called **nets**.

Here is the net of an open cube.

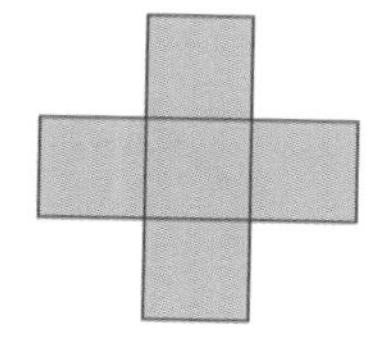

It folds to make a cube.

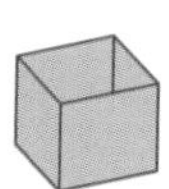

▶ **The area of the net of a 3-D shape is called the surface area.**

You find the surface area by adding the area of each surface together.

example

This net makes a cuboid.
Write down the dimensions of the cuboid and the surface area of the cuboid.

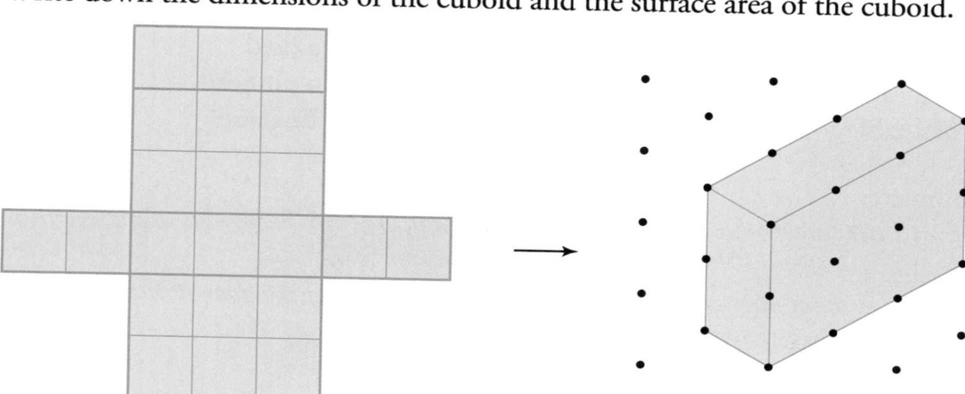

Dimensions of cuboid are 1 cm by 2 cm by 3 cm.
Surface area of cuboid is $6 + 2 + 3 + 2 + 6 + 3 = 22\text{ cm}^2$

Exercise S1.4

1 Copy these nets for an open cube (onto squared paper).
Colour the square which forms the base of the cube.

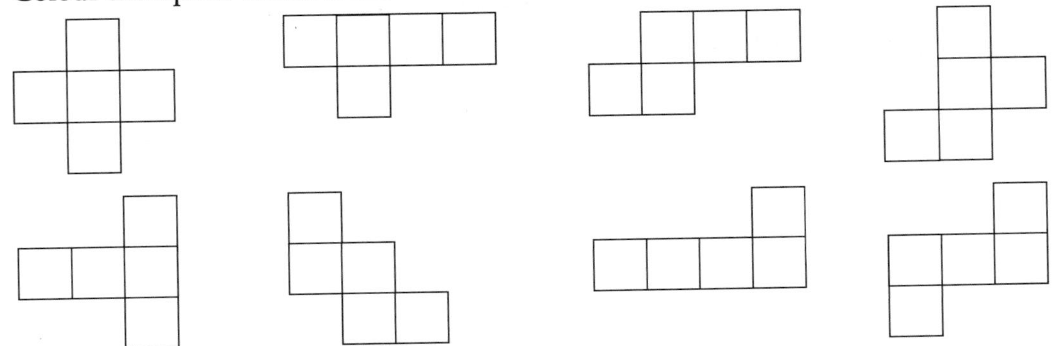

2 These nets make a cuboid.
Write down the dimensions of the cuboid and find the surface area.

a

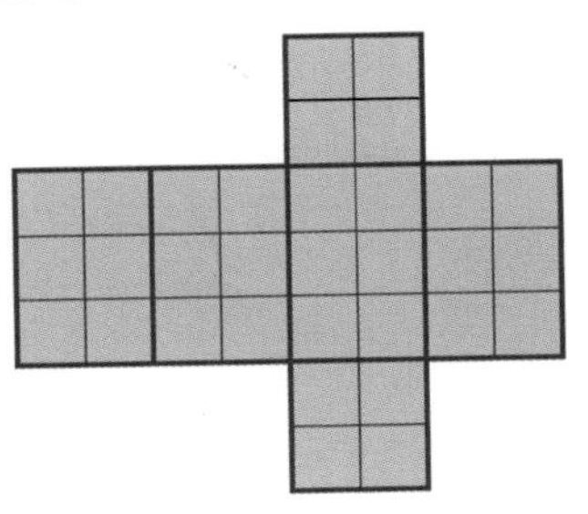

b

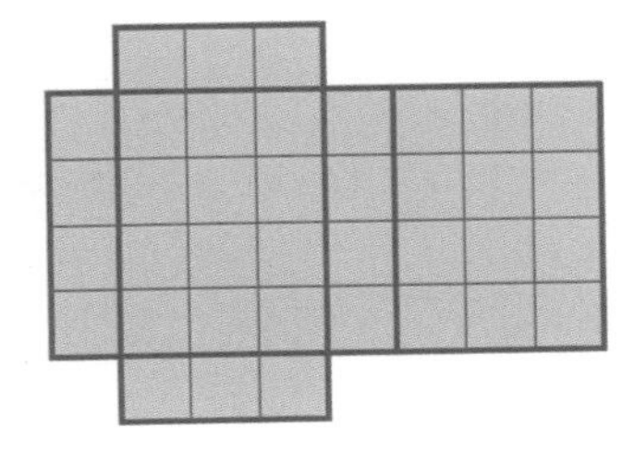

3 These nets make open cuboids. The length of each small square represents 1 cm.
For each cuboid give the length, the width and the height.

a

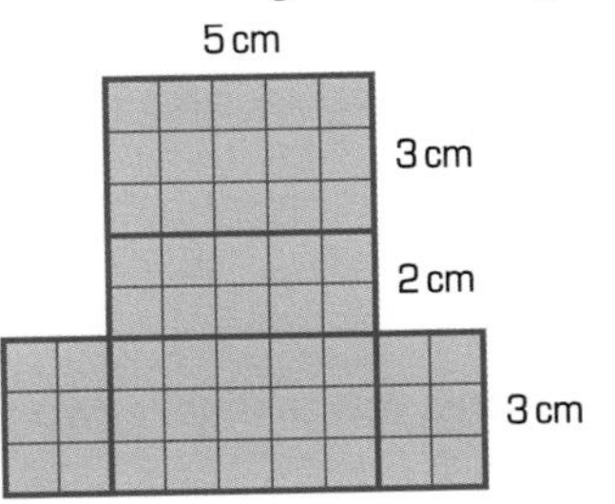

b

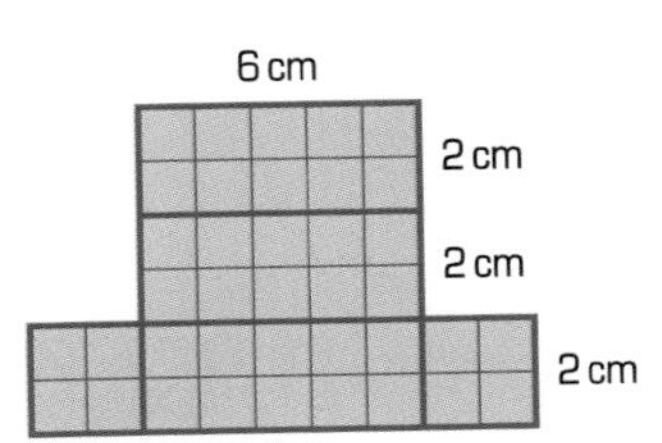

4 This closed box is a cuboid.

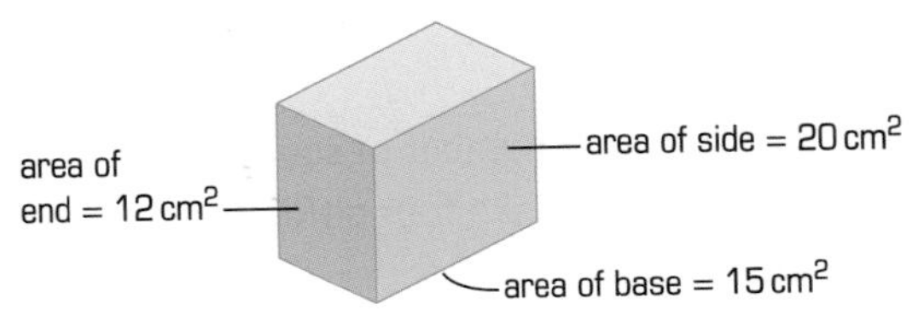

a Work out the surface area of the cuboid.
b Work out the dimensions of the cuboid.
c Draw a net for the cuboid.

S1 Summary

You should know how to ...

1 Understand and use the formula
area of rectangle = length × width
and that area is measured in square centimetres (cm^2).

2 Calculate the perimeter and area of simple compound shapes that can be split into rectangles.

3 Solve mathematical problems or puzzles, recognise and explain patterns and relationships, generalise and predict.

Check out

1 **a** What is the area of this rectangle? State the units of your answer.

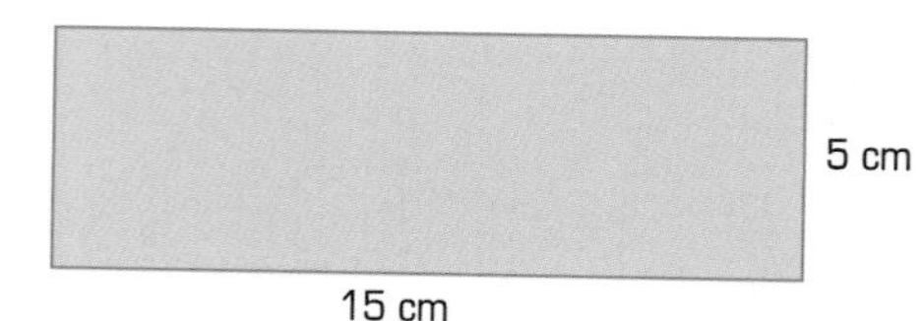

b Write down possible lengths and widths for a rectangle, if the area is 48 cm^2.

2 Find the perimeter and area of this shape.

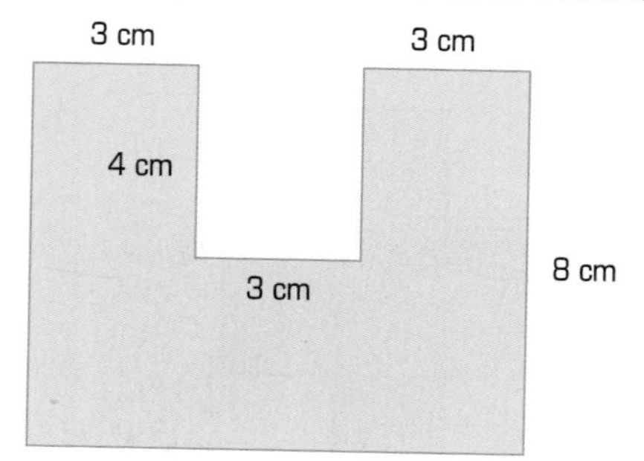

3 These rectangles are the same:

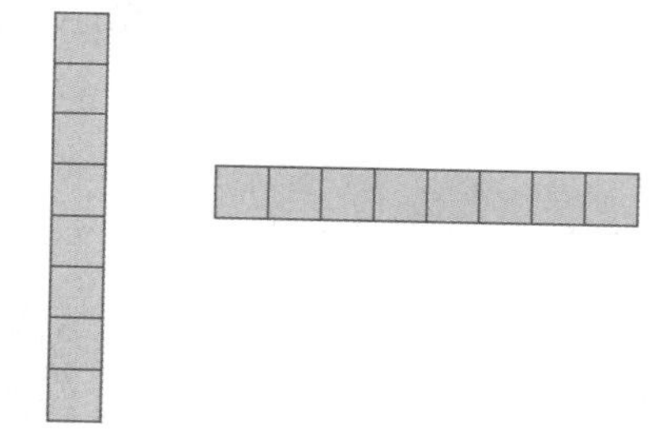

With eight squares you can make two different rectangles:

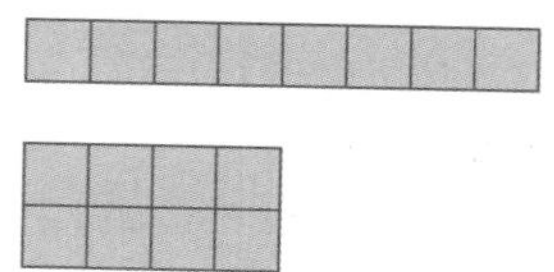

What is the smallest number of squares that can be arranged to make four different rectangles?

N2 Fractions, decimals and percentages

This unit will show you how to:

- Use fraction notation, including mixed numbers, and the vocabulary numerator and denominator.
- Change an improper fraction to a mixed number.
- Recognise relationships between fractions, including relating hundredths to tenths.
- Reduce a fraction to its simplest form by cancelling.
- Begin to add and subtract simple fractions.
- Relate fractions to division, and use division to find simple fractions, including tenths and hundredths, of numbers and quantities.
- Use a fraction as an 'operator' to find fractions.
- Use decimal notation for tenths and hundredths.
- Relate fractions to their decimal representations.
- Understand percentage as the number of parts in every 100.
- Express one half, one quarter, three quarters, and tenths and hundredths, as percentages.

This cheese is cut into twelve equal pieces, or twelfths.

Before you start

You should know how to …

1 Mentally multiply and divide numbers through recall of the 10 × 10 tables.

2 Add and subtract mentally.

Check in

1 Write down the answers to these:

a 3 × 4 **b** 15 ÷ 3 **c** 5 × 5
d 6 × 3 **e** 28 ÷ 4 **f** 7 × 6

2 Work out:

a 21 + 9 + 5 **b** 35 − 7 **c** 22 + 25
d 37 + 16 **e** 63 − 15

N2.1 Understanding fractions

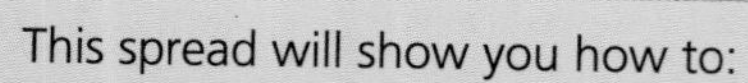

This spread will show you how to:

- Use fraction notation and the vocabulary numerator and denominator.
- Use fraction notation to describe parts of shapes.

KEYWORDS

Denominator Numerator
Divide
Fraction

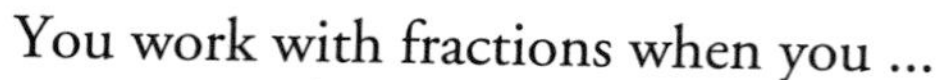

You work with fractions when you ...

... share a pizza

This is 1 out of the 6 pieces.

... colour part of a shape blue

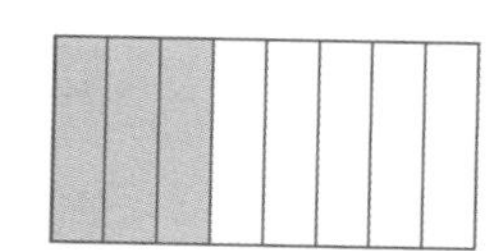

3 out of 8 parts are blue.

... tell the time.

It's half past 5.

- **A fraction describes part of a whole. The whole must be divided into equal parts.**

example

What fraction of this shape is shaded?

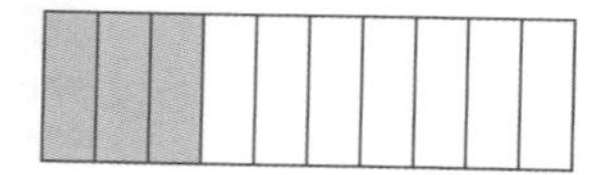

The rectangle is divided into 10 equal parts
3 parts are shaded.
$\frac{3}{10}$ of the shape is shaded.

- In a fraction:

You can write one number as a fraction of another when the numbers use the same units.

example

What fraction of £1 is 32p?

Make the units the same: £1 = 100p
The fraction is $\frac{32}{100}$.

Exercise N2.1

1 Write down the fraction of each shape that is shaded.

a 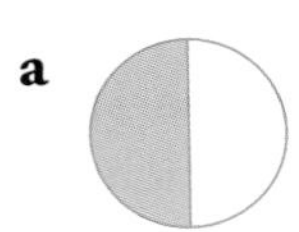 **b** **c** **d**  **e** 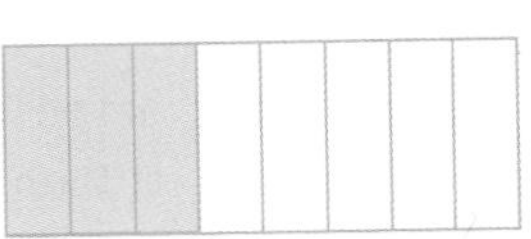 **f** 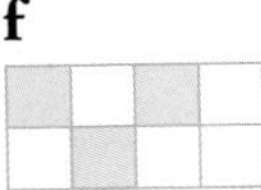

2 Copy these shapes and shade in the fraction given.

a 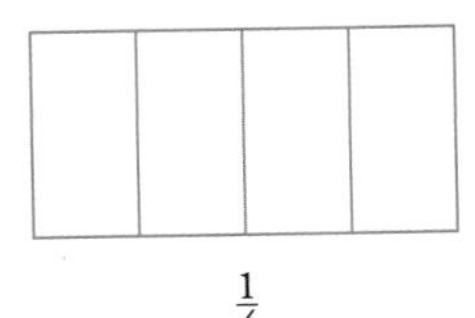
$\frac{1}{4}$

b 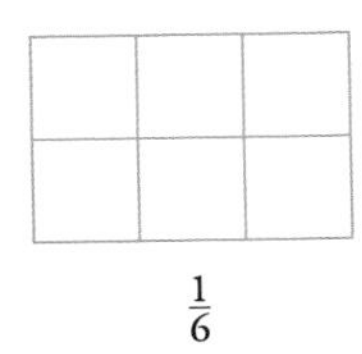
$\frac{1}{6}$

c 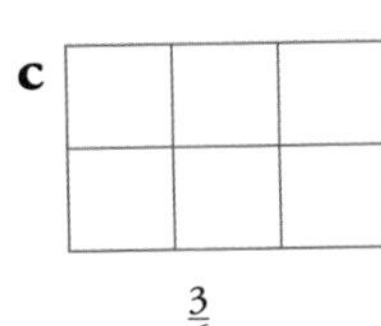
$\frac{3}{6}$

3 Draw a 4 × 4 square grid 20 times.
Find as many ways as you can to shade:
a $\frac{1}{2}$ of the squares **b** $\frac{3}{4}$ of the squares
Compare your results with your classmates.

4 Write these amounts as fractions of 10p:
a 1p **b** 7p **c** 5p **d** 4p **e** 6p **f** 8p

5 Write these distances as fractions of 25 cm:
a 1 cm **b** 11 cm **c** 24 cm **d** 15 cm **e** 20 cm **f** 5 cm

6 Write these amounts as fractions of £1:
a 71p **b** 32p **c** 9p **d** 99p **e** 80p **f** 50p

7 Write these amounts as fractions of 1 hour:
a 30 mins **b** 15 mins **c** 45 mins **d** 20 mins **e** 50 mins **f** 80 mins

8 Copy this shape four times and shade in the fraction given.
a $\frac{1}{2}$ **b** $\frac{1}{4}$ **c** $\frac{1}{8}$ **d** $\frac{3}{8}$

9 For each of these shapes:
- Make sure the shape is split into **equal sized parts.**
- Write down how many equal sized parts there are.
- Write down how many equal sized parts are shaded.
- Write down the fraction of each shape that is shaded.

a **b** **c**

d 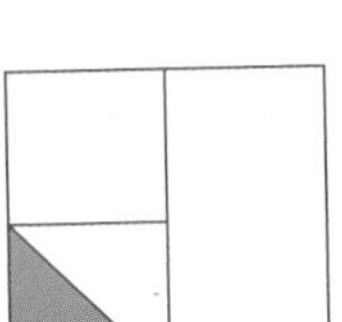 **e**  **f** 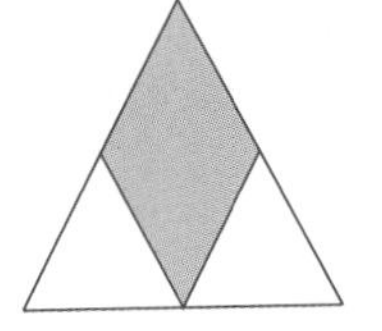

N2.2 Equivalent fractions

This spread will show you how to:

- Recognise when two simple fractions are equivalent.
- Reduce a fraction to its simplest form by cancelling.
- Compare and order simple fractions by writing them with the same denominator.

KEYWORDS

Denominator Numerator
Equivalent Fraction
Simplest form

In this shape, 2 out of the 4 squares are shaded.

You can say $\frac{2}{4}$ is shaded.

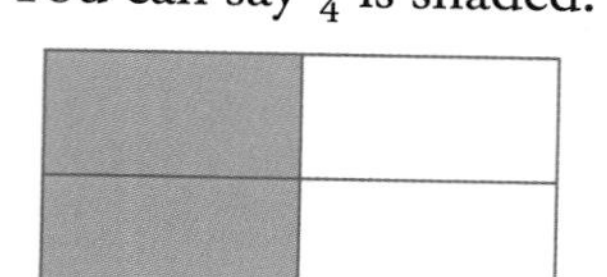

You can also say that $\frac{1}{2}$ is shaded.

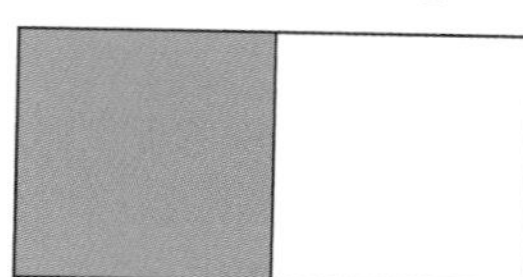

$\frac{1}{2}$ and $\frac{2}{4}$ show the same fraction of the shape. The fractions $\frac{1}{2}$ and $\frac{2}{4}$ are equivalent.
You can write: $\frac{1}{2} = \frac{2}{4}$

▸ **You find equivalent fractions by multiplying the numerator and denominator by the same number.**

example

Find three equivalent fractions for $\frac{1}{3}$:

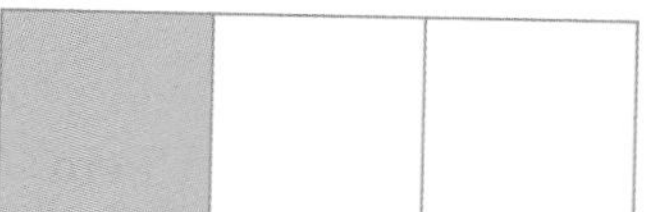

Multiply by 2: $\frac{1}{3} \overset{\times 2}{=} \frac{2}{6}$

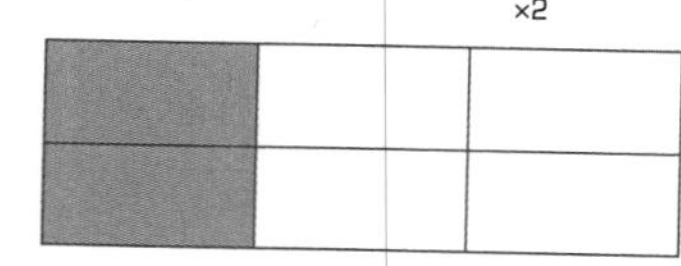

Multiply by 3: $\frac{1}{3} \overset{\times 3}{=} \frac{3}{9}$

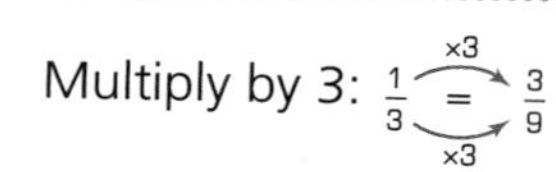

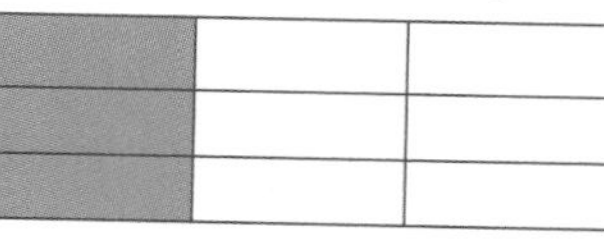

Multiply by 4: $\frac{1}{3} \overset{\times 4}{=} \frac{4}{12}$

▸ **You simplify fractions by dividing the numerator and denominator by the same number.**

example

Simplify:

a $\frac{3}{12}$

b $\frac{12}{18}$

a Divide by 3:

$\frac{3}{12} \overset{\div 3}{=} \frac{1}{4}$

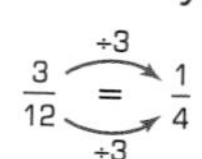

The fraction is now in its simplest form.

b Divide by 2: $\frac{12}{18} \overset{\div 2}{=} \frac{6}{9}$

Divide by 3: $\frac{6}{9} \overset{\div 3}{=} \frac{2}{3}$

$\frac{12}{18} = \frac{6}{9} = \frac{2}{3}$.

Exercise N2.2

1 For each of these shapes write down:

- the fraction that is shaded
- the fraction that is unshaded

a 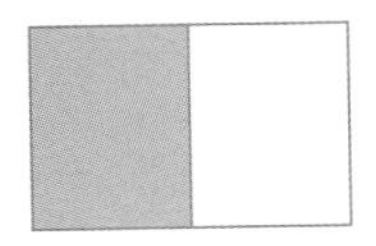

b 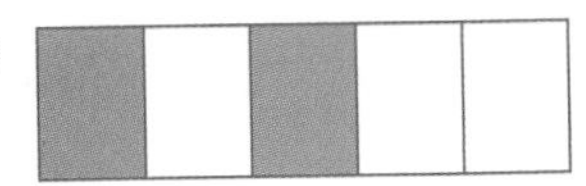

c 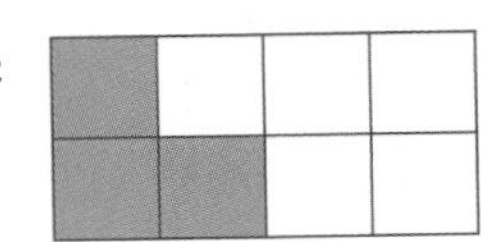

d 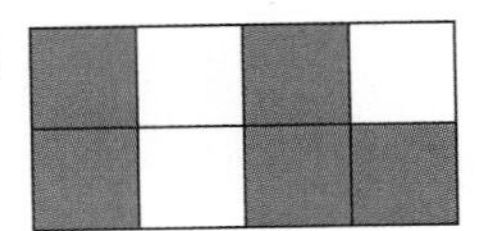

2 Copy and complete the equivalent fractions for the shaded parts.

a 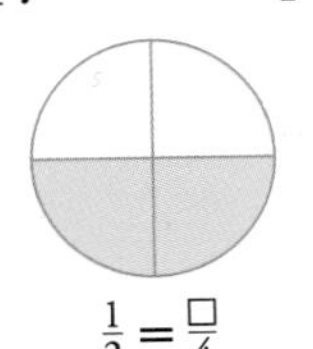

$\frac{1}{2} = \frac{\square}{4}$

b 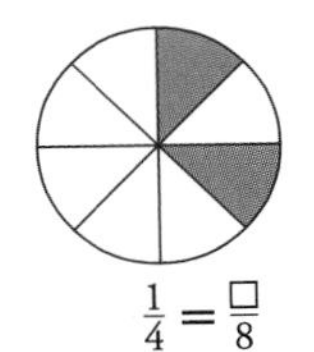

$\frac{1}{4} = \frac{\square}{8}$

c 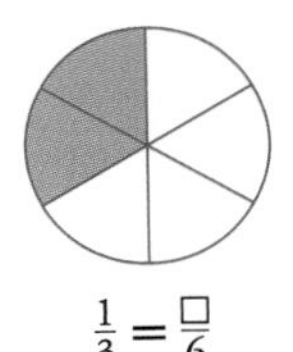

$\frac{1}{3} = \frac{\square}{6}$

d 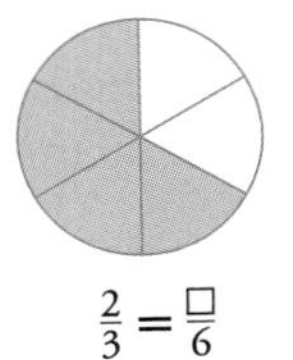

$\frac{2}{3} = \frac{\square}{6}$

3 Use two equivalent fractions to describe the shaded parts of these shapes.

a 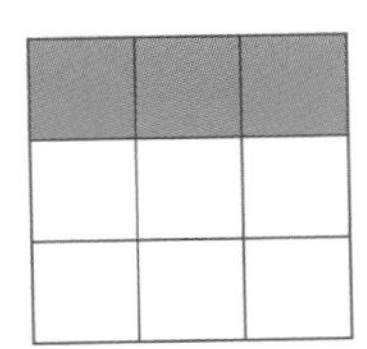

b 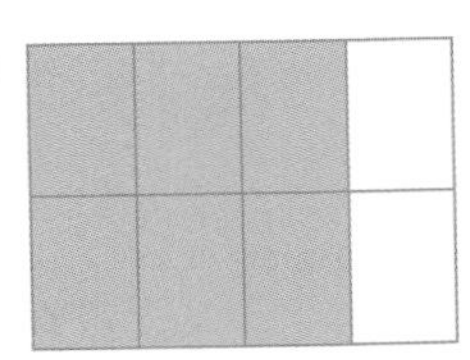

c 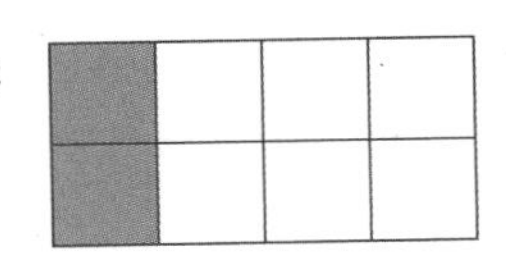

d 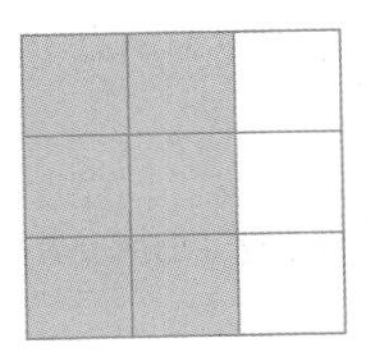

4 Use two equivalent fractions to describe the unshaded parts in question 3.

5 Write these fractions in order, starting with the smallest.

$\frac{7}{8}$ $\frac{5}{8}$ $\frac{1}{8}$ $\frac{2}{8}$ $\frac{6}{8}$

6 Copy and complete these equivalent fraction calculations:

a 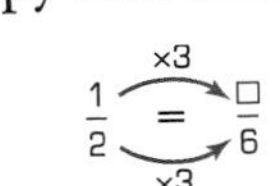

$\frac{1}{2} \xrightarrow{\times 3} \frac{\square}{6}$

b 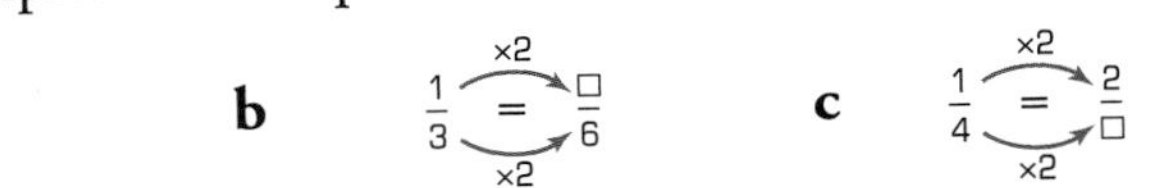

$\frac{1}{3} \xrightarrow{\times 2} \frac{\square}{6}$

c 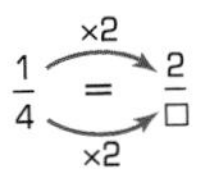

$\frac{1}{4} \xrightarrow{\times 2} \frac{2}{\square}$

d 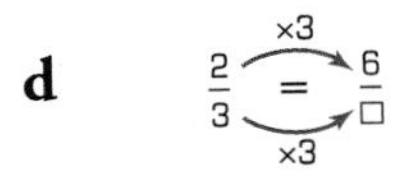

$\frac{2}{3} \xrightarrow{\times 3} \frac{6}{\square}$

e 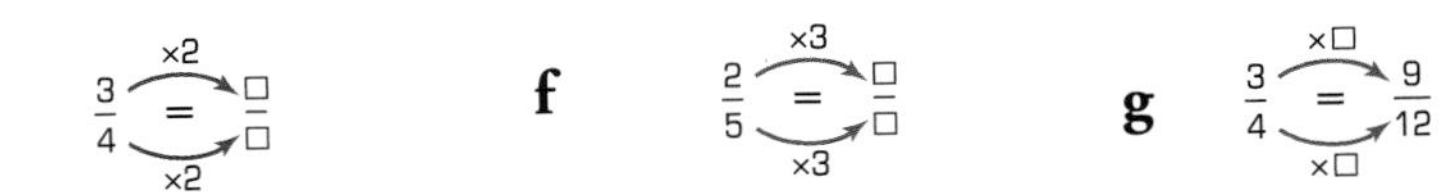

$\frac{3}{4} \xrightarrow{\times 2} \frac{\square}{\square}$

f 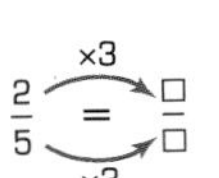

$\frac{2}{5} \xrightarrow{\times 3} \frac{\square}{\square}$

g 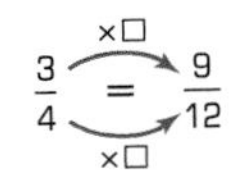

$\frac{3}{4} \xrightarrow{\times \square} \frac{9}{12}$

h 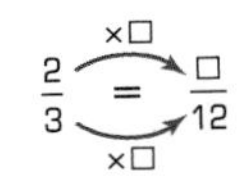

$\frac{2}{3} \xrightarrow{\times \square} \frac{\square}{12}$

7 Find three fractions equivalent to each of these fractions:

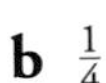
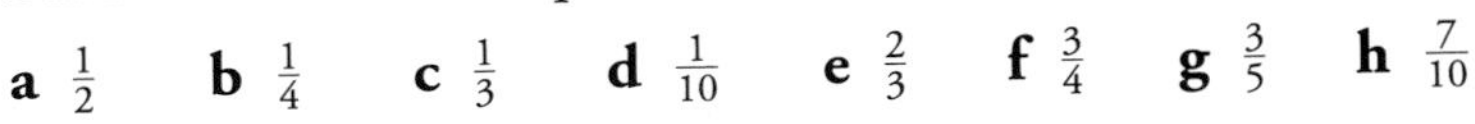

a $\frac{1}{2}$ **b** $\frac{1}{4}$ **c** $\frac{1}{3}$ **d** $\frac{1}{10}$ **e** $\frac{2}{3}$ **f** $\frac{3}{4}$ **g** $\frac{3}{5}$ **h** $\frac{7}{10}$

8 Use your answers to question **7e** and **7f** to say which is larger: $\frac{2}{3}$ or $\frac{3}{4}$?

9 Write these fractions in order, starting with the smallest.

$\frac{1}{3}$ $\frac{1}{5}$ $\frac{1}{2}$ $\frac{1}{4}$ $\frac{1}{6}$

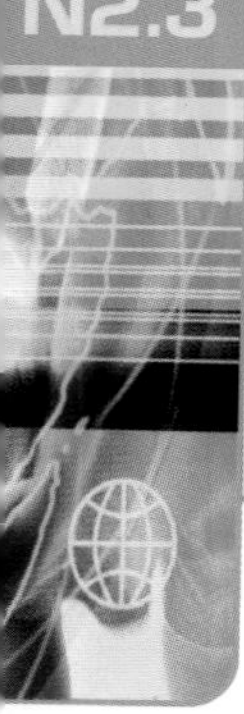

N2.3 Adding and subtracting fractions

This spread will show you how to:

- Begin to add and subtract simple fractions.
- Change improper fractions to mixed numbers.

KEYWORDS
Denominator Improper
Numerator
Mixed number

- **You name a fraction by its denominator, that is its fraction family:**

$\frac{1}{2}$s are halves — $\frac{1}{2}$ is one half
$\frac{1}{3}$s are thirds — $\frac{2}{3}$ is two thirds
$\frac{1}{4}$s are quarters — $\frac{3}{4}$ is three quarters

- **You add fractions with the same denominator by adding the numerators.**

For example, $\frac{3}{6} + \frac{2}{6} = \frac{5}{6}$.

example

Add these fractions and simplify where possible.

a $\frac{1}{8} + \frac{1}{8} + \frac{1}{8}$ **b** $\frac{1}{6} + \frac{1}{6} + \frac{1}{6}$

a Imagine a bar with 8 parts:

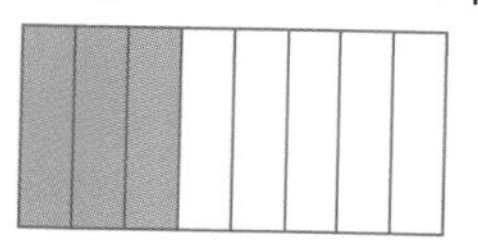

$\frac{1}{8} + \frac{1}{8} + \frac{1}{8} = \frac{3}{8}$

b Imagine a bar with 6 parts:

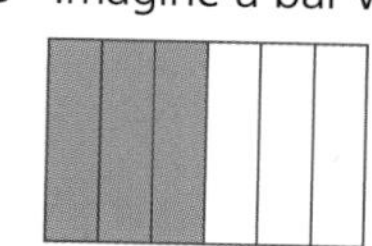

$\frac{1}{6} + \frac{1}{6} + \frac{1}{6} = \frac{3}{6}$

You can simplify: $\frac{3}{6} = \frac{1}{2}$

You subtract fractions in the same way:

$\frac{3}{6} - \frac{1}{6} = \frac{2}{6}$

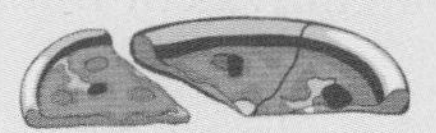

Sometimes you see fractions and whole numbers mixed ...

- **$2\frac{1}{2}$ is a mixed number.**
  You can change it into a single fraction like this:

$2\frac{1}{2} = 1 + 1 + \frac{1}{2}$
$2\frac{1}{2} = \frac{2}{2} + \frac{2}{2} + \frac{1}{2}$
$2\frac{1}{2} = \frac{5}{2}$

1000 m is $2\frac{1}{2}$ times round the running track.

$\frac{5}{2}$ is an improper or top heavy fraction.

Exercise N2.3

1 Write these fractions using words:

a $\frac{1}{2}$ **b** $\frac{1}{4}$ **c** $\frac{1}{3}$ **d** $\frac{1}{6}$ **e** $\frac{2}{3}$ **f** $\frac{3}{4}$

g $\frac{5}{8}$ **h** $\frac{11}{12}$ **i** $1\frac{1}{4}$ **j** $2\frac{2}{3}$ **k** $3\frac{5}{16}$ **l** $13\frac{2}{9}$

2 Write these fractions using numbers:

a three fifths **b** nine tenths **c** seven eighths

d two sevenths **e** one twelfth **f** seven twentieths

3 Work out the answer to each of these fraction problems.

a $\frac{1}{3}+\frac{1}{3}$ **b** $\frac{1}{5}+\frac{1}{5}+\frac{1}{5}$ **c** $\frac{1}{7}+\frac{1}{7}+\frac{1}{7}+\frac{1}{7}$

d $\frac{1}{8}+\frac{1}{8}+\frac{1}{8}$ **e** $\frac{1}{5}+\frac{1}{5}+\frac{1}{5}-\frac{1}{5}$ **f** $\frac{1}{4}+\frac{1}{4}+\frac{1}{4}-\frac{1}{4}-\frac{1}{4}$

g $\frac{1}{2}-\frac{1}{2}+\frac{1}{2}-\frac{1}{2}$ **h** $\frac{1}{3}-\frac{1}{3}+\frac{1}{3}+\frac{1}{3}$ **i** $\frac{3}{4}-\frac{1}{4}-\frac{1}{4}$

4 Work out the answer to each of these fraction problems.

a $\frac{3}{8}+\frac{2}{8}$ **b** $\frac{1}{9}+\frac{3}{9}$ **c** $\frac{2}{7}+\frac{2}{7}+\frac{1}{7}$

d $\frac{7}{9}-\frac{5}{9}$ **e** $\frac{4}{5}-\frac{1}{5}$ **f** $\frac{7}{8}-\frac{3}{8}+\frac{1}{8}$

g $\frac{3}{5}+\frac{2}{5}-\frac{4}{5}$ **h** $\frac{3}{6}+\frac{1}{6}-\frac{3}{6}$ **i** $\frac{2}{10}+\frac{5}{10}-\frac{3}{10}+\frac{5}{10}$

5 Work out the answer to each of these fraction problems.
Simplify your answers if possible.

a $\frac{1}{8}+\frac{1}{8}+\frac{1}{8}+\frac{1}{8}$ **b** $\frac{1}{6}+\frac{1}{6}+\frac{2}{6}$ **c** $\frac{7}{8}-\frac{5}{8}$

d $\frac{9}{10}-\frac{7}{10}$ **e** $\frac{7}{12}-\frac{3}{12}$ **f** $\frac{3}{12}+\frac{3}{12}+\frac{4}{12}$

g $\frac{6}{9}-\frac{3}{9}$ **h** $\frac{3}{16}+\frac{5}{16}-\frac{4}{16}$ **i** $\frac{11}{14}-\frac{4}{14}$

6 Change each of these mixed numbers into single fractions.

a $1\frac{1}{2}$ **b** $1\frac{1}{3}$ **c** $1\frac{3}{4}$ **d** $1\frac{5}{6}$ **e** $1\frac{1}{8}$ **f** $2\frac{1}{3}$

g $2\frac{1}{2}$ **h** $3\frac{1}{4}$ **i** $2\frac{7}{8}$ **j** $2\frac{5}{9}$ **k** $2\frac{1}{10}$ **l** $3\frac{5}{7}$

7 Change these fractions into mixed numbers.

a $\frac{6}{5}$ **b** $\frac{7}{4}$ **c** $\frac{9}{2}$ **d** $\frac{11}{3}$ **e** $\frac{16}{5}$ **f** $\frac{21}{5}$

8 Add these mixed numbers.
Simplify your answers if possible.

a $1\frac{1}{3}+1\frac{1}{3}$ **b** $2\frac{1}{5}+2\frac{1}{5}$ **c** $2\frac{3}{8}+1\frac{3}{8}$ **d** $4\frac{1}{4}+1\frac{3}{4}$ **e** $1\frac{1}{2}+2\frac{1}{2}$ **f** $4\frac{1}{6}+1\frac{5}{6}$

9 Copy and complete each of these sums. The first one is done for you.

a $\frac{1}{3}+\frac{1}{6}$
$=\frac{2}{6}+\frac{1}{6}$
$=\frac{3}{6}=\frac{1}{2}$

b $\frac{1}{2}+\frac{1}{4}$
$=\frac{\square}{4}+\frac{\square}{4}$
$=\frac{\square}{4}$

c $\frac{1}{4}+\frac{1}{8}$
$=\frac{\square}{8}+\frac{1}{8}$
$=\frac{\square}{8}$

N2.4 Fractions and decimals

This spread will show you how to:

- Use decimal notation for tenths and hundredths.
- Recognise the equivalence between decimal and fractional forms.

KEYWORDS
Convert
Equivalent
Decimal fraction

A fraction describes part of a whole.
A decimal also describes part of a whole.

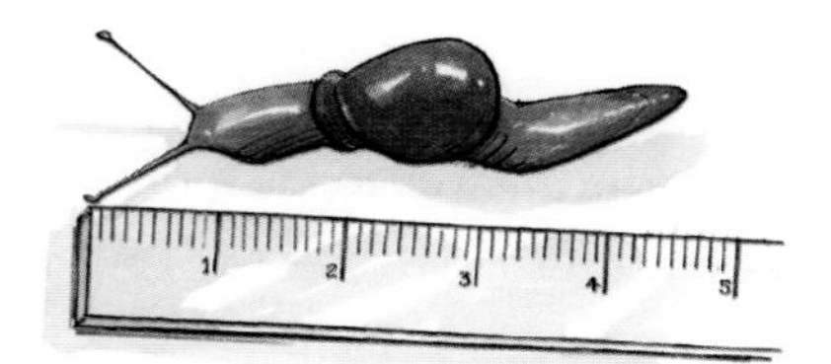

The snail is $4\frac{1}{2}$ cm or 4.5 cm long

You can write decimals as fractions.

0.6 and 0.25 are decimal fractions.

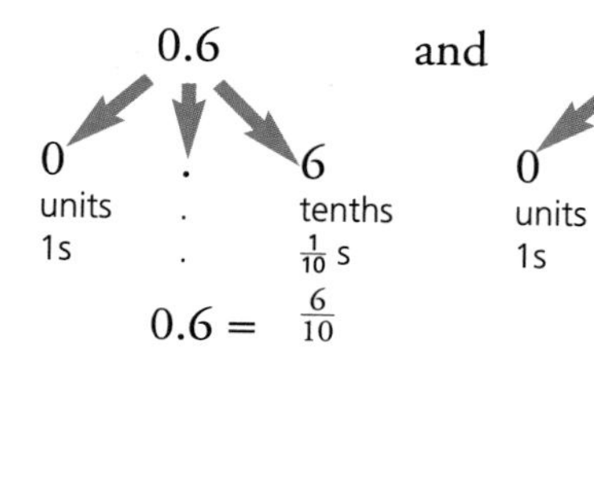

0.25: 0 units (1s) . 2 tenths ($\frac{1}{10}$ s) 5 hundredths ($\frac{1}{100}$ s)

$0.25 = \frac{2}{10} + \frac{5}{100}$

$0.25 = \frac{20}{100} + \frac{5}{100}$

$0.25 = \frac{25}{100} = \frac{1}{4}$

$\frac{2}{10} \overset{\times 10}{=} \frac{20}{100}$ (×10)

You can convert a decimal to a fraction on a number line.

▶ **You should know these common fractions and their decimal equivalents:**

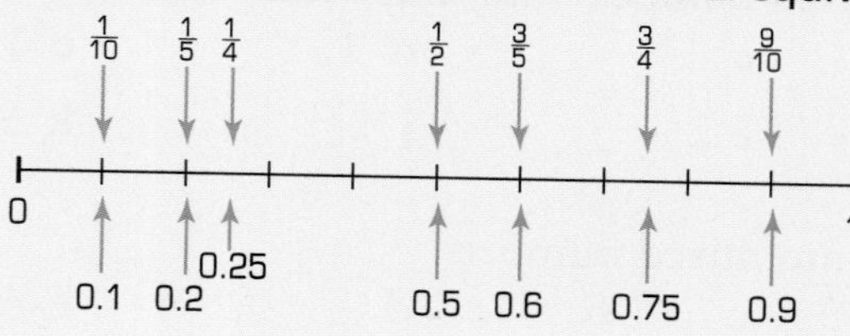

You can use these facts to convert other fractions to their decimal equivalents.

example

Convert to decimals:

a $\frac{2}{5}$ **b** $\frac{7}{10}$ **c** $\frac{23}{100}$

a $\frac{1}{5} = 0.2$
so $\frac{2}{5} = 0.2 + 0.2 = 0.4$

b $\frac{7}{10}$ is 0.7

c $\frac{23}{100} = \frac{20}{100} + \frac{3}{100}$
so $\frac{23}{100} = \frac{2}{10} + \frac{3}{100}$
so $\frac{23}{100} = 0.23$

Exercise N2.4

1 Convert these decimals into fractions.

a 0.5 **b** 0.9 **c** 0.2 **d** 0.7 **e** 0.25 **f** 0.75

2 Convert these fractions into their decimal equivalents.

a $\frac{1}{10}$ **b** $\frac{2}{10}$ **c** $\frac{3}{10}$ **d** $\frac{6}{10}$ **e** $\frac{1}{5}$ **f** $\frac{2}{5}$ **g** $\frac{3}{5}$ **h** $\frac{3}{4}$

3 This number line is split into tenths.

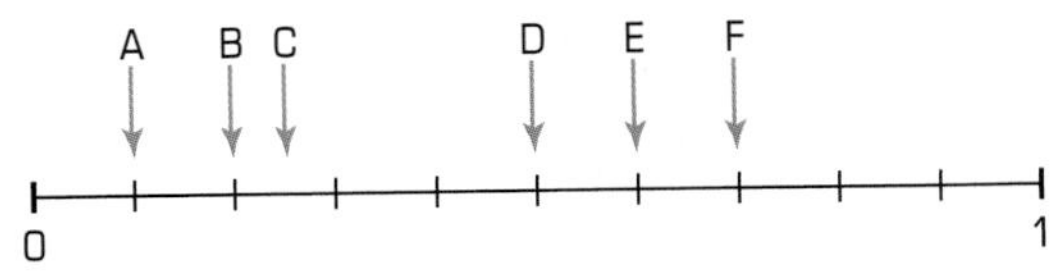

a Match each of these fractions and decimals to the points indicated on the number line:

0.5 $\frac{1}{10}$ 0.2 $\frac{7}{10}$ 0.6 $\frac{1}{4}$

b Write each fraction as its decimal equivalent, and each decimal as its fraction equivalent.

4 Put these fractions and decimals in order from lowest to highest.

$\frac{1}{2}$, $\frac{9}{10}$, 0.4, $\frac{3}{5}$, $\frac{7}{10}$, 0.25

5 Complete these sums. Here is an example:

$0.15 = \frac{1}{10} + \frac{5}{100}$

a $0.35 = \frac{3}{10} + \frac{\square}{100}$ **b** $0.24 = \frac{2}{10} + \frac{\square}{100}$ **c** $0.73 = \frac{\square}{10} + \frac{3}{100}$ **d** $0.81 = \frac{\square}{10} + \frac{\square}{100}$

e $0.9 = \frac{\square}{10}$ **f** $0.08 = \frac{\square}{100}$ **g** $0.07 = \frac{7}{\square}$ **h** $0.99 = \frac{9}{\square} + \frac{9}{\square}$

6 Use the numbers in the boxes to make up fractions and decimals that are equal to each other.

a $\frac{\square}{\square} = \square.\square$ [1 10 / 0 1]

b $\frac{\square}{\square} = \square.\square$ [3 3 / 0 10]

c $\frac{\square}{\square} = \square.\square$ [2 4 / 0 5]

d $\frac{\square}{\square} = \square.\square\square$ [1 0 5 / 2 4]

e $\frac{\square}{\square} = \square.\square\square$ [0 4 / 3 5 7]

f $\frac{\square}{\square\square\square} = \square.\square\square$ [7 0 0 7 / 1 0 0]

7 Put the correct sign between each pair of numbers. Choose from: > < =
The first one is done for you:

a $\frac{1}{2} > 0.25$ **b** 0.5 $\frac{3}{4}$ **c** 0.7 $\frac{3}{10}$ **d** $\frac{1}{5}$ 0.15

e $\frac{1}{4}$ 0.4 **f** 0.1 $\frac{1}{100}$ **g** 0.2 $\frac{1}{5}$ **h** 0.1 $\frac{1}{10}$

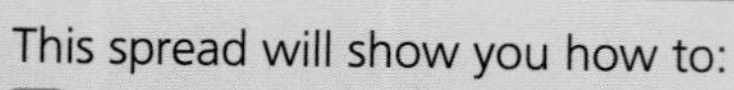

Fractions of an amount

This spread will show you how to:

- Relate fractions to division.
- Use a fraction as an operator to find fractions of quantities.

KEYWORDS

Fraction
Integer
Equivalent
Whole number

Liam has 10 pieces of chocolate. He eats half of them.
$\frac{1}{2}$ of 10 is 5 pieces.
He eats 5 pieces.

- Finding $\frac{1}{2}$ of an amount is the same as dividing the amount into 2 equal parts. $\frac{1}{2}$ of an amount = amount ÷ 2.
- $\frac{1}{5}$ of an amount = amount ÷ 5.

example

a Find $\frac{1}{10}$ of £420

b Calculate $\frac{1}{6}$ of 2.4 m

a $\frac{1}{10}$ of £420 is £420 ÷ 10
= £42

b $\frac{1}{6}$ of 2.4 m is 2.4 m ÷ 6
= 0.4 m

Stella needs to work out $\frac{3}{10}$ of £320.

She knows that $\frac{3}{10}$ is $\frac{1}{10} + \frac{1}{10} + \frac{1}{10}$

This means that $\frac{3}{10}$ is 3 lots of $\frac{1}{10}$ or $3 \times \frac{1}{10}$

$\frac{3}{10}$ of £320 is $3 \times \frac{1}{10}$ of £320
= 3 × £320 ÷ 10
= 3 × £32
= £96

example

Calculate

a $\frac{2}{3}$ of £60

b $\frac{5}{8}$ of 32 m

c $\frac{9}{10}$ of 40 kg

a $\frac{1}{3}$ of £60 is £60 ÷ 3 = £20
$\frac{2}{3}$ of £60 is $2 \times \frac{1}{3}$ of £60

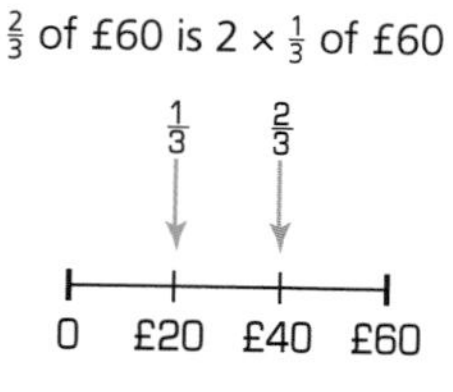

b $\frac{1}{8}$ of 32 m is 32 m ÷ 8 = 4 m
$\frac{5}{8}$ of 32 m is $5 \times \frac{1}{8}$ of 32 m

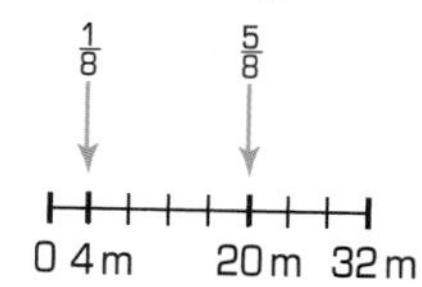

c $\frac{1}{10}$ of 40 kg is 40 kg ÷ 10 = 4 kg
so $\frac{9}{10}$ of 40 kg is $9 \times \frac{1}{10}$ of 40 kg

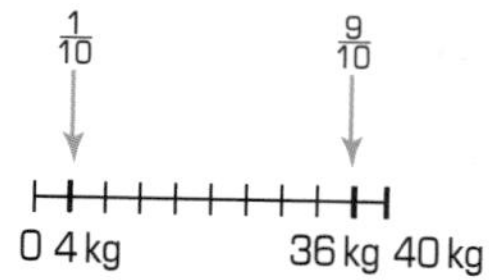

Exercise N2.5

1 Write down the answers to these questions.

a $\frac{1}{2}$ of 20 **b** $\frac{1}{2}$ of 8 **c** $\frac{1}{2}$ of 16 **d** $\frac{1}{2}$ of 100 **e** $\frac{1}{2}$ of 7 **f** $\frac{1}{2}$ of 23

2 Write down the answers to these questions.

a $\frac{1}{3}$ of 9 **b** $\frac{1}{4}$ of 8 **c** $\frac{1}{3}$ of 15 **d** $\frac{1}{3}$ of 30 **e** $\frac{1}{4}$ of 12 **f** $\frac{1}{4}$ of 20

3 Work out the answers to these problems.

a $\frac{1}{2}$ of 30 **b** $\frac{1}{3}$ of 45 **c** $\frac{1}{6}$ of 18 **d** $\frac{1}{7}$ of 21 **e** $\frac{1}{8}$ of 24 **f** $\frac{1}{10}$ of 60

4 There are sixty minutes in an hour. Work out how many minutes there are in these fractions of one hour.

a $\frac{1}{2}$ **b** $\frac{1}{4}$ **c** $\frac{1}{3}$ **d** $\frac{1}{5}$ **e** $\frac{1}{10}$ **f** $\frac{1}{20}$

5 There are 100 cm in a metre. Work out how many centimetres there are in these fractions of one metre.

a $\frac{1}{2}$ **b** $\frac{1}{4}$ **c** $\frac{1}{10}$ **d** $\frac{1}{5}$ **e** $\frac{1}{100}$ **f** $\frac{1}{20}$

6 Work out the answers to these problems.

a $\frac{2}{3}$ of 9 **b** $\frac{3}{4}$ of 12 **c** $\frac{3}{4}$ of 20 **d** $\frac{2}{3}$ of 30 **e** $\frac{2}{5}$ of 30 **f** $\frac{3}{5}$ of 25

7 Work out how many minutes are in these fractions of one hour.

a $\frac{3}{4}$ **b** $\frac{2}{3}$ **c** $\frac{3}{10}$ **d** $\frac{5}{6}$ **e** $\frac{7}{20}$ **f** $\frac{5}{12}$

8 Work out how many centimetres are in these fractions of one metre.

a $\frac{3}{10}$ **b** $\frac{3}{4}$ **c** $\frac{4}{5}$ **d** $\frac{7}{20}$ **e** $\frac{21}{50}$ **f** $\frac{4}{25}$

9 Work out the following amounts of money.

a $\frac{2}{3}$ of £6 **b** $\frac{3}{4}$ of 80p **c** $\frac{5}{7}$ of 35p **d** $\frac{2}{3}$ of £90 **e** $\frac{4}{5}$ of £250 **f** $\frac{7}{10}$ of £2000

10 Jenny has £5. She spends two fifths of this on magazines.
How much money does she spend?

11 Jim has £2. He spends three fifths of it on sweets.
How much does he spend?

12 Sherry has £1.80. She spends $\frac{2}{9}$ of it on a comb.

a How much does she spend?

b How much does she have left?

13 Which is the bigger amount of money?

$\frac{3}{4}$ of £20 or $\frac{2}{5}$ of £30

Show how you worked it out.

Fractions, decimals and percentages

This spread will show you how to:

- Understand percentage as the number of parts in every 100.
- Express simple fractions as percentages.

KEYWORDS

Convert
Percentage
Equivalent
Decimal fraction

Another way of describing part of a whole is by using a percentage.

You can use the symbol % to mean per cent.

Per cent means per 100.

- A percentage is a fraction written as a number of parts per 100.
For example: 60% = $\frac{60}{100}$ and 25% is $\frac{25}{100}$.

example

Write as a fraction:

a 60%

b 25%

a 60% means $\frac{60}{100}$

60%

20% 40% 80% 100%

0 1

$\frac{60}{100} = \frac{6}{10} = \frac{3}{5}$

$60\% = \frac{3}{5}$

b 25% means $\frac{25}{100}$

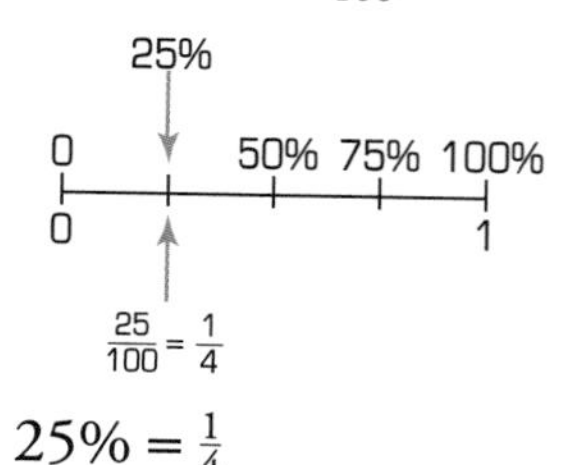

$25\% = \frac{1}{4}$

- You can convert a fraction to a percentage by writing it as an equivalent fraction with a denominator of 100.
For example: $\frac{3}{4} = \frac{75}{100} = 75\%$

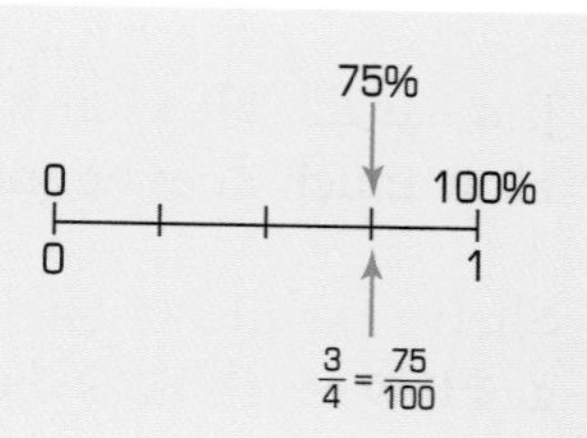

These number lines show fractions, decimals and percentages:

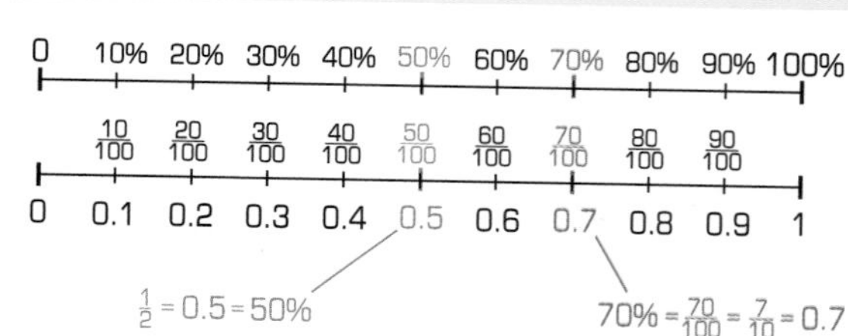

Exercise N2.6

1 Write the following percentages as fractions out of 100.

a 7% **b** 17% **c** 23% **d** 59% **e** 31% **f** 89% **g** 97% **h** 10%

2 Write these percentages as fractions, and cancel them down to their simplest form.

a 50% **b** 10% **c** 20% **d** 60% **e** 25% **f** 75% **g** 80% **h** 15%

3 Write the following fractions as percentages.

a $\frac{13}{100}$ **b** $\frac{20}{100}$ **c** $\frac{99}{100}$ **d** $\frac{50}{100}$ **e** $\frac{1}{2}$ **f** $\frac{1}{10}$ **g** $\frac{3}{10}$ **h** $\frac{1}{5}$

4 Convert these percentages into decimals.

a 20% **b** 25% **c** 50% **d** 80% **e** 10% **f** 75% **g** 17% **h** 37%

5 Convert these decimals into percentages.

a 0.25 **b** 0.75 **c** 0.32 **d** 0.81 **e** 0.4 **f** 0.9 **g** 0.01 **h** 0.07

6 This number line is split into 10 equal parts.

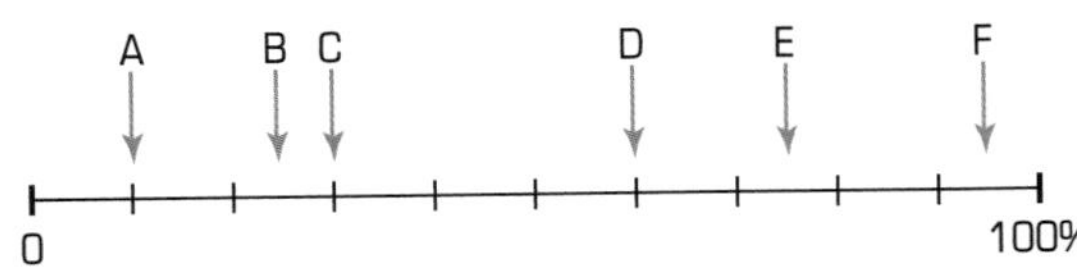

Match each of these fractions, decimals and percentages to the letters on the number line.

$\frac{3}{5}$ 10% $\frac{3}{4}$ 0.3 95% 0.25

7 This diagram shows the equivalences of 25%:

$25\% = \frac{25}{100} =$ → 0.25 ; → $\frac{1}{4}$

Complete these equivalences.

a $18\% = \frac{\square}{100} =$ → $\square.\square\square$; → $\frac{\square}{50}$

b $20\% = \frac{20}{\square} =$ → $\square.\square$; → $\frac{\square}{5}$

c $75\% = \frac{75}{\square} =$ → $\square.\square\square$; → $\frac{3}{\square}$

d $70\% = \frac{\square}{100} =$ → $\square.\square$; → $\frac{\square}{10}$

8 Use the numbers in the boxes to find equivalent fractions, decimals and percentages.

a $\square.\square\square = \square\square\%$ [7 7 3 / 0 3]

b $\square.\square\square = \square\%$ [8 8 / 0 0]

c $\square\square\% = \frac{\square}{\square\square}$ [3 7 / 2 0 5]

d $\square.\square = \frac{\square}{\square}$ [0 2 / 5 4]

N2 Summary

You should know how to ...

1 Use fractions as an operator to find fractions.

2 Simplify fractions by cancelling all common factors.

3 Use decimal notation for tenths and hundredths.

4 Understand % as parts per 100.

Check out

1 a Write down the fraction of each shape that is shaded.

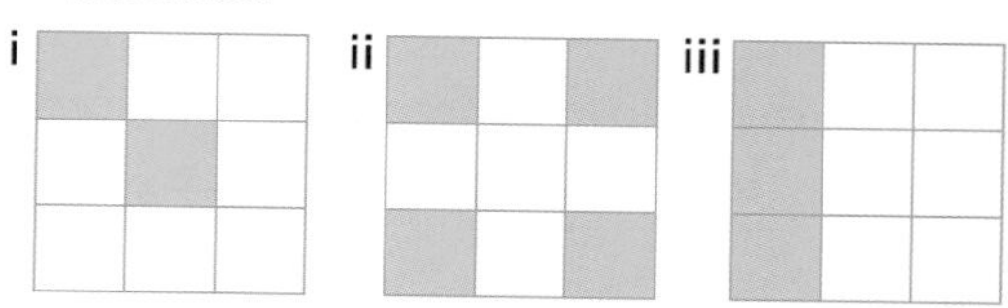

b Write these amounts as fractions of 20p:

i 7p ii 13p iii 11p
iv 5p v 10p vi 4p

c Write these distances as fractions of 2 metres:

i 1 m ii 50 cm iii 25 cm iv 40 cm

d Work these out:

i $\frac{1}{2}$ of 86p iv $\frac{1}{10}$ of 40p
ii $\frac{1}{4}$ of 60 cm v $\frac{3}{10}$ of 60p
iii $\frac{3}{4}$ of 60 cm vi $\frac{3}{5}$ of 60 minutes

2 a Copy and complete these equivalent fractions:

i $\frac{1}{2} = \frac{\square}{14}$ ii $\frac{1}{4} = \frac{\square}{12}$ iii $\frac{2}{3} = \frac{10}{\square}$

b Simplify each of these fractions:

i $\frac{3}{6} = \frac{\square}{2}$ ii $\frac{4}{8}$ iii $\frac{2}{6}$ iv $\frac{3}{9}$ v $\frac{4}{10}$

3 a Work out these fraction problems:

i $\frac{1}{7} + \frac{1}{7} + \frac{1}{7}$ iv $\frac{1}{8} + \frac{6}{8} - \frac{3}{8}$
ii $\frac{3}{8} + \frac{2}{8}$ v $\frac{2}{11} + \frac{5}{11} + \frac{3}{11} - \frac{6}{11}$
iii $\frac{5}{9} - \frac{3}{9}$ vi $\frac{3}{11} - \frac{4}{11} + \frac{6}{11}$

b Change these fractions into decimals:

i $\frac{1}{2}$ ii $\frac{1}{4}$ iii $\frac{1}{10}$ iv $\frac{3}{4}$ v $\frac{7}{10}$ vi $\frac{1}{5}$

4 Copy and complete this % decimal fraction table.

30%	0.3	
	0.4	$\frac{2}{5}$
	0.25	
7%		$\frac{7}{100}$

1 Statistics and probability

This unit will show you how to:

- Use the language associated with probability to discuss events, including those with equally likely outcomes.
- Solve a problem by interpreting data in tables, graphs, charts and diagrams.
- Find the mode of a set of data.
- Begin to find the median and mean of a set of data.

People are all different but the average person represents us all.

Before you start

You should know how to ...

1 Plot points on a coordinate grid.

2 Read scales.

Check in

1 Plot these points on a grid:
(2, 1) (2, 3) (4, 3) (4, 1)
Join them in order.
What shape do you make?

2 What reading do the arrows show on this scale?

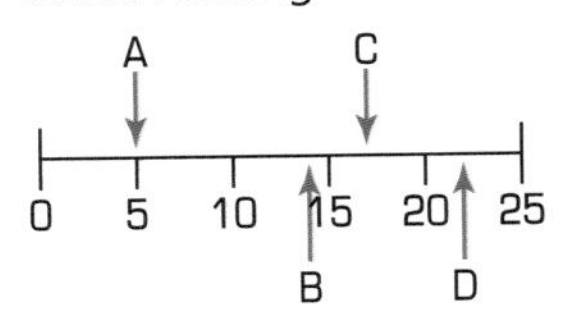

Finding the average

This spread will show you how to:
- Find the mode of a set of data.
- Begin to find the median of a set of data.

KEYWORDS
Average Median
Data Mode

You often see the word 'average' used:

An **average** is a single value that describes a set of data.
There are three different measures of average.
Here are two of them:

The mode

▶ **The mode is the value in the data that occurs most often.**

example

Woolly Warmers sell 12 jumpers one day.
The sizes sold are:

S	M	S	M	M	S
L	M	M	S	L	M

Key:
S = Small
M = Medium
L = Large

What is the modal size?

The modal size is the most common size.
They sell 4 small, 6 medium and 2 large jumpers.
Medium is the most common.
The modal size is Medium.

The median

▶ **The median is the value in the middle of the data.**
The data must be arranged in order first.

example

Find the median of these:

a shoe sizes: 3, 7, 4, 7, 5, 6, 4, 5, 6

b temperatures in °C: 21, 19, 24, 18, 18, 22

a arrange in size order: 3, 4, 4, 5, (5), 6, 6, 7, 7
The middle value is 5.
The median shoe size is 5.

b arrange in size order: 18, 18, (19, 21), 22, 24
There are two middle values: 19 and 21
The median is halfway between: 20
The median temperature is 20°C.

Exercise D1.1

1 The registration letters of cars for sale in a garage were:

T R V P W Y Y X V W V

What was the modal letter?

2 Find the mode for each of these sets of data:

a Shoe sizes: 5, 4, 10, 3, 3, 4, 7, 4, 6, 5

b Dress sizes: 10, 12, 10, 18, 12, 14, 12

c Hours of sunshine one week: 4, 5, 3, 4, 5, 0, 1

d Daily temperatures one week in °C: 17, 25, 22, 18, 22, 16, 19

3 Nine students wrote their shoe size on the board:

a Arrange the shoe sizes in order.

b What is the median shoe size?

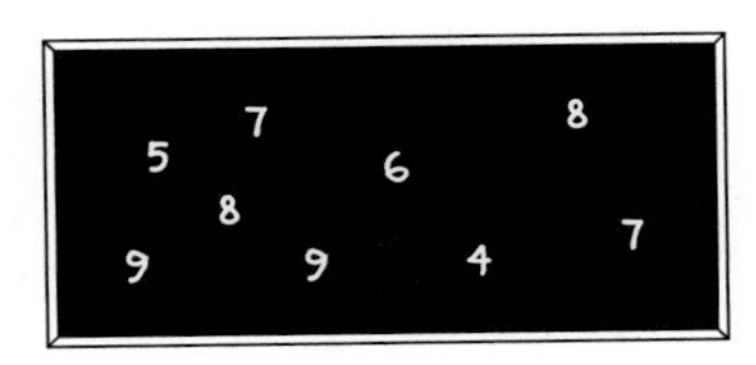

4 Arrange each of these sets of data in order and find the median:

a Shoe sizes: 3, 6, 9, 1, 2, 5, 8, 4, 7

b Dress sizes: 12, 10, 8, 14, 20, 16, 14, 16, 18

c Daily temperatures one week in °C: 18, 20, 16, 17, 22, 25, 24

5 Six students wrote their shoe size on the board:

a Arrange the shoe sizes in order.

b Find the halfway value.

c What is the median shoe size?

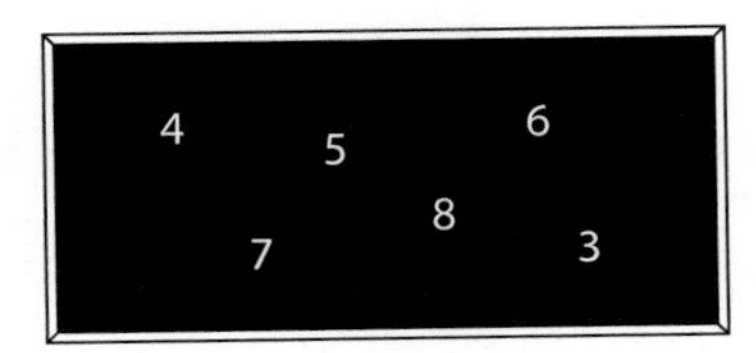

6 Find the median of each of these sets of data:

a Shoe sizes: 3, 6, 9, 1, 2, 5, 8, 4, 7, 3

b Temperatures one fortnight in °C: 18, 20, 16, 17, 22, 25, 24, 19, 22, 21, 23, 17, 20, 21

7 This table shows the length of five rivers:

a Write the rivers in order of size starting with the smallest.

b Write down the median length of the five rivers.

River	length (km)
Danube	2858
Mekong	4186
Nile	6695
Volga	3685
Yukon	3185

8 The median of this set of six lengths is 4.3 m.

4 m   4.7 m   ☐   4.1 m   4.2 m   4.5 m

What is the missing length?

The mean

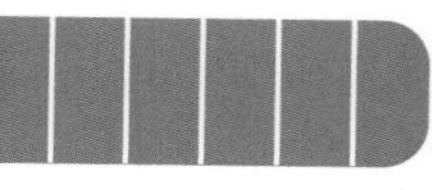

This spread will show you how to:

- Begin to find the mean of a set of data.

KEYWORDS

Average Mean
Data Value

There is a third average you can use – the **mean**.
It is what most people mean when they say average!

It is mean because you have to work it out!

Here are three towers:

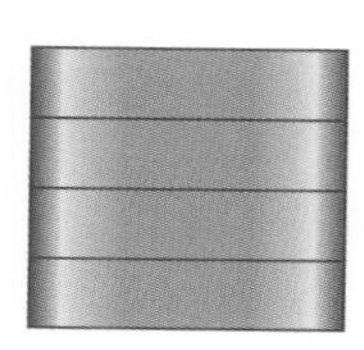
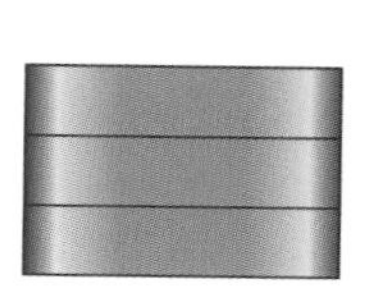
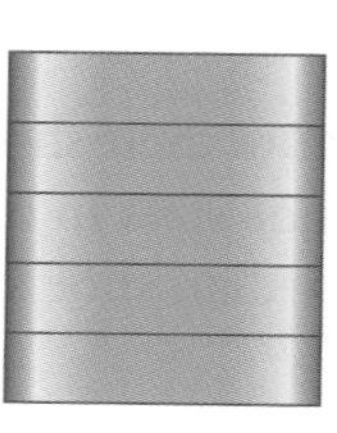

You can level out the towers so that they are all the same height:

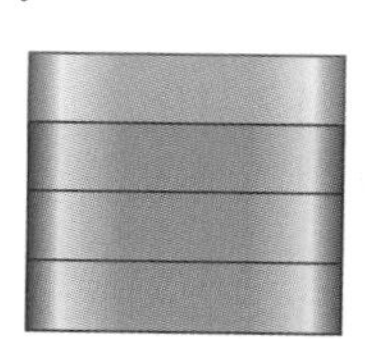
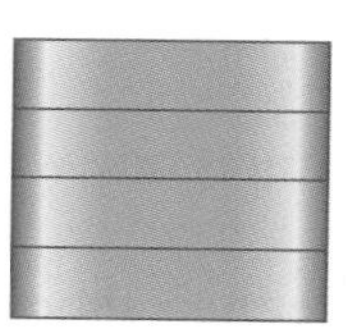

This is the mean height. The mean height is 4.

- **The mean of a set of data is the sum of all the values divided by the number of values of data there are.**

Unlike the median and the mode, the mean uses every piece of data.

example

Find the mean average of each set of data:

a Shoe sizes: 3, 6, 5, 4, 6, 4, 7

b Peas in a pod: 7, 9, 10, 4, 6, 3

a The number of values is: 7
The sum of the values is:
3 + 6 + 5 + 4 + 6 + 4 + 7 = 35

The mean = 35 ÷ 7
= 5

b The number of values is: 6
The sum of the values is:
7 + 9 + 10 + 4 + 6 + 3 = 39

The mean = 39 ÷ 6
= 6.5

Exercise D1.2

1 **a** Find the mean number of blocks for this set of towers:

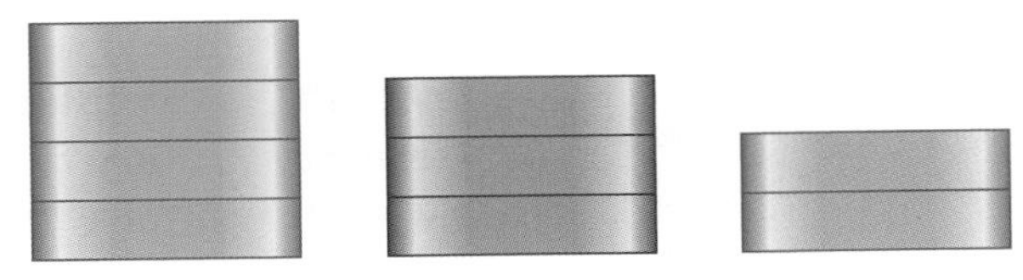

b Find the mean number of pencils in a group:

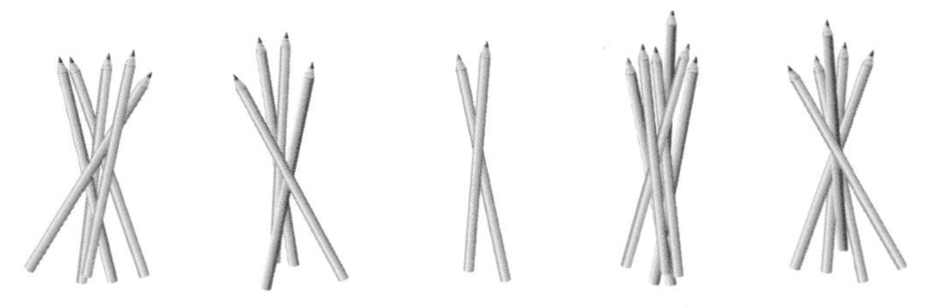

c Find the mean number of matches per box:

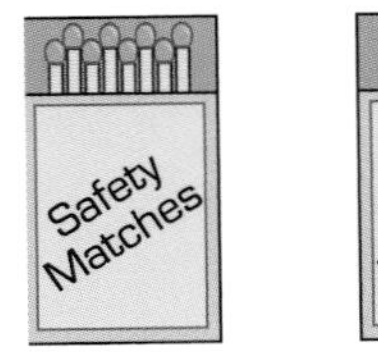

d Find the mean number of blocks:

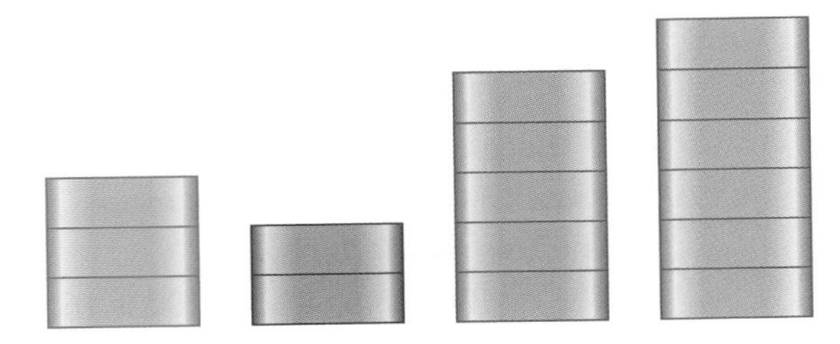

2 Here are four shapes made of cubes:

a What is the total number of cubes used?

b What is the mean number of cubes used?

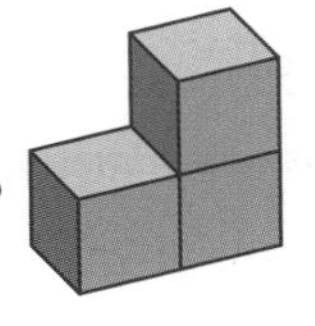
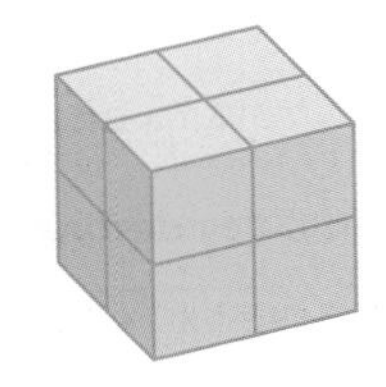
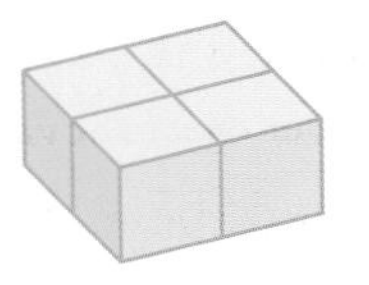
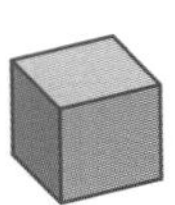

3 Find the mean average:

a number of chocolates in boxes containing: 10, 14, 9, 15

b length of fabric on rolls containing: 4 m, 5 m, 4 m, 2 m, 5 m

c price per kilogram of apples bought from shops charging:
98p, £1.02, 96p, £1.05, 95p, £1.04

4 Find the mean of each set of data. Use a calculator to help you.

a number of sweets in a packet: 29, 34, 24, 26, 27

b number of matches in a box: 49, 56, 39, 52, 47, 42, 29, 44, 54, 28

c money spent on food in a week: £78, £84, £86, £93, £49

D1.3 Interpreting diagrams

This spread will show you how to:

- Extract and interpret data in tables, graphs and charts.

KEYWORDS

Bar chart, Statistics, Frequency, Tally, Pie chart, Table

You often find statistics presented in diagrams.
This frequency table summarises the colour of students' hair in a class:

Colour of hair	Tally	Frequency
Brown	𝍸 𝍸 III	13
Black	IIII I	6
Blonde	𝍸 𝍸	10
Red	𝍸	5
Dyed	I	1

- Brown is the modal colour because it has the highest frequency.
- There are 35 students in the class – you find the total by adding the frequencies: $13 + 6 + 10 + 5 + 1 = 35$

You can show data from a frequency table on a chart or diagram:

- **A pie chart uses a circle to display data.**

Pie chart showing number of children in a family.

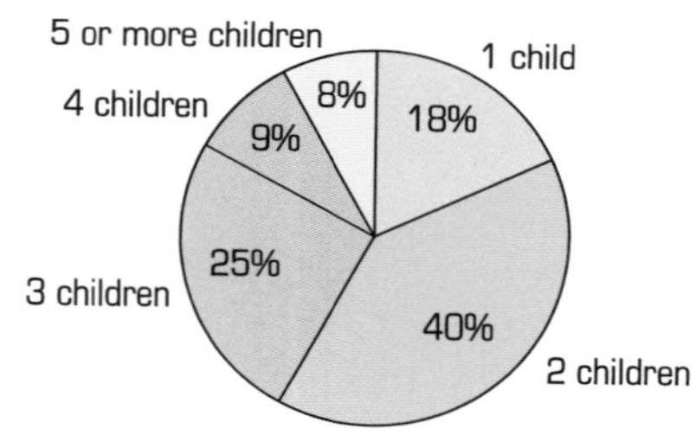

- The modal number is 2 children as this is the largest category.
- If there were 200 people in the survey then 16 of them had 5 or more children in the family: 8% of 100 is 8 so 8% of 200 is 16.

- **A bar chart uses bars to display data:**

Bar chart showing holiday destinations one summer:

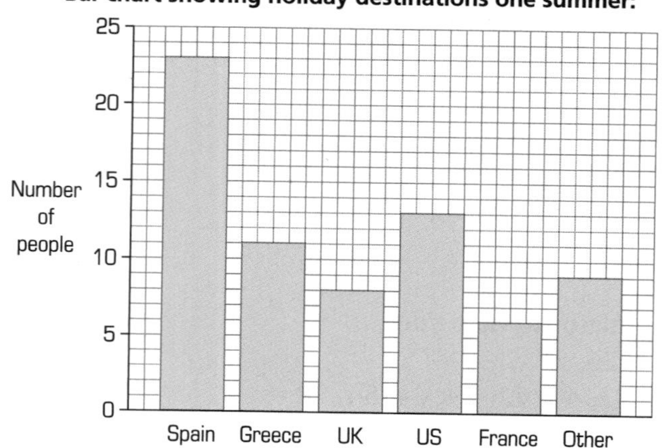

- The bar chart shows that most people in the survey went to Spain – that is the modal destination.
- To find the total number of people in the survey you add the frequencies, so add the heights of the bars: $23+11+8+13+6+9 = 70$.

Exercise D1.3

1 The pie chart shows the percentages of different film ratings in the *Movie and Video Guide*.

☆☆☆☆ 5%
No Stars 20%
☆☆☆ 20%
☆ 25%
☆☆ 30%

a What is the modal rating in the Guide?

b The Guide rated 100 films.
How many 2* films are there?

c How many of each rating are there if 1000 films are reviewed in the Guide?

d As a fraction, 20% is $\frac{1}{5}$.

20% of the films in the Guide were given no stars.

e What fraction of the films were given a rating of 3 stars or more?

2 This bar chart shows the number of mobile phone users in four secondary schools.

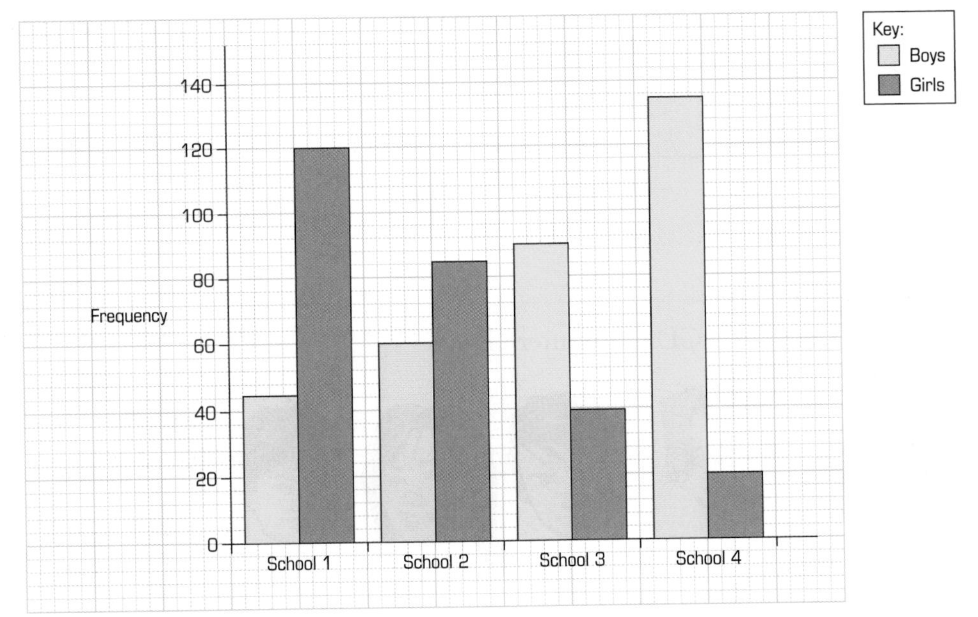

a Which school has the largest number of girls with mobile phones?

b Which school has the smallest total number of mobile phone users?

c In which school is there the biggest difference between the numbers of boys and girls who own mobile phones?

3 The pie chart shows the proportion of different types of room in two hotels.

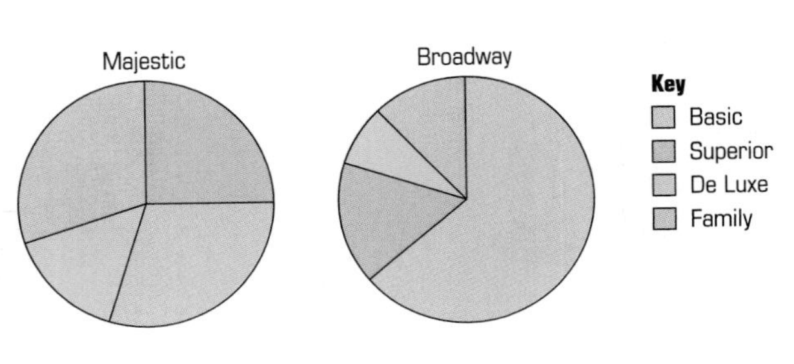

a Roughly what percentage of the rooms at the Majestic Hotel are *Basic*?

b Hanif says 'You can see that there must be more *Superior* rooms at the Majestic Hotel than at the Broadway.' Explain why Hanif is wrong.

D1.4 Introducing probability

This spread will show you how to:

- Use the language of probability to discuss events.

KEYWORDS

Certain, Likely, Chance, Probability, Impossible

The chance of an event happening is somewhere between certain and impossible:

Impossible ——→ Certain

Getting more likely

▶ **Probability** is a measure of the chance of an event happening. It describes how likely something is to happen.

The probability of pigs flying is 0 — The probability of the sun rising tomorrow is 1

Impossible — Certain

example

Six pupils enter a lucky draw for a CD Computer Game.

Judy

Jack

Fred

Sheila

Kemal

Steven

Their names are written on pieces of paper which are then put in a hat.
The name picked out of the hat wins the CD.
Who is more likely to win a prize – a girl, or Fred?

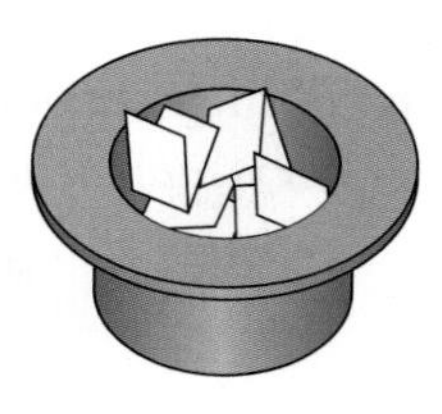

Fred has a 1 in 6 chance of being picked.

There are 2 girls.
There is a 2 in 6 chance of a girl being picked.

You can say a girl is more likely to be picked than Fred.

Exercise D1.4

1 Copy this probability line: Impossible ———————— Certain

Show these events on your line:

a You will have a birthday next year.

b A dog will have kittens.

c There will be snow next Christmas.

d It will be sunny next week.

2 **a** Write down the names of four animals.
Label them A, B, C and D.
Which do you think are more likely to hurt you?

b Put the animals in the order you think are the most likely to hurt you.

Least likely ——→ Most

3 You may get hurt playing these sports:
A: Soccer B: Rugby C: Cricket D: Tennis

a Are you more likely to get hurt playing A or B?

b Compare the other events in pairs. Write:
Getting hurt playing A is more likely than...

c Put the events in order, least likely to most likely.

4 Use the information in the example to answer these questions:

a What is the probability of Kemal being picked?
Copy and complete:
Kemal has a __ in __ chance of being picked.

b Order these events on a probability line:
A: Kemal gets picked.
B: A boy gets picked.
C: Either Jack or Sheila gets picked.
D: No-one wins the CD.

5 Five friends are trying to decide where to go on Friday.
They each write down their preferred choice on a card:

a Copy and complete this sentence:
There is a __ in __ chance they will go to the cinema.

b Write the probability they will go to the cinema as a fraction: $\frac{\square}{\square}$

c Order these events on a probability line:
A They go to the Club
B They to to the Cinema
C They go to the Theatre
D They do not go Bowling

D1.5 Calculating probabilities

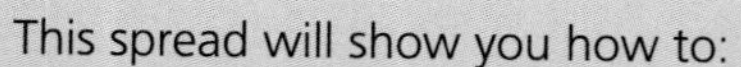

This spread will show you how to:

- Use the language of probability to discuss events, including those with equally likely outcomes.

KEYWORDS

Equally likely
Probability scale
Outcome
Random

- All probabilities are measured on a scale of 0 to 1:

Probabilities can be written as fractions.

This box contains the numbers 1, 2, 3, 4 and 5.
There are 5 possible outcomes.
Each of the outcomes 1, 2, 3, 4 and 5 is **equally likely**.
Each outcome has a 1 in 5 probability of being picked.

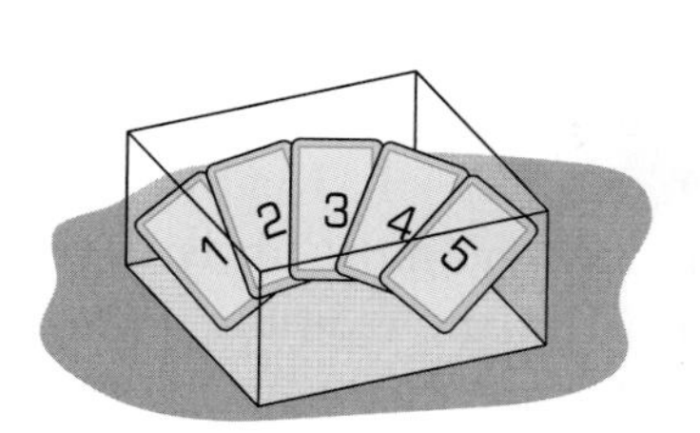

As a fraction:

- Probability of an event happening = $\dfrac{\text{Number of ways the event can happen}}{\text{Total number of possible outcomes}}$

example

Work out the probability of picking from the box:

a a 3 **b** an even number **c** a number greater than 1.

a There is only one 3.

There is a 1 in 5 probability of picking the 3.

The probability of picking a 3 is $\frac{1}{5}$.

b There are two even numbers.

There is a 2 in 5 probability of picking an even number.

The probability of picking an even number is $\frac{2}{5}$.

c There are 4 numbers greater than 1.

There is a 4 in 5 probability of picking a number greater than 1.

The probability of picking a number greater than 1 is $\frac{4}{5}$.

Exercise D1.5

1 The letters in the word APRIL are placed in a bag and one letter is picked out.

Copy and complete:

There are _____ possible outcomes.
Each outcome has a 1 in _____ chance of being picked.

What is the probability of picking out:

a an A? **b** an R? **c** a vowel? **d** a consonant?

Show your answers on a probability line.

2 An 8-sided spinner is used to play a game.

a How many possible outcomes are there from one spin?
b How many ways are there of getting an even number?
c What is the probability of getting an even number.

3 Joe shuts his eyes and presses one of the options on this drinks machine:

What is the probability of him getting:

a coffee **b** milk **c** hot chocolate **d** tea?

4 There are 10 tracks on a CD:
5 tracks have female singers,
3 have male singers,
the rest are dance tracks with no singing.
Martine presses the 'Random' button, which plays the tracks in a random order.
What is the probability that:

a The first track to play features a male singer?
b The first track does *not* have a female singer?
c The first track has no singing at all?

Experimental probability

This spread will show you how to:

- Use the language of probability to discuss events, including those with equally likely outcomes.
- Interpret data in tables.

KEYWORDS
Frequency chart
Experiment
Fair
Outcome
Tally

You can **calculate** the probability of an event happening:

- Probability of an event happening = $\frac{\text{Number of ways the event can happen}}{\text{Total number of possible outcomes}}$

You can **estimate** the probability of an event happening by carrying out an experiment.

example

Rachel and Paul think the dice they have been using is biased.
To check they throw the dice 50 times and record the outcome.

Here are their results:

Dice score	Tally	Frequency
1	~~IIII~~ IIII	9
2	~~IIII~~ III	8
3	~~IIII~~ III	8
4	~~IIII~~ II	7
5	~~IIII~~ III	8
6	~~IIII~~ ~~IIII~~	10

a What is the most frequent score on the dice?
b What is the least frequent score?
c Estimate the probability of getting a 1 with this dice.
d Do you think the dice is biased? Explain.

---

a The table of results shows 6 is the most frequent outcome.
b The least frequent outcome is a 4.
c There were 9 scores of 1 out of a total of 50 scores so the estimated probability of getting a 1 is $\frac{9}{50}$.
d Even with a fair dice you would expect some differences in the scores. The differences are small so it is probably a fair dice.

Exercise D1.6

1 **Experiment 1: Test a drawing pin**

- Take a drawing pin
- Drop it on your table 50 times
- Record whether it lands 'point up' or 'point down'
- Record your results in a copy of this table:

Outcome	Tally	Frequency
Point up		
Point down		

Use your table to answer these questions:

a Which outcome seems to be the most likely?

b Estimate the probability of the pin landing point down

How could you get a more reliable estimate for the probability that the pin lands point down?

Experiment 2: Test a spinner

Make a rectangular spinner with the design shown:

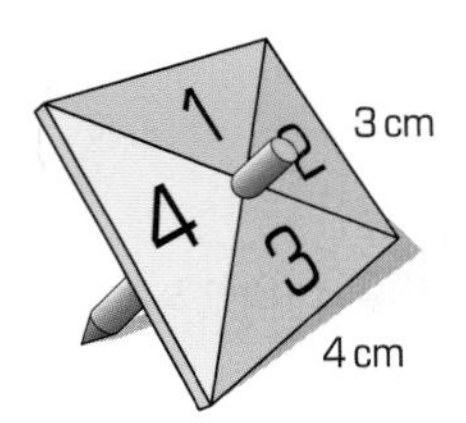

Now test your spinner:

- Spin it 100 times
- Record the results in a copy of this table:

Spinner score	Tally	Frequency
1		
2		
3		
4		

a Which score seems to be most likely?

b Which score seems to be least likely?

c Use your table to estimate the probability of each score.

d Do you think your spinner is fair? Explain.

Summary

You should know how to ...

Solve a problem by representing, extracting and interpreting data in tables, charts and diagrams.

Check out

Here are the ages of 10 people at a club:

18, 17, 17, 15, 16, 15, 18, 17, 15, 17

a Find the median age.

b Find the mode of the ages.

c Find the mean age.

d Show the data on a bar chart.

These 10 members of the club all enter a prize draw.

They each write their name on a ticket, and the winning ticket is picked out of a hat.

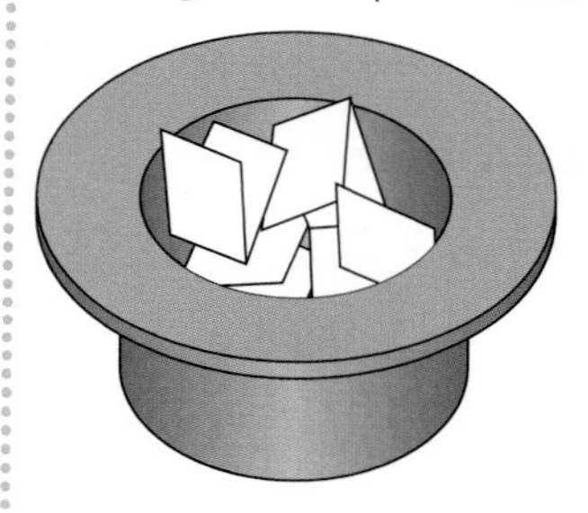

e What is the probability that the winner's age will be:

i 15

ii 18

iii 16 or 17

iv More than 14?

f Mark the answers on a probability scale. Explain in words how likely each of the four outcomes is. Choose the words from this list:

- Certain
- Evens
- Likely
- Highly likely
- Impossible
- Unlikely
- Highly unlikely

A2 Expressions and formulae

This unit will show you how to:

- Relate fractions to division.
- Use the relationship between addition and subtraction.
- Understand and use the relationships between the four operations, and the principles (not the names) of the arithmetic laws.
- Develop from explaining a generalised relationship in words to expressing it in a formula using letters as symbols.

A formula can help convert between measures.

Before you start

You should know how to ...

1 Know how to use the four rules of number $+ - \div \times$.

Check in

1 Answer the following:

a $\frac{20}{4} = \_$ **b** $2 \times 9 = \_$

c $\_ \times 6 = 30$ **d** $4 \times \_ = 32$

e $\frac{10}{5} = \_$ **f** $\frac{15}{?} = 3$

g $4 \times 2 = \_$ **h** $2 + 2 + 2 + 2 = \_$

i $3 + 3 + 3 = \_$ **j** $3 \times 5 = \_$

k $4 + 4 + 4 + 4 + 4 + 4 + 4 = \_$

l $7 \times 4 = \_$

m $9 + 9 + 9 + 9 + 9 + 9 + 9 + 9 + 9 + 9 = \_$

n $10 \times 9 = \_$

o $1 + 1 + 1 + 1 + 1 + 1 + 1 + 1 + 1 + 1 + 1 = \_$

p $11 \times 1 = \_$

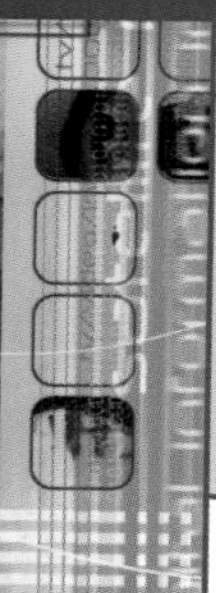

A2.1 Using letter symbols

This spread will show you how to:

- Understand the operation of multiplication and its relationship to addition.
- Develop from explaining a generalised relationship in words to expressing it in a formula using letters as symbols.

KEYWORDS

Algebra, Variable, Rule, Unknown, Value

You use a letter to stand for a value when ...

... the value is variable or ... the value is unknown.

There are c days to your birthday – c depends on when you read this.

There are g grains of sand on the beach – there are too many to count!

Using letters to represent values is called **algebra**.

Algebra is a shorthand way of writing unknowns or variables.

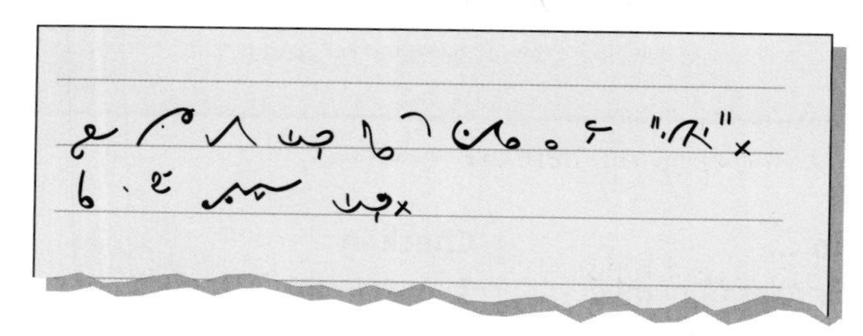

Like shorthand you must learn how to read and write algebra. Luckily there are only a few rules.

example

a Johnny eats d donuts a day for 5 days. How many does he eat altogether?

b There are b bricks in a wall. How many bricks are there in 6 identical walls?

a Johnny has 5 lots of d donuts. You could write $d + d + d + d + d$. This is $5 \times d$. In shorthand you write $5d$.

b There are 6 lots of b bricks. In shorthand this is $6b$.

Exercise A2.1

1 Use algebra shorthand to represent these sentences.
The first one is done for you.

a The number of cars in the world.
There are n cars in the world.

b The number of hairs on your head.
There are ___ hairs on your head.

c The number of words in a newspaper.

d The number of burgers in America.

e The number of fish in the sea.

f The number of stars in the sky.

g The number of blades of grass in a field.

h The number of plants in a garden.

i The number of books in the library.

j The number of internet users in the UK.

2 Use algebra shorthand to represent these sentences.
The first one is done for you.

a The number of hairs on 10 heads.
There are t hairs on 1 head, so there are $10t$ hairs on 10 heads

b The number of words in 5 newspapers.
There are ___ words in 1 newspaper, so there are ___ in 5 newspapers.

c The number of blades of grass in 5 fields.

d The number of plants in 6 gardens.

e The number of books in 20 libraries.

f The number of sweets in 20 tubes.

g The number of coins in 12 moneyboxes.

h The number of bricks in 8 houses.

i The number of houses in 35 streets.

j The number of leaves on 100 trees.

k The number of pairs of trainers in 5 sports shops.

3 If there are x leaves on a tree, and y trees in a forest, how many leaves are in the forest?

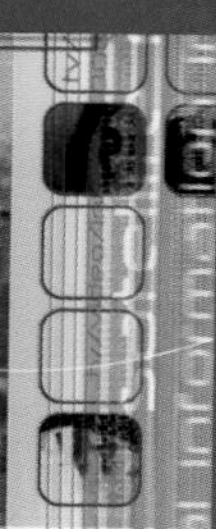

A2.2 Rules of algebra

This spread will show you how to:

- Develop from explaining a generalised relationship in words to expressing it in a formula using letters as symbols.

KEYWORDS

Algebra, Unknown, Expression, Value, Rule

Johnny usually eats d donuts a day.

He eats d donuts.

One Tuesday he gets an extra donut.

He eats $d + 1$ donuts!

You can write one more or one less than an unknown like this:

One less		Unknown		One more
$n - 1$	← -1	n	$+1$ →	$n + 1$

example

a There are s sweets in a tube.
You eat 3 sweets. How many are left?

b There are p pieces of gum left in a pack.
You buy another pack of 10. How many do you have now?

---

a You start with s and take away 3.
You have 3 less than s.
You write $s - 3$.

b You start with p then add 10 to it.
You have 10 more than p.
You write $p + 10$.

$s - 3$ and $p + 10$ are expressions.

- An **expression** is a sentence that uses letters and numbers.

Exercise A2.2

1 Write an expression for each of these sentences.
Use n to stand for the number of counters.

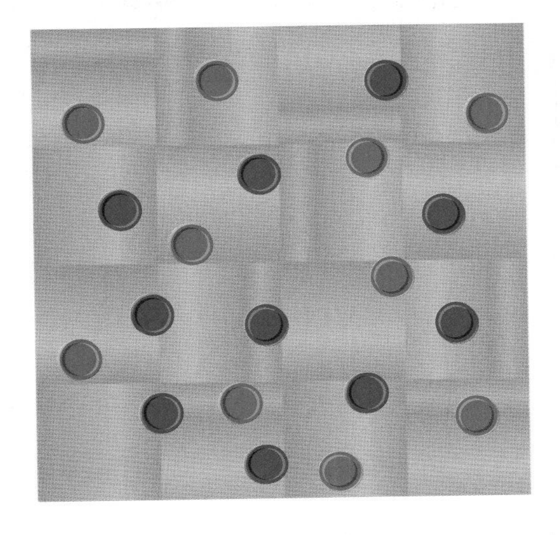

a I have some counters.
I drop 4 on the floor.
I now have __ counters.

b I have some counters.
I find 6 counters on the floor.
I now have __ counters.

c I have some counters.
I give 20 counters to a teacher.
I now have __ counters.

d I have some counters.
I find 3 more in the cupboard.
I now have __ counters.

e I have some counters.
I drop five counters.
I now have __ counters.

f I have some counters.
I find 10 counters on the floor.
I now have __ counters.

g I had some counters.
I buy 3 more.
I now have __ counters.

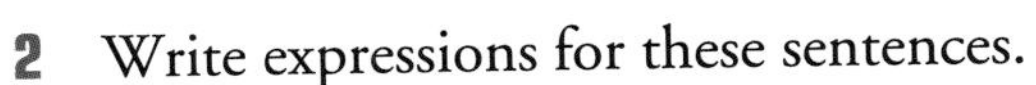

2 Write expressions for these sentences.

a Add 5 to an unknown number.
b Take 5 away from an unknown number.
c Take an unknown number away from 10.
d 5 minus an unknown number.
e Subtract an unknown number from 20.
f Add an unknown number to 10.
g Add an unknown number to 3.
h Add 7 to an unknown number.

3 The cost of a chocolate bar is c pence.
The cost of a can of pop is d pence.

a Which of these expressions give the total cost of 4 chocolate bars and 3 cans of pop?

c + 3	4d − c	4c + 3d	4 + 3

b Write an expression for the cost of 6 cans of pop and 1 bar of chocolate.

c Write a sentence to describe the expression $2c + d$.

A2.3 Simplifying expressions

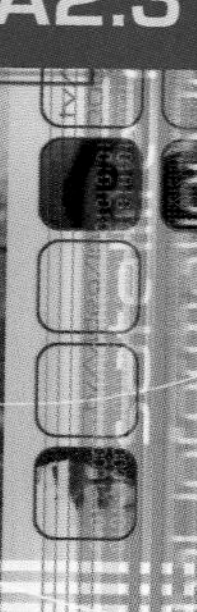

This spread will show you how to:

- Develop from explaining a generalised relationship in words to expressing it in a formula using letters as symbols.

KEYWORDS

Consecutive, Simplify, Expression, Term

A 100 square is organised in rows of 10.

1	2	3	4	5	6	7	8	9	10
11	12	13	14	15	16	17	18	19	20
21	22	23	24	25	26	27	28	29	30
31	32	33	34	35	36	37	38	39	40
41	42	43	44	45	46	47	48	49	50
51	52	53	54	55	56	57	58	59	60
61	62	63	64	65	66	67	68	69	70
71	72	73	74	75	76	77	78	79	80
81	82	83	84	85	86	87	88	89	90
91	92	93	94	95	96	97	98	99	100

$58 - 10 = 48$

$58 - 1 = 57$

$58 + 1 = 59$

$58 + 10 = 68$

Here is part of a 100 square. No one knows which part it is.
You can fix a place using a letter (x).
Once you fix one place in the 100 square you can fix them all.

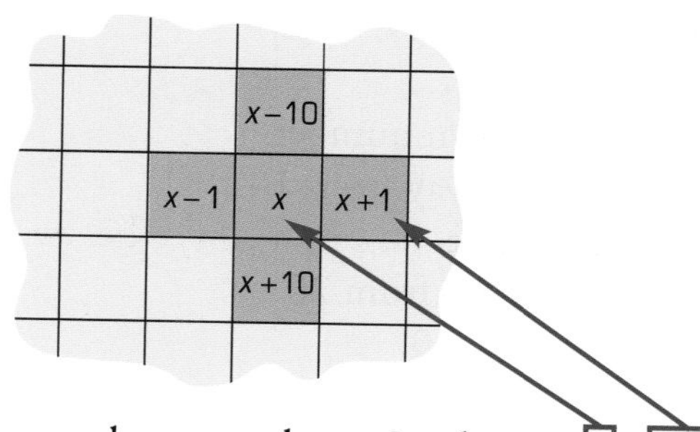

You can add consecutive unknown numbers together using letters: $x + x + 1$

Each part of the expression is called a **term**.
In $x + x + 1$ there are three terms: x, x and 1.

The x terms can be collected together:
They are called like terms because they both use x.

$x + x + 1$
$= 2x + 1$

- To simplify an expression you **collect like terms** together. Terms are **like** when they use the same letters.

Exercise A2.3

1 Simplify these expressions:

a $5 + 4 + 2 - 3$ **b** $12 - 3 + 4 - 2 + 10$

c $10 - 2 + 1 + 7$ **d** $10 - 1 + 6 + 5 - 3$

e $3 - 1 + 5$ **f** $2 + 3 + 4 - 9$

2 Simplify these expressions

a $x + x + x$ **b** $b + b + b + b$

c $y + y + 2y$ **d** $d + 2d + d + 2d$

e $t + t + t - t$ **f** $2f + 3f - f$

g $x + 4x + 3x + 6x$ **h** $2x + 5x + 2x - 4x + 6x$

i $7y - 3y + 5y + 3y$ **j** $12r - 3r - 2r - r$

k $10n - 4n + 2n$ **l** $5x + 5x + 5x + 5x + 5x + 5x$

m $5y + 3y - 10y$ **n** $2y + 2y + 5y - 2y$

3 Simplify these expressions by collecting like terms.

a $x + x + 3 + 2$

b $w + 2w + 1 + 2$

c $10y + 3y + 4 + 2$

d $17 + 3 + 13x - 2x$

e $3t + 5t - 2t + 10 + 3 - 2 - 1$

f $5x + 2x + 6 - 1$

g $5y - 2y + 3y + 10 - 3 + 6$

h $y + 2y + 3y + 1 + 3$

i $3x + 4x - 2 + 5$

j $w + 3 + 3w + 4 + 2w$

k $3t + 3 + 2t + 2$

l $3 + 2 + 4t + 10 + 6t - 2t$

4 Write an expression for the perimeter of the shape below.
All sides are cm.

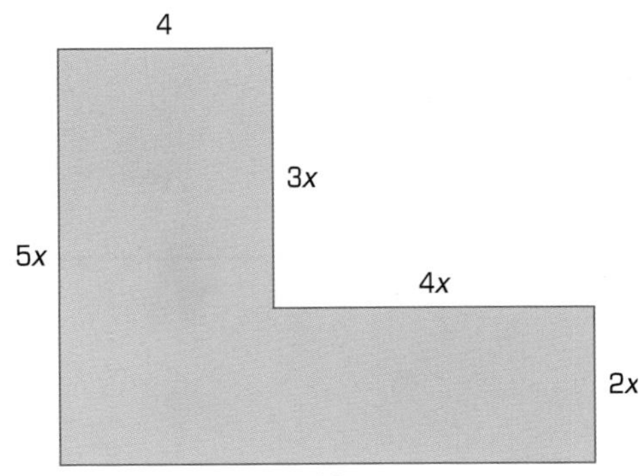

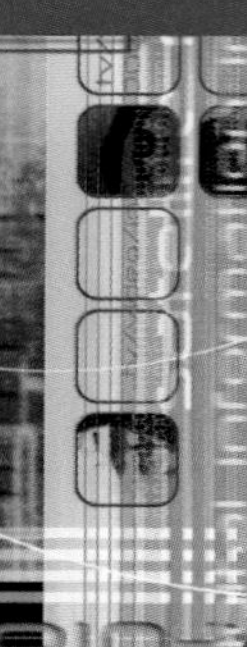

A2.4 Substitution

This spread will show you how to:

- Relate fractions to division.
- Understand the operation of multiplication and its relationship with addition.

KEYWORDS

Expression Value
Substitute Variable

These are the rules of algebra:

- $3d$ means $3 \times d$ or 3 lots of d
- In number you multiply before you add:

  $2 \times 5 + 1$
  $= 10 + 1 = 11$
- In algebra you multiply before you add:

  $2d + 1$ means double d then add 1.

$2d + 1$, $d - 3$ and $5d$ are expressions.
The value of d can vary.

If you know the value of d you can find the value of the expression.

- To **evaluate** an expression you **substitute** the value of the variable into the expression.

example

The letter d stands for the number showing after the throw of a dice.
Evaluate the expressions when the number on the dice (d) is 5:

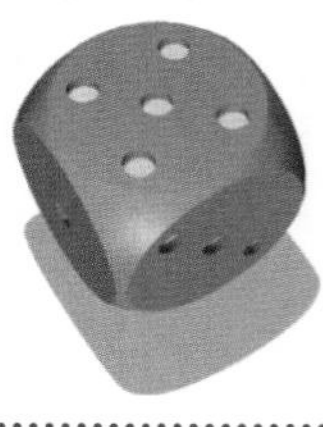

a $d + 3$ **b** $3d + 2$ **c** $2d - 1$ **d** $11 - d$

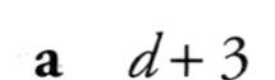

a $d + 3$
$= 5 + 3$
$= 8$

b $3d + 2$
$= 3 \times 5 + 2$
$= 15 + 2$
$= 17$

c $2d - 1$
$= 2 \times 5 - 1$
$= 10 - 1$
$= 9$

d $11 - d$
$= 11 - 5$
$= 6$

You need to use this rule to play the game in the exercise.

- $\frac{10}{2}$ means $10 \div 2$ so $\frac{d}{2}$ means $d \div 2$.

When d is 5, $\frac{d}{2} = d \div 2$
$= 5 \div 2 = 2.5$

Exercise A2.4

Play this game. The rules are in the centre of the grid.

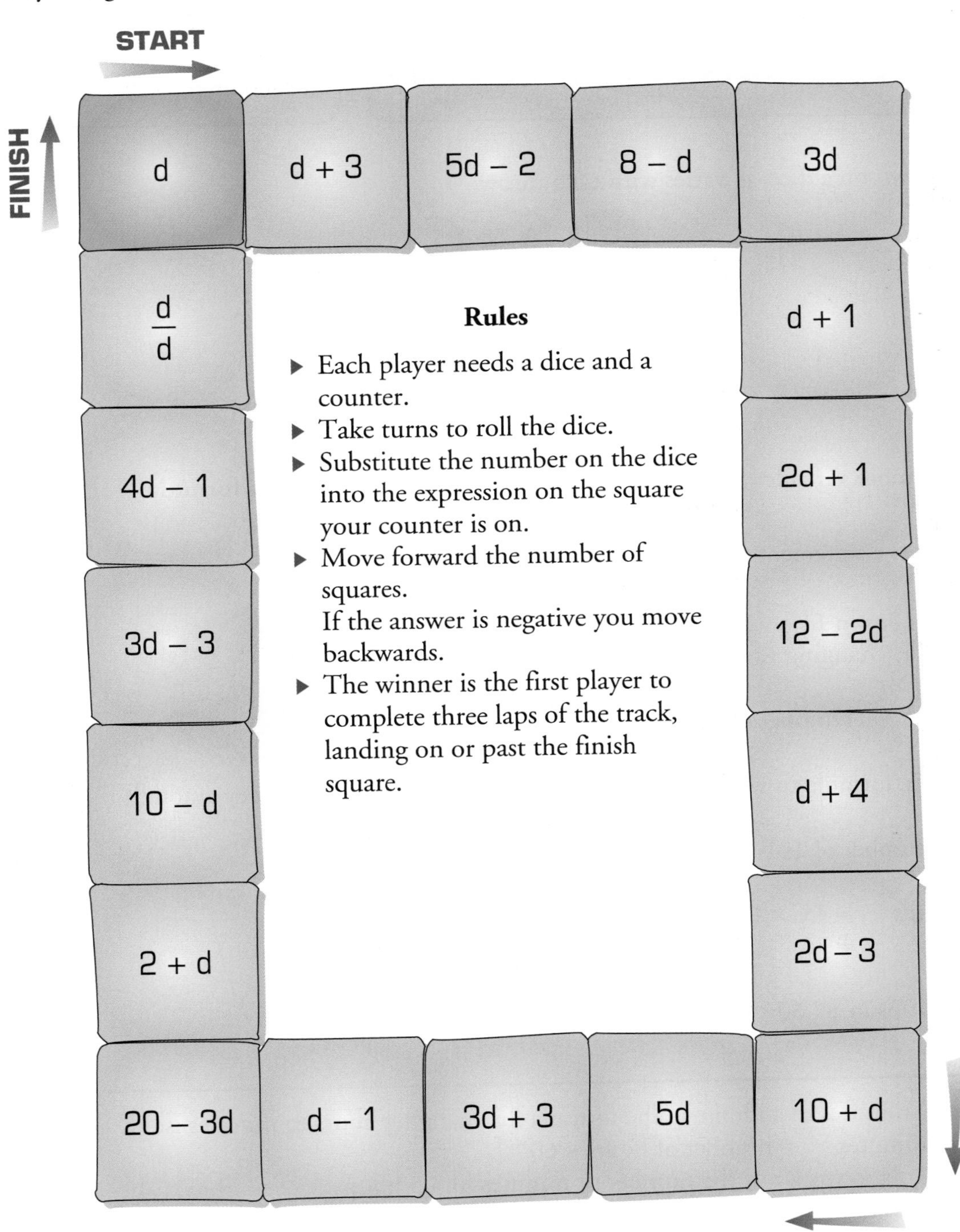

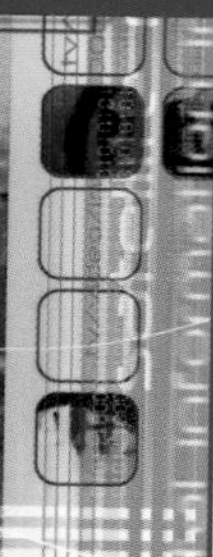

A2.5 Using formulae

This spread will show you how to:

- Develop from explaining a relationship in words to expressing it in a formula using letters as symbols.

KEYWORDS

Value
Variable
Relationship

The value shown on a dice, *d*, varies with each throw:

d is a **variable**.

When two or more variables are linked, you can write the relationship in a **formula**.

▶ A **formula** is a statement that links variables.

There are ten 10ps in £1.
The formula for changing £s to 10ps is:

Number of 10ps = number of £s × 10

You can use the formula to work out how many 10ps there are in any money amount.

In £7.60 the number of £s is 7.60 or 7.6

Number of 10ps $= 7.6 \times 10$
$= 76$

There are 76 10ps in £7.60

example

There are 60 minutes in an hour so the formula for changing hours to minutes is:
Number of minutes = number of hours × 60
Use the formula to work out the number of minutes in 3.5 hours.

The number of hours is 3.5 so:
The number of minutes $= 3.5 \times 60$
$= 210$ minutes

Exercise A2.5

1 Use the formula

Number of 10ps = number of £s × 10

to work out the number of:

a 10ps in £5 **b** 10ps in £2.30 **c** 10ps in £21.20

2 Use the formula

Number of minutes = number of hours × 60

to work out the number of minutes in:

a 3 hours **b** 6 hours **c** 2.5 hours
d $1\frac{1}{2}$ hours **e** $2\frac{1}{4}$ hours **f** a day

3 Use the formula in question 1 to work out the number of:

a 10ps in 90p **b** £s in 120p **c** £s in 2700p

4 Use the formula in question 2 to work out the number of hours in:

a 120 minutes **b** 240 minutes **c** 90 minutes
d 135 minutes **e** 100 minutes **f** 2 hours 20 minutes

5 Use the formula

Total Cost = number of items × price of one item

to solve these problems:

a Joe buys five pencils. The price of each pencil is 20p.
b Daniel buys three cans of pop. Each can cost 40p.
c Arif pays in total 55p. He bought five erasers. What is the price of one?
d Luke paid 99p for some collector cards. Each card was 33p. How many did he buy?

6 Use the formula

Area = length × width

to find the missing values for these rectangles.

a length = 3 and width = 5 **b** area = 15 and length = 15
c width = 4 and length = 8 **d** area = 25 and width = 5
e length = 14 and width = 2 **f** width = 6 and length = 6

All lengths are measured in cm and areas in cm^2.

Summary

You should know how to ...

Develop from explaining a generalised relationship in words to expressing it in a formula using letters as symbols.

Check out

1 Write algebraic expressions for the following:

a There are d daffodils in a field. Amy picks 5. How many are left in the field?

b There are n kittens in a basket. Two more climb in. How many are there now?

c Karen has y keyrings. Her friend buys her one, her mother another one and she buys herself three more. How many are there now altogether?

d There are x cars in a show room. How many cars are there in three identical garages?

e A lot of money is won by four friends. Write an expression to show how much each friend gets.

2 Write the following as expressions, using algebraic conventions:

a 6 less than x

b 3 more than y

c z more than 4

d f less than 7

e x multiplied by 4

f 7 multiplied by y

g 26 divided by z

h h divided by 5

i 5 multiplied by y

j 12 divided by x

k 8 subtract x

l x subtract 8

3 Write a formula for:

a the cost of n packets of crips at 29 pence each

b the cost of m pints of milk at 35 pence each

c the number of months, m, in y years.

2 Angles and shapes

This unit will show you how to:

- Use units of time; read the time on a 24-hour digital clock and use 24-hour clock notation, such as 19:53.
- Use timetables.
- Recognise positions and directions: read and plot coordinates in the first quadrant.
- Classify triangles and quadrilaterals.
- Find coordinates of points determined by geometric information.
- Recognise perpendicular and parallel lines.
- Recognise and estimate angles.
- Use all four operations to solve word problems involving time.

Each sector of a dartboard makes the same angle at the centre.

Before you start

You should know how to ...

1 Recognise clockwise and anticlockwise turns.

2 Read points on the first quadrant.

Check in

1 Describe these turns as clockwise or anticlockwise.

a 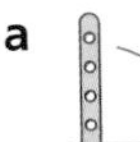 b 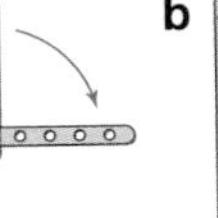 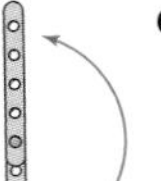 c 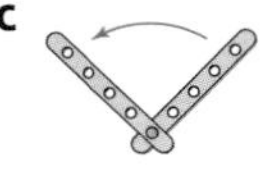 d 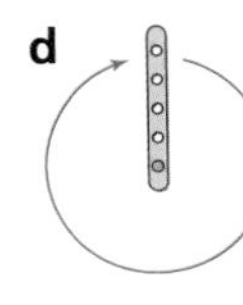

2 Give the coordinates of the points A and B.

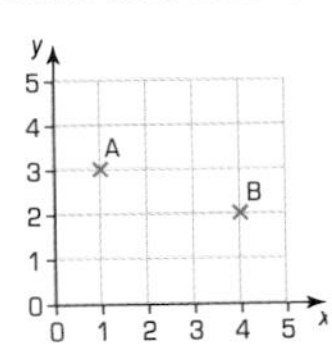

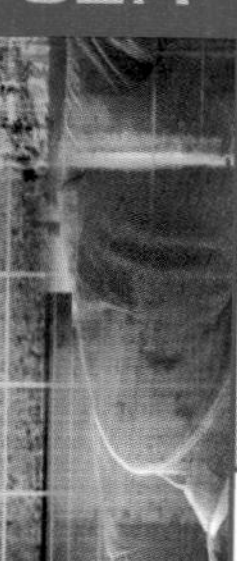

S2.1 Time for a change

This spread will show you how to:

- Read clocks and use timetables.
- Recognise angles.

KEYWORDS
Angle
Degree °
Time

There are 24 hours in a day, but an analogue clock is numbered from 1 to 12.
You use am for times before midday and pm for times after midday.

The clock shows 30 minutes past 8 in the morning.

8.30 am
am means before midday or noon

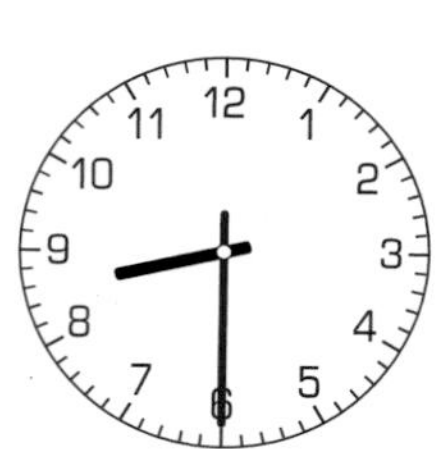

The clock shows 30 minutes past 8 in the evening.

8.30 pm
means after midday or noon

A digital clock uses the 24 hour clock.
You add 12 hours onto am times to find pm times.

8 30 pm
20:30

An analogue clock has hands that turn:

The hand has turned $\frac{1}{4}$ of the way round.

You can measure turn in degrees, ° for short.

A full turn is 360°.

A half turn is 180°.

A quarter turn is 90°.

Exercise S2.1

1 These clocks all show pm times. Write the times using the 24 hour clock.

a **b** **c** **d**

Describe what you were doing at each of these times yesterday.

2 Copy and complete this table.

Event	12 hour clock	24 hour clock
Breakfast	8.00 am	08.00
Morning Break	11.00 am	
Lunch	12.30 pm	
Afternoon Break	3.00 pm	
Tea	5.30 pm	
Supper	9.30 pm	

3 Write these times using am or pm:

a 14.30 **b** 10.25 **c** 08.45 **d** 22.25

4 Write these times using the 24 hour clock:

a 8.30 pm **b** 9.45 am **c** 10.20 pm **d** midday

5 Add 30 minutes to each of these times:

a 8.15 am **b** 6.45 pm **c** 7.05 am **d** 7.55 pm
e 08.30 **f** 16.10 **g** 17.45 **h** 10.50

6 Here is part of a tram timetable:

High Street	19.40
Arena	19.50
Meadowhall	19.55

How long is the journey from:

a High Street to Meadowhall **b** High Street to Arena?

7 What angle is shown on each clockface?

a **b** **c**

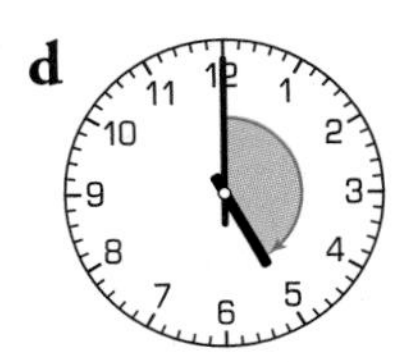

d

S2.2 Angles and lines

This spread will show you how to:

- Recognise parallel and perpendicular lines.
- Recognise a point of intersection.
- Recognise properties of rectangles.

KEYWORDS

Intersect, Opposite, Parallel, Perpendicular, Shape, Diagonal

You can describe a shape using its properties.
Here are some useful terms:

- **Parallel** lines are always the same distance apart.
  You can write // as shorthand for the word parallel.
- **Perpendicular** lines meet at a corner or 90° (at a right angle).
  You can write ⊥ as shorthand for the word perpendicular.

The sides of a ladder are parallel and they are perpendicular to the rungs.

You mark parallel and perpendicular lines on shapes like this:

The small square shows the sides are perpendicular.

This rectangle has four pairs of perpendicular sides.

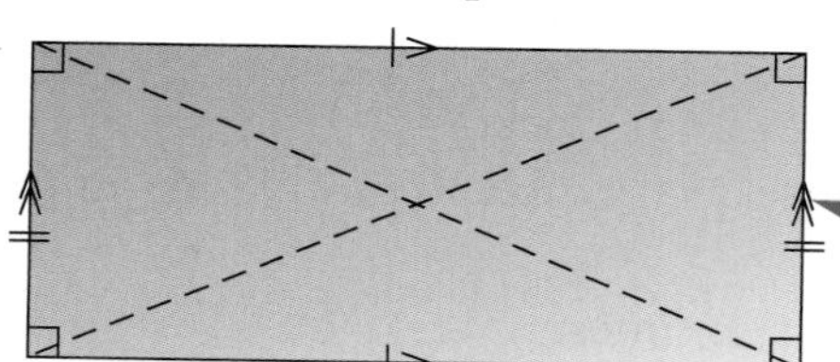

The arrows show the lines are parallel.

This rectangle has two pairs of parallel lines.

The dashed lines are diagonals.

Opposite sides are equal in length.
The marks show the lengths that are equal.

- Lines **intersect** when they meet.

Parallel lines never intersect.
They are equidistant.

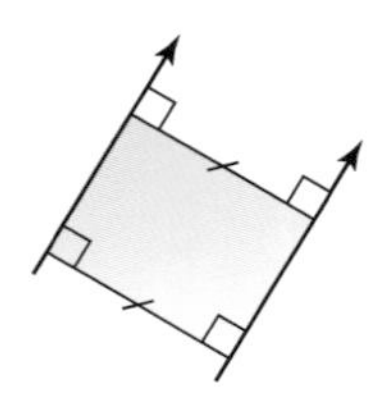

Perpendicular lines intersect at right angles.

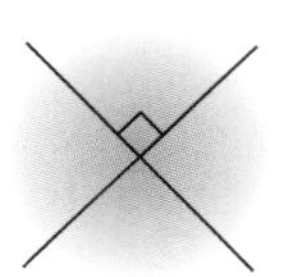

Other straight lines intersect at an angle.
The opposite angles are equal.

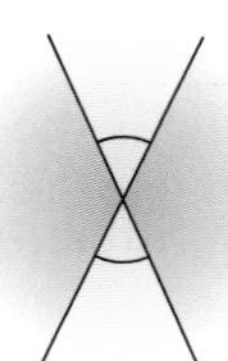

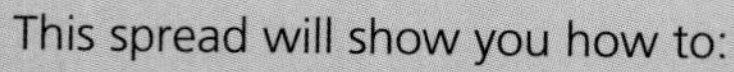

Exercise S2.2

1 The two vertical sides of a door frame are parallel.
Make a sketch of all the pairs of parallel and perpendicular lines you can see. Mark the parallel and perpendicular lines on your sketch.

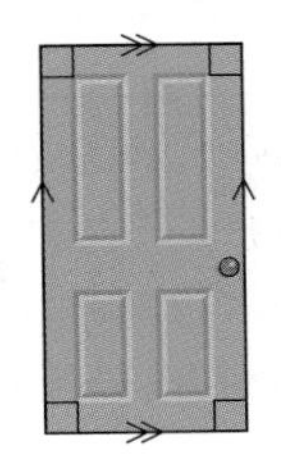

2 On a copy of each of these shapes:

- Show the pairs of sides that are parallel. Use > and >>
- Show the pairs of sides that are perpendicular. Use ⊾

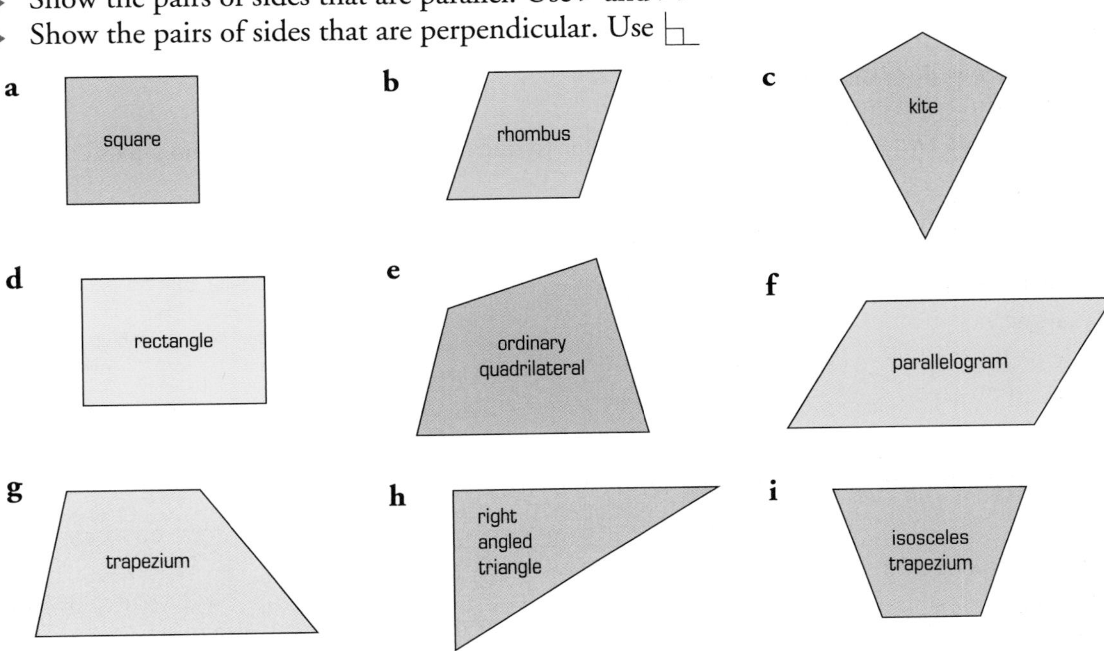

3 Copy and complete the table putting each shape in question 2 in the right place:

	At least one pair of parallel sides	No parallel sides
At least one pair of perpendicular sides		
No perpendicular sides		

4 Describe each of the shapes in question 2 as fully as possible.
Choose the words to use from this list:

Parallel	Perpendicular	Intersect
Equal angles	Equal sides	Opposite

S2.3 Coordinates and shapes

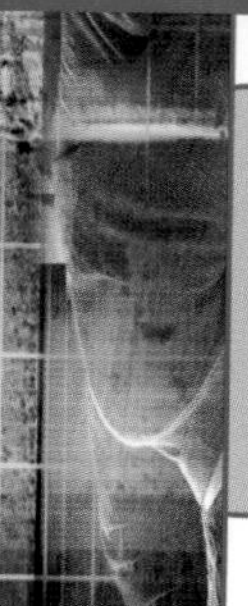

This spread will show you how to:

- Recognise positions.
- Read and plot points using coordinates in the first quadrant.
- Find coordinates determined by geometric information.
- Classify triangles.

KEYWORDS

Coordinates	*x*-axis
Grid	*y*-axis
Position	Vertex
Triangle	Vertices

You can plot points on a grid.

▶ **A grid has two axes that are perpendicular to each other: the *x*-axis and the *y*-axis.**

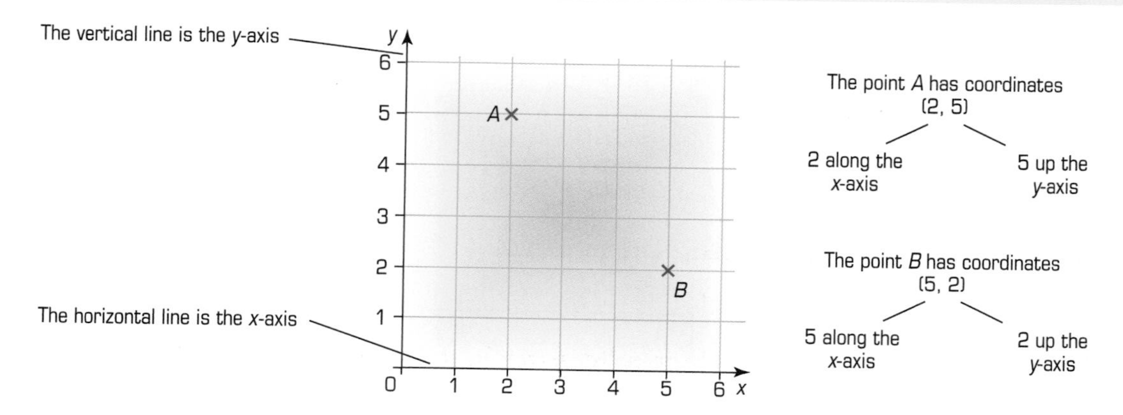

You can draw shapes on a grid. These are the types of triangles you can draw:

Scalene
No sides equal

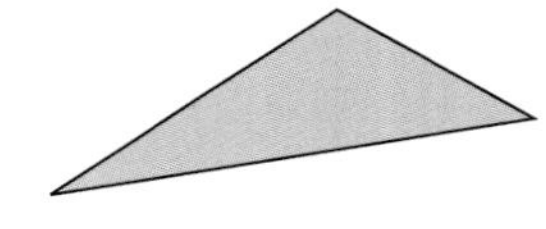

Isosceles
2 sides equal

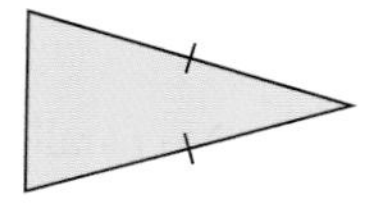

Equilateral
All 3 sides equal

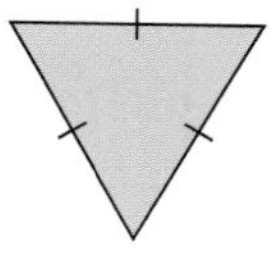

Right-angled

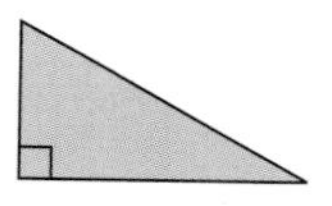

example

Two vertices of an isosceles triangle are $P(4, 4)$ and $Q(3, 2)$.
What are the coordinates of the third vertex?

Plot the points *P* and *Q*.
The third point could take either of the positions shown and more.

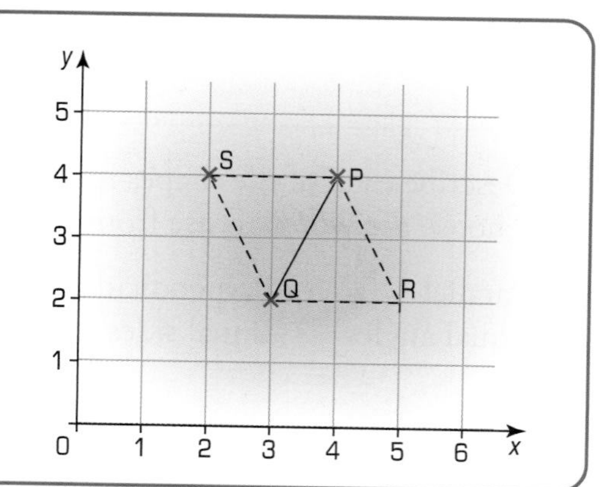

Exercise S2.3

1 Write down the coordinates of the points A → J on the grid.

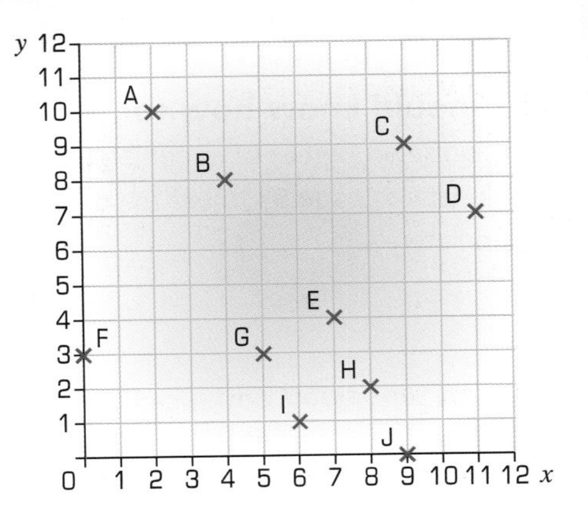

2 Plot the following points and complete the shapes. Each shape has one vertex missing. Write down the missing coordinate for each shape.

a Right-angled triangle (2, 3) (2, 9) (,)
b Isosceles right-angled △ (2, 3) (2, 9) (,)
c Isosceles triangle (2, 3) (2, 9) (,)
d Scalene triangle (2, 3) (2, 9) (,)
e Scalene right-angled triangle (2, 3) (2, 9) (,)

3 The points (1, 4) (5, 4) (5, 2) are three vertices of a rectangle.
Write down the coordinates of the fourth vertex.
Find the perimeter and area of the rectangle.

4 Plot these three points:
(3, 3) (8, 8) and (8, 3)
What fourth point will make a square?
Find the perimeter and area of the square.

5 Find two other possible positions of the third vertex in the Example on page 84.

6 On squared paper draw axes as shown and plot the points *A* (5, 6) and *B* (7, 8).
A (5, 6) and *B* (7, 8) are two vertices of a triangle.
Plot point C to make different size triangles. Record your results in a table.
Which type of triangle couldn't you get?

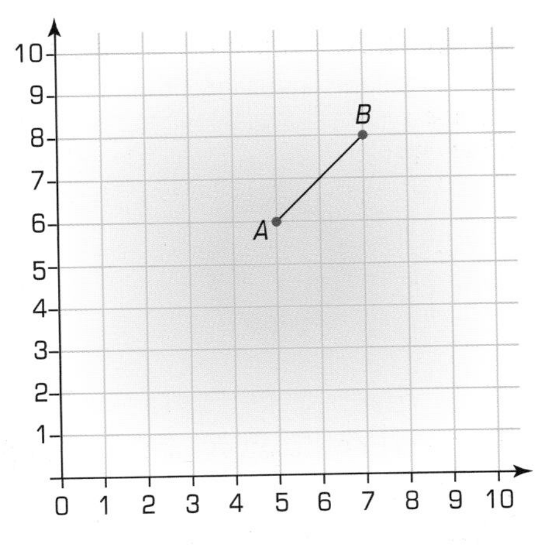

Co-ordinates of point C	Type of triangle
(4, 2)	Scalene right-angled triangle

Try to explain why.

Summary

You should know how to …

1 Recognise properties of quadrilaterals.
Look back to page 83 to see the different quadrilaterals.

2 Recognise perpendicular and parallel lines.

3 Identify and use appropriate operations to solve word problems involving time.

Check out

1 Find these shapes in this diagram:
kite
trapezium
square
rhombus

2 Find lines that are:
a parallel
b perpendicular
in the diagram in question 1.

3 This is part of a train timetable:

Stevenage	0745
Welwyn Garden City	0755
Finsbury Park	0813
London Kings Cross	0821

a How long is the journey from Stevenage to London Kings Cross?

Jordan arrives at Finsbury Park at 0808.
b How long must she wait for the train?
c How long will her train journey take to Kings Cross?

2 Handling data

This unit will show you how to:

- Order a set of numbers.
- Solve a problem by representing, extracting and interpreting data in tables, graphs, charts and diagrams.

This is part of a poll by an environmental action group:

15. Are you...?

- Vegetarian ☐
- Looking to become Vegetarian ☐
- Vegan ☐
- An animal eater ☐

The way the question is phrased shows the bias of the group.

The question could annoy people who eat meat.

Before you start

You should know how to ...

1 Represent data in tables and charts.

2 Use information from tables and diagrams to solve problems.

Check in

1 Make a bar chart for the information in this table.

Day	Absences
Mon	4
Tues	5
Wed	3
Thu	1
Fri	3

2 How could you *see* which day had most absences, using the bar chart from question 1?

Discussing statistical methods

This spread will show you how to:

- Solve a problem by collecting data.

KEYWORDS
Survey

The governors of Maypole High School are concerned about students' safety at the beginning and end of the school day because of parked cars outside the school gates.

They hold a meeting to decide what action to take.

They decide that the most important questions to ask are:

- How do students travel to school?
- How long does it take students to travel to school?

You can turn these questions into statements.

- Most students travel to school by car.
- Most students take over 30 minutes to get to school.

- You can conduct a **survey** to test a statement.

Exercise D2.1

1 The Governors of Maypole High School ask Year 7 to identify the issues affecting the traffic at the beginning and the end of the day.

Here are their suggestions:

a The Governors need to turn each of these suggestions into a statement. Match each of these statements to one of the Year 7 suggestions.

A: The school car park is usually full.
B: Most students live more than a mile from school.
C: The buses to school are usually full.
D: More than half of the parked cars do not belong to a parent or a teacher.

b Which two suggestions are not matched by one of the statements in **a**? Try to write a statement for each of these.

2 Imagine you need to identify the issues affecting the traffic outside school. Make a list of all the questions you would ask.

3 At Maypole High School, class 7W decide to find out if:

- the number of people that travel in each car is only two
- most students travel to school by bus
- students who walk to school have less distance to travel
- most cars bring only one student to school
- students living furthest away take the longest time to travel to school

Compare these questions with your own.
Comment on which questions are the most relevant and why.

Collecting data

This spread will show you how to:

- Solve a problem by collecting and organising data.

KEYWORDS
Questionnaire
Survey

To decide whether traffic outside school can be reduced, the Maypole High Governors want to ask drivers:

- *How far is it from your house to school?*
- *How many people do you take to school?*
- *What buses go from near your house?*
- *Why do you drive to school?*

These are **closed** questions. They have definite answers.

These are **open** questions. They can include answers you haven't thought of.

The Governors develop a questionnaire:

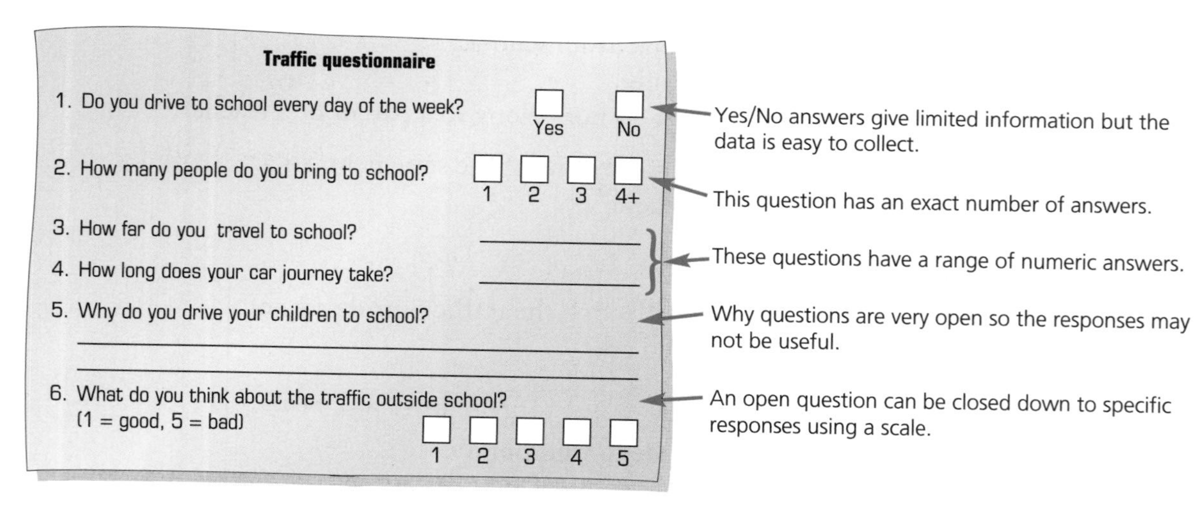
Traffic questionnaire

1. Do you drive to school every day of the week? ☐ Yes ☐ No
2. How many people do you bring to school? ☐ 1 ☐ 2 ☐ 3 ☐ 4+
3. How far do you travel to school? ________
4. How long does your car journey take? ________
5. Why do you drive your children to school? ________
6. What do you think about the traffic outside school? (1 = good, 5 = bad) ☐ 1 ☐ 2 ☐ 3 ☐ 4 ☐ 5

- You can use a questionnaire to conduct a survey.
  Open questions invite any response.
  Closed questions invite choice.

You can close down an open question using a scale.

Exercise D2.2

1 Rakhi and Jem are investigating people's journeys to school. They are designing a questionnaire. Here are the first three questions.

Ways to school questionnaire

1. How do you travel to school?

2. How long is your journey to school?
   Short Medium Long

3. Do you listen to the radio on the way to school?

a For each question, decide:

- Does it ask for important information?
- Is it easy to answer?
- Will the answers be easy for Rakhi and Jem to use?

b Suggest improvements to each question.

2 Write down one more question for a school journey survey.
Try to make sure that it will be easy to answer, and will give clear information.

3 Improve each of these questions by giving some options for people to choose from.
a What is your favourite sort of television programme?
b How good is your local bus service?
c How many hours a week do you spend playing computer games?
d What time do you go to bed on a school day?

4 Rewrite each of these questions to make them fairer.
a How much time do you waste each week sending pointless text messages?
b Why don't people like sprouts?
c Do you agree that there are far too many sports programmes on television, and sometimes you can't even find a good film or drama?
d Which shampoo would you rather use – one of the boring old ones, or the new super enriched GlamLite shampoo, which is rich in vitamins and came top in a recent survey?

5 Sheila works in a school library. She wants to find out what young people want to read.
Design a short questionnaire to help her decide.
Your questionnaire should have about 5 good questions.

Organising the data

This spread will show you how to:

- Solve problems by organising data into tables and charts.

KEYWORDS
Frequency chart
Range
Tally
Interval

The Maypole High governors have conducted their survey and want to organise the data that they have collected.

This question has exactly four responses:

How many people do you take to school?
1 ☐ 2 ☐ 3 ☐ 4 ☐

This sort of data is called **discrete** data.

- **Discrete data can only take exact values.**

You can organise the data in a frequency chart.

Number of people	Tally	Frequency
1	𝍸 I	6
2	𝍸 𝍸 I	11

Other questions can have a range of responses:

This sort of data is called **continuous** data.

- **Continuous data can take any value in a given range.**

You can organise continuous data in a frequency table, grouped into **intervals**.

Journey time (minutes)	0–10	11–20	21–30	31–40	41–50	51–60
Number of pupils	5	12	9	4	2	1

- **You can use a tally or frequency chart to organise data as you collect it.**

Exercise D2.3

1 Mandy collected data about how many pets people have.
Here are her results:
0, 3, 1, 2, 1, 0, 2, 2, 5, 6, 4, 7, 3, 2, 4, 2, 1, 0, 2, 2.

Copy and complete this frequency table for the data.

Number of pets	0–1	2–3	4–5	6–7
Tally				
Number of people				

2 Alison counted the number of matches in 20 boxes.
Here are her results:
48, 50, 49, 53, 61, 45, 48, 51, 52, 49, 44, 48, 54, 43, 51,
60, 50, 49, 47, 39.

Copy and complete this frequency table:

Number of matches	30–39	40–49	50–59	60–69
Tally				
Number of people				

3 Complete a frequency table for each of these sets of data.

a Heights of members of class 7Y (to the nearest cm):
152, 161, 148, 139, 158, 163, 160, 155, 153, 147, 141,
162, 154, 160, 153, 151, 159, 152, 155, 150

Height (cm)	135–139	140–144	145–149	150–154	155–159	160–164
Tally						
Frequency						

b Number of brothers or sisters of class 7Y:
2, 0, 1, 1, 2, 0, 0, 3, 0, 1, 2, 1, 1, 1, 0, 2, 4, 1, 0, 2

4 Say whether each of these sets of data is discrete or continuous.

a The age of teachers in your school.
b The time taken to run 100 metres.
c The number of rooms in each house in a street.
d The height of the trees in a forest.

Displaying your results

This spread will show you how to:

- Represent data in graphs and charts.

KEYWORDS

Bar chart
Data
Bar-line graph
Pie chart

The Governors of Maypole High want to illustrate their survey by using diagrams.

They must choose the most appropriate diagram to get the point across.

Some of their data is **numerical**...

- *How many people do you bring to school?*
- *How far do you travel to school?*
- *How long does your journey take?*

and some is **non-numerical**.

- *How do you travel to school?*
- *Do you live near a bus route?*
- *Why do you drive your children to school?*

You represent different types of data in different ways.

A **pie chart** uses a circle to show data.

How students travel to school

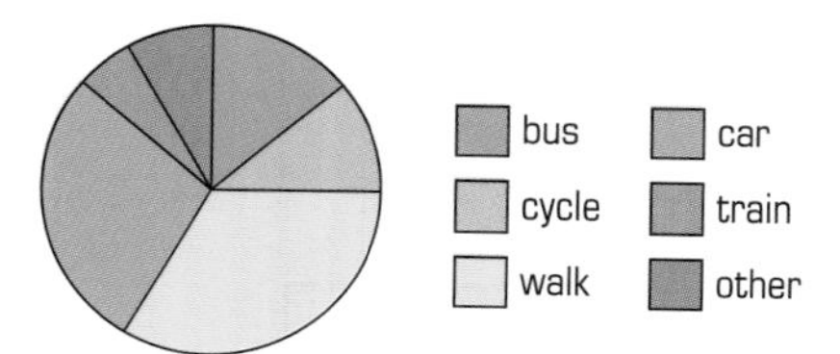

The size of the sector represents the proportion in the category.

Pie charts are useful when you want to compare a few categories.

A **bar chart** uses bars to show data.

Number of people in a car

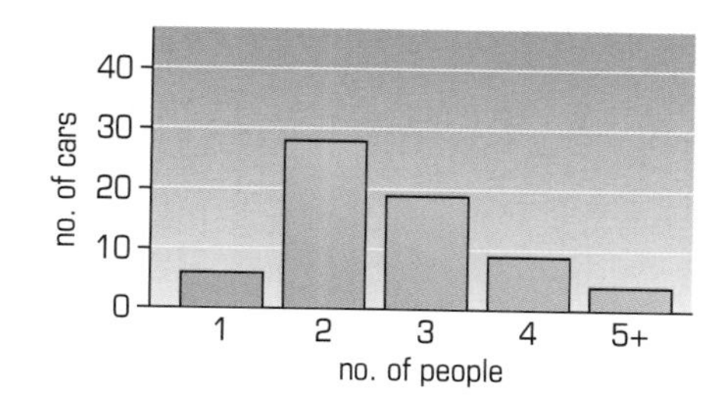

The bars are equal width.
The height of the bar represents the frequency.

Bar charts show the actual values and are useful for displaying more categories.

Exercise D2.4

1 For each set of data:
- ▶ Organise the data in a frequency table.
- ▶ Draw a bar chart on squared paper.

a Number of goals scored by a football team:
2, 3, 0, 0, 1, 5, 2, 0, 1, 1, 6, 2, 1

b Number of hours of sunshine one fortnight:
7, 2, 0, 1, 8, 9, 5, 4, 6, 0, 0, 0, 1, 4

c Favourite colour of a group of students:
green, purple, pink, green, pink, red, green, blue, orange, red, blue, red, green, blue, purple, orange, purple

2 Use your bar charts to write down the mode of each set of data in question 1.

3 The bar-line chart shows the results of a survey about the number of cats people own.

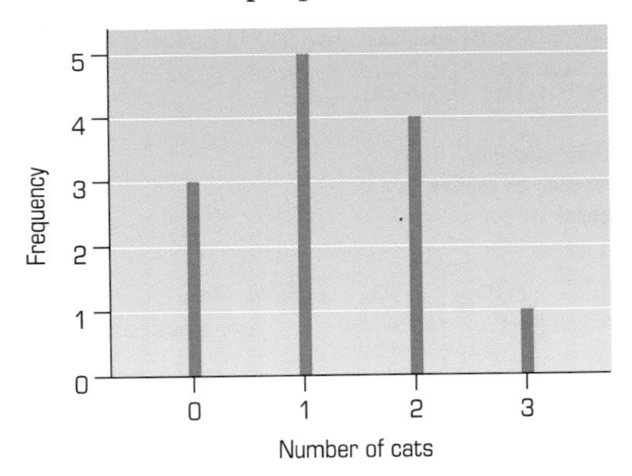

Use the chart to answer these questions:

a What was the most number of cats anyone owned?
b How many people owned no cats?
c How many people took part in the survey?
d What was the most common number of cats owned?

4 **a** Use this data to make a frequency table:

Number of goals scored in 20 football matches
1, 2, 0, 0, 1, 2, 5, 3, 0, 0, 1, 3, 3, 3, 2, 4, 3, 2, 1, 2.

b Now use the frequency table to produce a bar-line chart.

D2.5 Interpreting your diagrams

This spread will show you how to:

- Interpret data in graphs and charts.

KEYWORDS

Bar chart Pie chart
Interpret

Different people may find different conclusions from the same diagram ...

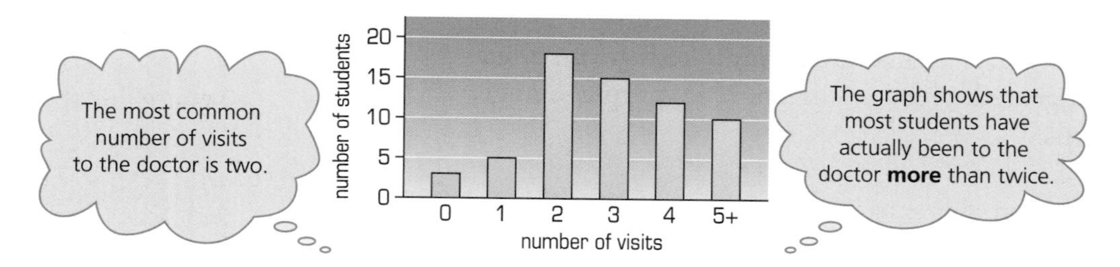

- Once you have drawn a diagram you should interpret it. To interpret a diagram you make a conclusion about what it shows.

The Maypole High Governors interpret the diagrams for the drivers.

Bar chart showing the number of students taken to school by car each day

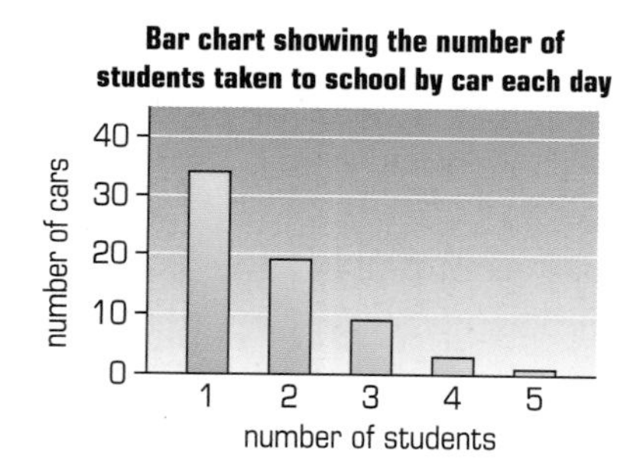

The bar chart shows that most people bring only one student to school.

Pie chart showing information on drivers to school

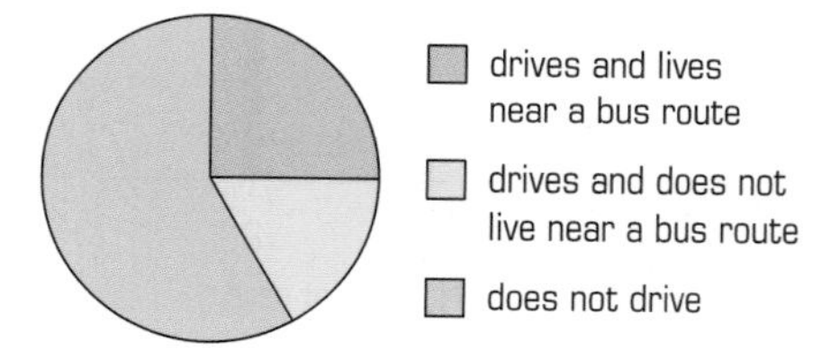

The pie chart shows that more than half the people who drive live near a bus route.

From these graphs, the Maypole High Governors conclude that:

- half the traffic can be stopped if those students living near a bus route use the bus.
- parents should consider sharing cars where possible.

When you finish a survey you may find further issues to explore, such as:
Why do students near a bus route not use the bus?

Exercise D2.5

1 Kim is a photographer. She records the number of sucessful photographs that she gets from each film she develops. The chart shows her results.

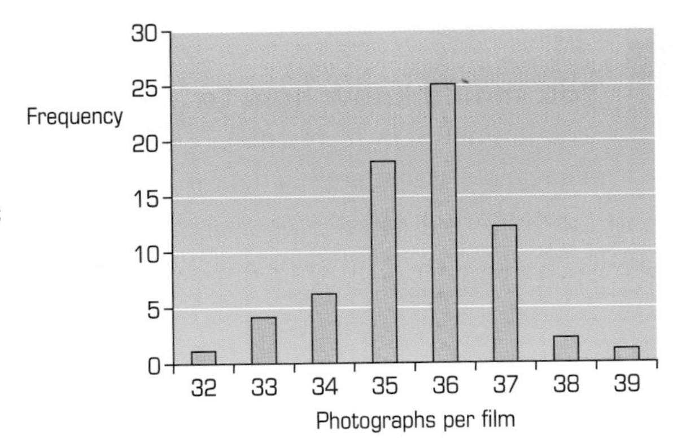

Use the chart to answer these questions:

a What was the highest number of photographs that Kim got from a film?

b What was the mode of the number of photographs?

c How many films does the chart show data for?

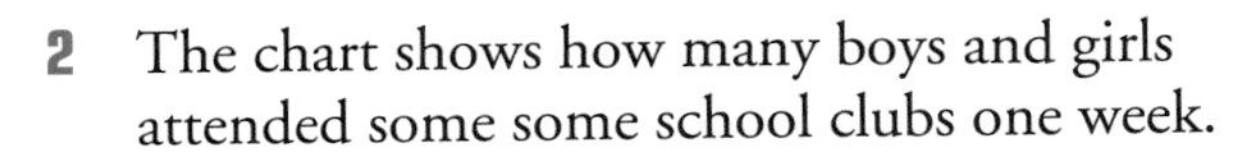

2 The chart shows how many boys and girls attended some some school clubs one week.

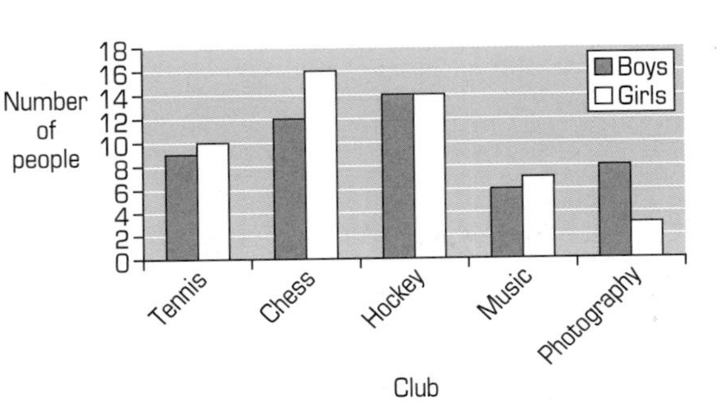

Use the chart to answer these questions:

a How many people went to chess club?

b Which club had equal numbers of boys and girls?

c Which club had more boys than girls?

d What was the total attendance at all of the clubs?

3 A shop sells five flavours of crisps. The pie chart shows information about how many of each flavour were sold.

Flavours of crisps sold

Plain 16%
Cheese 30%
Bacon 12%
Pickle 18%
Tomato 24%

a Which of these statements are true:

A: The most popular flavour was cheese.

B: The shop sold twice as many packets of tomato flavour as bacon flavour.

C: More than a quarter of the packets sold were tomato flavour.

b Which two flavours of crisps made up more than 50% of the total amount sold?

4 Jo is a school technician. She keeps a record of how many times she has to fix the photocopier.

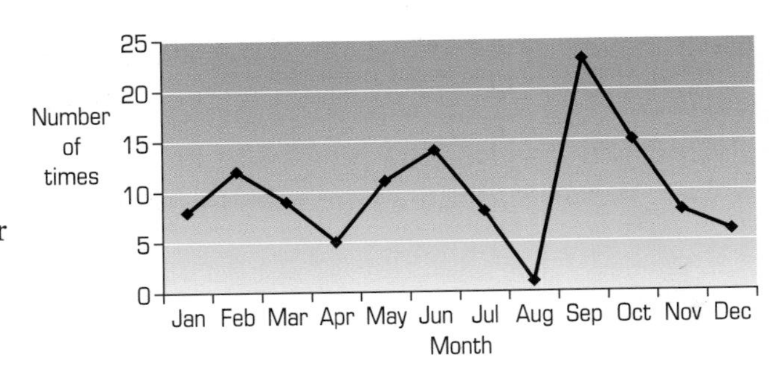

Use the chart to answer these questions.

a Which month had the biggest number of breakdowns?

b Which three months had the smallest number of breakdowns?

c Explain why you think the number of breakdowns has the pattern shown in the chart.

Summary

You should know how to ...

Solve a problem by representing, extracting and interpreting data in tables, charts and diagrams.

Check out

1 25 boys and 25 girls in Year 7 were asked to say how good they thought the school dinners were. The results are shown in this bar chart.

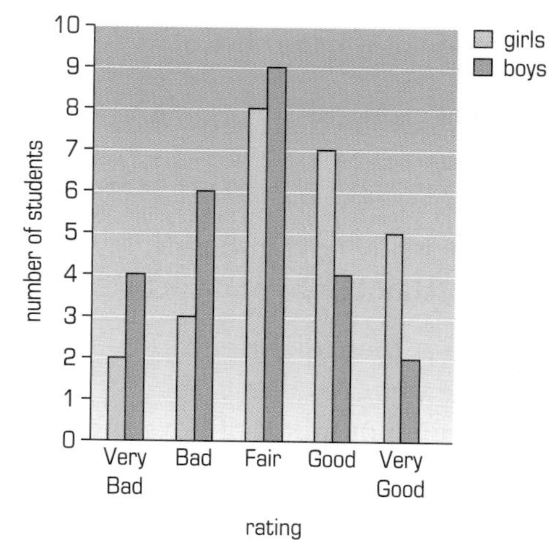

a How many students altogether thought the meals were 'Very Good'?

b How many girls said the meals were 'Fair'?

c Overall, who were happier with the school dinners – the boys or the girls?
Explain your answer.

2 The pie chart shows some information about the number of people using different facilities in a leisure centre.

Pie chart showing use of facilities in a leisure centre

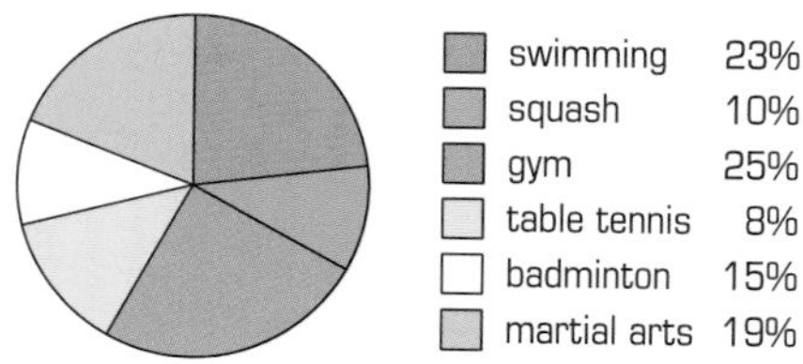

Which of these statements are true:

A: Less than a quarter of the people used the swimming pool.

B: More than half of the people were either playing badminton or squash.

C: The gym was over three times as popular as table tennis.

13 Multiplication and division

This unit will show you how to:

- Multiply and divide any positive integer by 10 or 100 and understand the effect.
- Understand the effect of and relationships between the four operations, and the principles (not the names) of the arithmetic laws as they apply to multiplication. Begin to use brackets.
- Know all multiplication facts up to 10 × 10 and corresponding division facts.
- Use doubling or halving, starting from known facts; partition.
- Recognise multiples up to 10 × 10.
- Factorise number up to 100.
- Extend written methods to:
  - short multiplication of HTU or U.t by U;
  - long multiplication of TU by TU;
  - short division of HTU by U.
- Use a calculator effectively.
- Estimate by approximating then check result.
- Use, read and write standard metric units (km, m, cm, mm), including their abbreviations, and relationships between them.
- Convert larger to smaller units.
- Identify and use appropriate operations to solve word problems.

Sometimes you need to be able to estimate quickly.

Before you start

You should know how to ...

1 Add and subtract integers.

2 Recall multiplication and division facts up to 10 × 10.

3 Use a calculator for basic calculations.

Check in

1 **a** Work out these mentally:

i 34 + 22 **ii** 63 – 21

iii 31 + 23 + 18 **iv** 61 – 19 + 2

b Work out these additions and subtractions using the column written method:

i 325 + 134 **ii** 768 – 253

iii 327 + 158 **iv** 625 – 371

2 Work these out:

a 3 × 7 **b** 24 ÷ 3 **c** 9 × 4 **d** 8 × 5 **e** 45 ÷ 9

3 Use a calculator to work out each of these:

a 1026 + 953 **b** 527 – 369

c 21 × 36 **d** 562 × 45

e 4707 ÷ 9 **f** 7981 ÷ 23

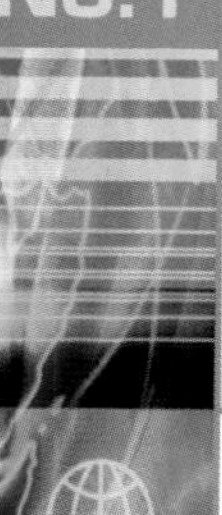

N3.1 Number and measures

This spread will show you how to:

- Multiply and divide whole numbers by 10 and 100.
- Convert between one metric unit and another.

KEYWORDS

Convert | Multiply
Divide | Place value

The numbers we use are based on the decimal system.
This makes it easy to multiply by 10 or 100.

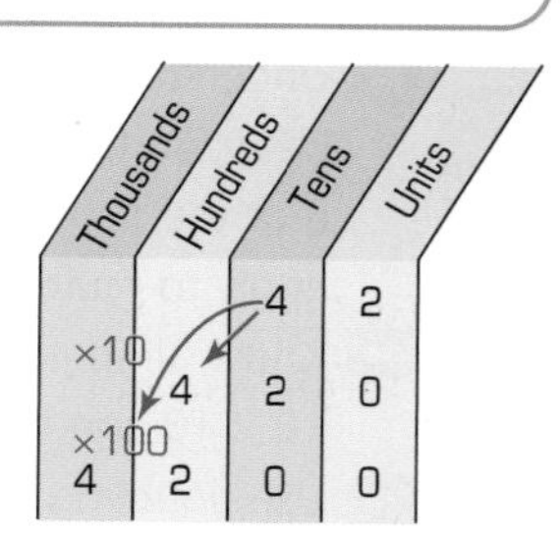

When you multiply by 10 each digit moves 1 place to the left: 42 × 100 = 420

When you multiply by 100 each digit moves 2 places to the left: 42 × 100 = 4200

Dividing is similar but the numbers get smaller.
Metric measures are based on the decimal system too.

When you divide the digits move to the right.

8500 ÷ 10 = 850
8500 ÷ 100 = 85

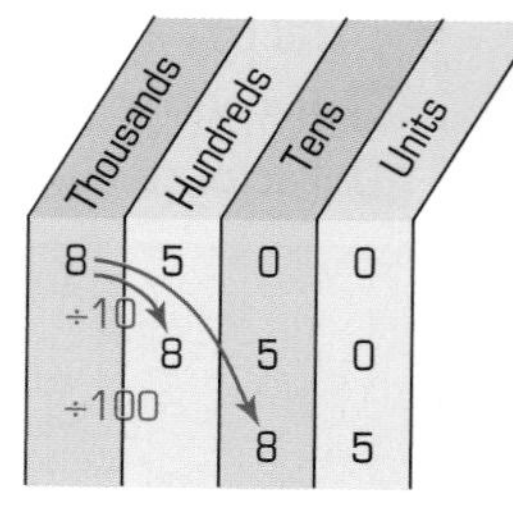

1 cm = 10 mm

1 km = 1000 m

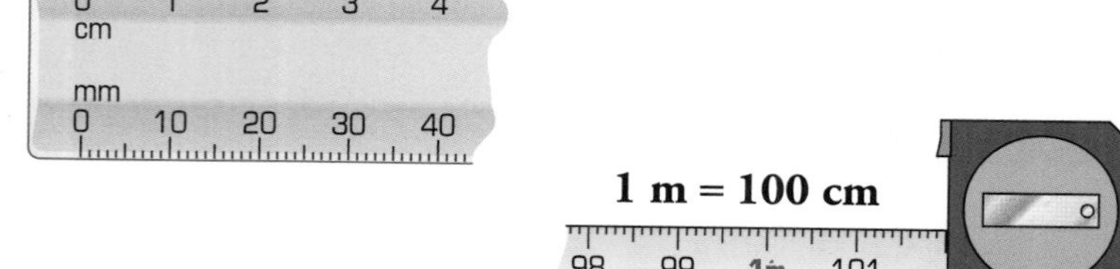

1 m = 100 cm

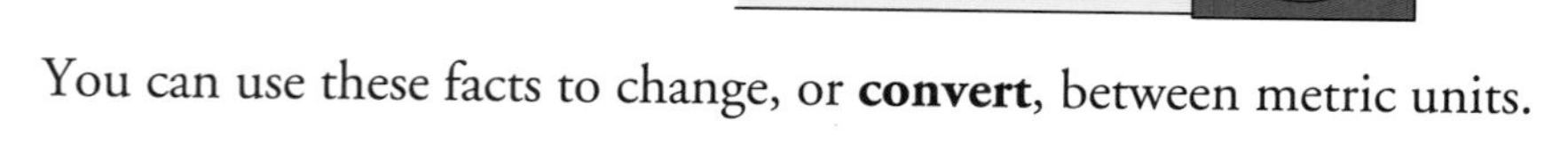

You can use these facts to change, or **convert**, between metric units.

example

Convert:

a 3 cm to mm **b** 5 m to cm **c** 4000 m to km

a 1 cm = 10 mm
3 cm = (3 × 10) mm
= 30 mm

b 1 m = 100 cm
5 m = (5 × 100) cm
= 500 cm

c 1000m = 1 km
4000 m =
(4000 ÷ 1000) km
= 4 km

- To convert from large units to small units you multiply.
- To convert from small units to large units you divide.

m —×100→ cm

m ←÷100— cm

Exercise N3.1

1 Multiply these numbers by 10.
a 8 **b** 9 **c** 10 **d** 12 **e** 15 **f** 18 **g** 100 **h** 220

2 Divided these numbers by 10.
a 70 **b** 60 **c** 200 **d** 500 **e** 300 **f** 180 **g** 230 **h** 3700

3 Multiply these numbers by 100.
a 3 **b** 8 **c** 20 **d** 70 **e** 68 **f** 130 **g** 155 **h** 178

4 Divide these numbers by 100.
a 500 **b** 400 **c** 900 **d** 2000 **e** 5000 **f** 2500 **g** 18 000

5 Jade buys 100 ice lollies at 20p each for a youth club picnic.
Work out how much she spends in total.
a Write your answer in pence.
b Write your answer in pounds.

6 100 plastic toy soldiers cost £4 in total.
a What is £4 in pence?
b Work out in pence how much each soldier costs.

7 Josie has 60 sweets. She and her nine friends share the sweets equally.
a Write down how many people share the sweets.
b Work out how many sweets each person gets.

8 Here are four metric units for measuring distance.
km mm m cm
Write them in order of size, starting with the smallest.

9 Change each of these measurements into millimetres (mm).
a 5 cm **b** 3 cm **c** 8 cm **d** 10 cm **e** 50 cm **f** 72 cm

10 Change each of these measurements into centimetres (cm).
a 3 m **b** 7 m **c** 11 m **d** 40 m **e** 23 m **f** 100 m

11 Joel buys 13 m of ribbon. How much is this in cm?

12 Change each of these measurements into metres (m).
a 2 km **b** 5 km **c** 8 km **d** 12 km **e** 28 km **f** 87 km

13 Nandini runs 5 km. How far has she run in metres?

14 Convert each of these measurements.
The first one is done for you.
a 3.7 cm into mm **b** 6.2 cm into mm **c** 12.5 cm into mm
d 3.85 m into cm **e** 8.15 m into cm **f** 12.7 km into m

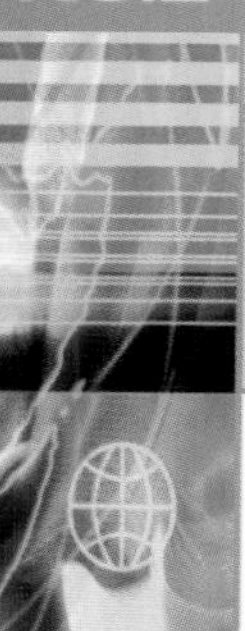

N3.2 Order of operations

This spread will show you how to:

- Know and use the order of operations.
- Use brackets.

KEYWORDS

Brackets, Multiply, Divide, Operations

In arithmetic, **operations** such as add and multiply can appear more than once.

The correct answer is 11. Multiplication comes before addition.

When there are a lot of different operations, you must do them in the correct order:

example

Work out: $3 + 4 \times (7 - 2)$

Brackets	$= 3 + 4 \times 5$
Division or multiplication	$= 3 + 20$
Addition or subtraction	$= 23$

- **If brackets appear in a problem, always do them first.**
- **Multiplication and division come before addition and subtraction.**

If more than one + or – appears, you should work from left to right.
The same rule applies with × and ÷.

example

Work out: **a** $3 \times 6 \div 2$ **b** $13 - 7 - 4$

a $3 \times 6 \div 2$
$= 18 \div 2 = 9$

b $13 - 7 - 4$
$= 6 - 4 = 2$

You would not get the same answer if you did the 7 – 4 first. Try it and see.

Exercise N3.2

Do all these questions in your head.

1 Work out the answers to each of these.

a $8 + 2 + 3$
b $7 + 5 - 3$
c $3 - 2 + 5$
d $10 - 6 - 3$
e $8 - 3 + 7$
f $6 + 5 - 8$
g $8 - 3 + 10$
h $12 - 7 + 3$
i $14 + 6 - 10$
j $3 + 9 - 7$

2 Work out the answers to each of these.

a $2 \times 4 \div 2$
b $8 \div 4 \times 2$
c $6 \div 3 \times 3$
d $3 \times 3 \times 2$
e $4 \times 4 \div 2$
f $8 \div 4 \div 2$
g $10 \div 2 \times 3$
h $12 \div 2 \div 2$
i $5 \times 4 \div 2$
j $10 \times 2 \div 5$

3 Work out the answers to these problems.
Remember to do the brackets first.

a $(8 + 4) - 6$
b $6 + (8 - 2)$
c $12 - (5 - 2)$
d $15 - (8 + 3)$
e $6 - (8 - 6)$
f $(12 + 5) - 3$

4 These are a little harder but you still do the brackets first.

a $6 + (5 - 3) - 2$
b $8 - (8 - 6) + 5$
c $7 + (8 - 3) + 2$
d $7 + 8 - (3 + 5)$
e $12 - 5 - (8 - 6)$
f $16 - (3 + 9) + 6$

5 Work out the answers to these problems.
Remember to think which operation to do first.

a $2 \times 4 + 3$
b $4 + 3 \times 2$
c $3 + 6 \div 2$
d $6 \div 2 + 3$
e $10 \div 2 + 3$
f $12 - 3 \times 2$
g $12 - 8 \div 2$
h $8 + 3 \times 4$
i $3 + 10 \div 2$
j $7 - 12 \div 4$

6 Put brackets into these problems if they are needed.

a $8 + 4 - 3 = 9$
b $6 - 3 \times 2 = 6$
c $8 + 4 \times 2 = 24$
d $8 \div 3 + 1 = 2$
e $9 \div 3 + 5 = 8$
f $6 + 3 \times 2 = 18$
g $6 \div 3 + 5 = 7$
h $7 - 5 \div 2 = 1$
i $14 \div 6 + 1 = 2$
j $20 - 10 \div 2 = 5$

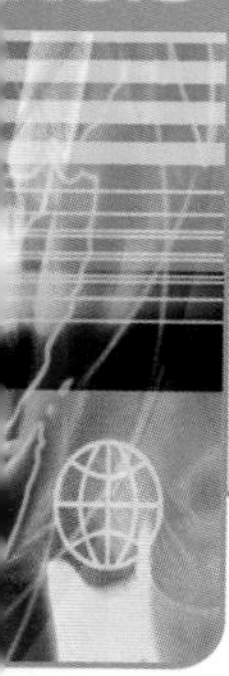

N3.3 Mental methods

This spread will show you how to:

- Know multiplication facts and corresponding division facts.
- Use doubling and halving.
- Understand the operation of division and its relationship to multiplication.

KEYWORDS

Double, Multiply, Halve, Partition

Maths is much easier if you can remember your times tables up to 10×10.

- 2 times table: Just double the number. → $21 \times 2 = 42$
- 3 times table: Double and add the number again. → 21×3; $21 \times 2 = 42$; $42 + 21 = 63$
- 4 times table: Just double the number twice. → 21×4; $21 \times 2 = 42$; $42 \times 2 = 84$
- 5 times table: Multiply by 10 and halve the answer. → 21×5; $21 \times 10 = 210$; $210 \div 2 = 105$

Once you have learned these, the rest are simple.

To times 21 by 8, I just double it 3 times in my head: 42, 84 and the answer is 168.

To times 21 by 6, I multiply by 3 to get 63 and then double it to give 126.

The 7 and 9 times tables are easier to work out if you split the numbers up.

example

Work out:

a 32×7 **b** 17×9

a $32 \times 7 = 32 \times 5 + 32 \times 2$
$= 160 + 64$
$= 224$

b $17 \times 9 = 17 \times 10 - 17 \times 1$
$= 170 - 17$
$= 153$

This method is called **partitioning**.

Your times tables help with dividing as well.

example

Work out $12 \div 3$

3 times ? makes 12.
The answer is 4. So $12 \div 3 = 4$

Exercise N3.3

Do all these questions in your head.

1 Work out these times table questions.

a 3×3 **b** 4×5 **c** 2×6 **d** 3×4 **e** 6×3 **f** 5×5
g 2×8 **h** 3×9 **i** 9×2 **j** 4×3 **k** 3×6 **l** 8×2

2 Work out these division questions.

a $16 \div 2$ **b** $16 \div 4$ **c** $15 \div 3$ **d** $15 \div 5$ **e** $12 \div 4$ **f** $20 \div 4$
g $25 \div 5$ **h** $18 \div 3$ **i** $30 \div 5$ **j** $28 \div 7$ **k** $24 \div 6$ **l** $27 \div 3$

3 Work out the answers to these 2 times table questions by doubling.

a 15×2 **b** 21×2 **c** 42×2 **d** 16×2 **e** 26×2 **f** 38×2

4 Work out the answers to these 4 times table questions by doubling twice.

a 15×4 **b** 21×4 **c** 31×4 **d** 13×4 **e** 17×4 **f** 28×4

5 Work out the answers to these 8 times table questions by doubling 3 times.

a 13×8 **b** 21×8 **c** 31×8 **d** 40×8 **e** 16×8 **f** 102×8

6 Work out the answers to these 5 times table questions by multiplying by 10 and then halving.

a 15×5 **b** 18×5 **c** 21×5 **d** 43×5 **e** 70×5 **f** 102×5

7 Work out the answers to these 3 times table questions.

a 11×3 **b** 21×3 **c** 25×3 **d** 17×3 **e** 31×3 **f** 43×3

8 Work out the answers to these 6 times table questions by multiplying by 3 and then doubling.

a 15×6 **b** 21×6 **c** 25×6 **d** 31×6 **e** 43×6 **f** 102×6

9 Try to use the methods you have learned.
Copy and complete these multiplication grids.

a

×	4	8	6
5			
3		24	
2			

b

×	7		
	28		
8		32	
5			15

N3.4 Multiplying by partitioning

This spread will show you how to:

- Partition to multiply mentally two-digit by one-digit numbers.
- Extend written methods to two-digit numbers.

KEYWORDS
Multiply
Partition

When you are multiplying large numbers, sometimes you can split one of them into smaller pieces.

For example, to find 21×7, this is just 7 lots of 21: $21 + 21 + 21 + 21 + 21 + 21 + 21$

You can split this into: 5 lots and 2 lots

$21 + 21 + 21 + 21 + 21$ $\quad$ $21 + 21$

$21 \times 5 = 105$ $\quad$ $21 \times 2 = 42$

$105 + 42 = 147$

This method is called **partitioning**.
You partition or split numbers to make the multiplication easier.

To work out 23 x 9, I can first say 23 x 10 equals 230.
Then I can subtract 23 to get 207.

If I want to work out 17 × 11, I can do 17 × 10, which equals 170.
Then I can add 17 to get 187

You can also use partitioning in written methods of multiplication.

example

Work out 23×24

This is:

$$
\begin{array}{rr}
 & 23 \times 10 = 230 \\
+ & 23 \times 10 = 230 \\
+ & 23 \times 4 = 92 \\
\hline
 & 552
\end{array}
$$

Exercise N3.4

Use the partitioning method to answer questions 1–4 in your head.

1 Work out:

a 7×9	**b** 7×11	**c** 8×9	**d** 8×11
e 12×9	**f** 13×11	**g** 18×9	**h** 18×11
i 23×9	**j** 31×11	**k** 26×9	**l** 35×11

2 A music cassette costs £3.

a How much would 10 cassettes cost?
b How much would 11 cassettes cost?
c How much would 9 cassettes cost?

3 A magazine costs 80p.

a How much would 10 magazines cost?
b How much would 11 magazines cost?
c How much would 9 magazines cost?

4 A bag of porridge weighs 3.5 kg.

a How much would 10 bags weigh?
b How much would 9 bags weigh?
c How much would 11 bags weigh?

Do questions 5 – 8 using a written method.

5 Work out:

a 22×11	**b** 21×21	**c** 31×21	**d** 50×19
e 32×31	**f** 32×19	**g** 35×29	**h** 42×41

6 John is holding a big barbeque.
A bag of sausages costs £4.50. The shop only has 21 bags.
How much will it cost John to buy all 21 bags?

7 Sally and 18 other friends are going to a nightclub.
Tickets cost £7.50 each.
How much do Sally and her friends pay in total?

8 Josh weighs a packet of crisps. It weighs 55 g.
How much will 29 identical bags weigh?

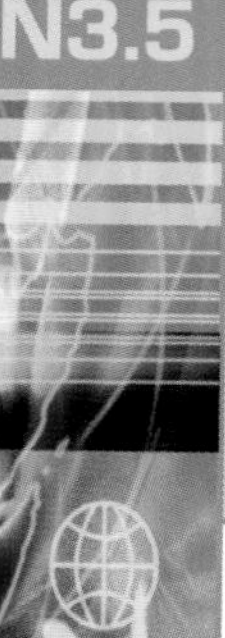

N3.5 Multiplying on paper

This spread will show you how to:
- Develop and refine written methods for multiplication.
- Estimate by approximating.

KEYWORDS
Estimate | Roughly
Round | Total
Decimal place

When you multiply large numbers, you need to use a written method. First you should **estimate** in your head.

To work out 28×43, first estimate:

- Round the numbers to the nearest 10: 30×40
- Multiply them together: 1200

Now work out the answer by using a written method:

- Split both numbers into tens and units and put them into a grid.

×	20	8
40		
3		

- Multiply the numbers together and write the answers in the grid.

×	20	8
40	800	320
3	60	24

- Add up the numbers inside the grid.

1204 is close to the estimate of 1200 so it is probably correct!

This method is called the **grid method**.

$$\begin{array}{r} 800 \\ 320 \\ 60 \\ +24 \\ \hline 1204 \end{array}$$

example

Nine planks of wood each measure 185 centimetres.
What is their total length?

First estimate:
9×185 is roughly 10×200.
$10 \times 200 = 2000$

Then calculate:

×	100	80	5
9	900	720	45

Add together:
$900 + 720 + 45 = 1665$

The answer is 1665 cm, or 16 m 65 cm

Exercise N3.5

1 Copy and complete these number patterns.

a
$4 \times 2 = 8$
$40 \times 2 = 80$
$4 \times 20 =$ ___
$40 \times 20 =$ ___
$400 \times 20 =$ ___
$40 \times 200 =$ ___

b
$5 \times 3 = 15$
$50 \times 3 = 150$
$5 \times 30 =$ ___
$50 \times 30 =$ ___
$500 \times 30 =$ ___
$50 \times 300 =$ ___

2 Work out the answers to each of these in your head.

a 30×3 **b** 3×20 **c** 50×5 **d** 80×4 **e** 30×30 **f** 30×20
g 50×50 **h** 80×40 **i** 300×30 **j** 30×200 **k** 500×50 **l** 800×400

3 For each of the following questions:
- ▶ Work out a rough estimate first.
- ▶ Use the grid method to work out the actual answer.

a 53×8 **b** 6×48 **c** 251×3 **d** 325×4 **e** 6×234 **f** 565×8
g 828×6 **h** 639×9 **i** 25×28 **j** 24×33 **k** 35×52 **l** 46×51

4 Four people each have £64.
a Work out a rough estimate for how much they have in total.
b Use the grid method to work out exactly how much they have.

For questions 5 to 8:
- ▶ Do a rough estimate first.
- ▶ Use the grid method of multiplication to work out the exact answer.

5 Janice runs 6 caravan parks. Each park has exactly 285 caravans.
How many caravans are there in total in the 6 parks?

6 Westdene High School puts on a play for 8 nights.
There are 175 tickets each night and each night is sold out.
How many tickets are sold altogether?

7 Stan buys some bricks. They are sold in packs of 75. He buys 25 packs.
How many bricks does he buy in total?

8 Plastic cups are sold in packs of 32.
How many cups are there in 36 packs?

9 Using the digits 2, 3, 4 and 5:
- ▶ Make a two-digit number. For example: 23.
- ▶ Make another number with the remaining digits. For example: 45.
- ▶ Multiply them together. 23×45
- ▶ What is the biggest answer you can make?
- ▶ Repeat with different two-digit numbers.
- ▶ What is the smallest answer you can make?

N3.6 Dividing on paper

This spread will show you how to:

- Understand division as it applies to whole numbers.
- Develop and refine written methods for division.
- Factorise numbers up to 100.
- Recognise multiples up to 10 × 10.

KEYWORDS

Divide, Subtract, Quotient, Multiple, Factor

This is a division: $20 \div 4 = 5$

These are the names for each part of the division: 20 — dividend, 4 — divisor, 5 — quotient

You use division when you share something.

Peter shares 12 presents equally between his 4 friends.
$12 \div 4 = 3$
They get 3 presents each.

You can also think of division as grouping.

12 people get into groups of 4.
$12 \div 4 = 3$
There are 3 groups altogether.

Repeated subtraction method

In this method you subtract multiples of the divisor until you can't subtract any more.

▸ Division is the opposite of multiplication: $5 \times 3 = 15$ so $15 \div 5 = 3$.

example

Work out:

a $20 \div 4$

b $161 \div 7$

a 4)20

20	
− 4	4 × 1
16	
− 4	4 × 1
12	
− 4	4 × 1
8	
− 4	4 × 1
4	
− 4	4 × 1
0	

You have subtracted 5 lots of 4.

So $4 \times 5 = 20$ and $20 \div 4 = 5$

b 7)161

161	
−70	7 × 10
91	
−70	7 × 10
21	
−21	7 × 3
0	

You have subtracted 23 lots of 7.

So $7 \times 23 = 161$
and $161 \div 7 = 23$

Exercise N3.6

1 Work out these divisions in your head.

> **Example**: $21 \div 3$
> $3 \times ? = 21$
> $3 \times 7 = 21$
> So $21 \div 3 = 7$

a $15 \div 5$ **b** $12 \div 3$ **c** $14 \div 7$ **d** $16 \div 4$ **e** $32 \div 4$ **f** $35 \div 5$

2 Work out these divisions on paper. Do a rough estimate first.
a $123 \div 3$ **b** $115 \div 5$ **c** $148 \div 4$ **d** $215 \div 5$ **e** $216 \div 3$ **f** $248 \div 8$

3 Work out which of the numbers in the grid are **multiples** of:
a 7 **b** 6 **c** 9

105	78	162	161	261	378	84	126	132	266

(Some are multiples of more than one number.)

4 Calculate:
a $184 \div 8$ **b** $468 \div 9$ **c** $217 \div 7$ **d** $210 \div 6$ **e** $360 \div 8$ **f** $648 \div 9$

5 **Investigation**
The number 12 has exactly six **factors** (1, 2, 3, 4, 6 and 12):
$1 \times 12 = 12$
$2 \times 6 = 12$
$3 \times 4 = 12$
a Find all the factors of 20, 60 and 180
b Find a number less than 100 with exactly seven factors.

6 **Investigation**
'All numbers that divide by 2 always end in a 0, 2, 4, 6 or 8 digit.'
For example, $124 \div 2 = 62$.
a Investigate numbers that divide by 5
What do you notice?
b Investigate other divisors.

7 Convert these fractions to mixed numbers.
(Turn to page 44 to remind yourself about improper fractions.)
a $\frac{24}{6}$ **b** $\frac{35}{7}$ **c** $\frac{112}{8}$ **d** $\frac{153}{9}$
Explain how you worked out your answers.

8 Calculate the following.
a $154 \div 11$ **b** $204 \div 12$ **c** $225 \div 15$ **d** $247 \div 13$
e $252 \div 21$ **f** $506 \div 22$ **g** $372 \div 31$ **h** $850 \div 25$

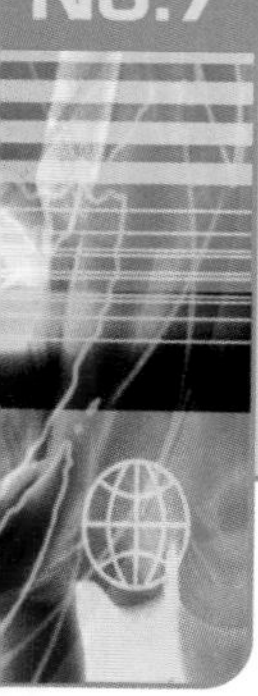

N3.7 Dividing with remainders

This spread will show you how to:

- Develop and refine written methods for division.
- Understand the operation of division and its relationship to multiplication.

KEYWORDS
Estimate
Roughly
Remainder
Decimal place

- **Division is the opposite of multiplication.**

Sometimes when you divide there is a remainder left over.

example

Work out $30 \div 7$

$$\begin{array}{r} 7\overline{)30} \\ -28 \\ \hline 2 \end{array} \qquad 7 \times 4$$

So $30 \div 7 = 4$ remainder 2

When the numbers are larger, you should estimate first.

example

Work out $333 \div 8$

First estimate: $320 \div 8 = 40$ so the answer should be roughly 40.

$$\begin{array}{rl} 8\overline{)333} & \\ -320 & \quad 8 \times 40 \\ 13 & \\ -\;\;8 & \quad 8 \times 1 \\ 5 & \end{array}$$

Check: $8 \times 41 + 5 = 333$

So $333 \div 8 = 41$ remainder 5

Exercise N3.7

1 Work out these divisions.
Be careful – they all have remainders!

a $18 \div 5$ **b** $21 \div 4$ **c** $13 \div 4$
d $17 \div 8$ **e** $26 \div 9$ **f** $35 \div 8$

2 **Remainder Max Game**
(2 players)

- The object of the game is to have the largest remainder.
- Each player takes it in turns to choose one number from Box A and one from Box B.
- Each number can only be chosen once.

Box A			
38	47	31	56
22	62	49	75

Box B			
3	5	7	9
4	6	8	10

- Divide the number in box A by the number in Box B.
  The player scores the value of the remainder.
- After four turns each the players add up the total.
- The winner is the player with the largest total remainder.

3 These divisions all have remainders.
Calculate:

a $130 \div 4$ **b** $155 \div 3$ **c** $241 \div 7$
d $493 \div 8$ **e** $310 \div 3$ **f** $263 \div 4$
g $430 \div 6$ **h** $452 \div 8$ **i** $845 \div 4$
j $913 \div 7$ **k** $513 \div 8$ **l** $823 \div 9$

4 Identify which of the numbers in the grid are multiples of:

a 12 **b** 15 **c** 13

180	195	228	312	156	104	240	169

5 Calculate the following.
Some of them have remainders.

a $130 \div 11$ **b** $155 \div 12$ **c** $238 \div 17$
d $263 \div 15$ **e** $420 \div 20$ **f** $323 \div 19$
g $208 \div 13$ **h** $270 \div 15$ **i** $188 \div 14$
j $548 \div 15$ **k** $936 \div 18$ **l** $517 \div 11$

6 Convert these improper fractions to mixed numbers.

a $\frac{23}{7}$ **b** $\frac{48}{5}$ **c** $\frac{92}{8}$
d $\frac{107}{4}$ **e** $\frac{128}{7}$ **f** $\frac{163}{10}$

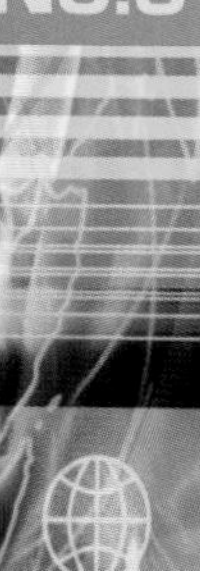

N3.8 Calculator methods

This spread will show you how to:
- Develop calculator skills and use a calculator effectively.
- Estimate by approximating and check the results.

KEYWORDS
Calculator Estimate
Interpret Brackets

You need to be able to decide how to solve problems in maths.
Here are three problems:

5+3×8
I would do this in my head because it's easy.

54×67
I would do this on paper, but estimate in my head first.

70÷34
This one's difficult! I would use a calculator, but estimate first.

You should use a calculator for problems that:

- are too difficult to work out in your head, or
- take too long to work out on paper.

example

Work out $(4.3 + 2.6) \times 4.8$

The example uses a Casio fx 82 scientific calculator.

First estimate: $(4 + 3) \times 5$
$= 7 \times 5 = 35$
Then use your calculator:

a Using brackets
Input: $(4.3 + 2.6) \times 4.8 =$
The output should be: 33.12

b Using the memory keys
Input: $4.3 + 2.6 =$
The answer should be: 6.9
Press: M+
Press: C
Press: MR × 4.8 =
The output should be: 33.12

M+ puts your answer in the memory.
C clears the screen.
MR recalls the memory.

- **When you use your calculator you must interpret the display to find the answer.**

example

£801 is to be shared between 45 people.
How much will each person receive?

First estimate: $800 \div 50 = 16$
Then use a calculator: $801 \div 45 = 17.8$
This is in pounds, so the answer is £17.80.

£17 and 8p is 17.08

Exercise N3.8

You will need a scientific calculator for this exercise.

1 This question refers to the keys on a Casio fx82.

a
- Work out $3 + 8 \div 2$ in your head. (Remember the order of operations.)
- Work out $8 \div 2$ on your calculator.
- Press the M+ key to store your answer
- Input 3 + MR
- You should have 7 as the answer!

b
- Work out $12 - 3 \times 3$ in your head.
- Input $3 \times 3 =$
- Press the M+ key to store your answer.
- Input 12 – MR.
- You should have 3 as the answer.

2 Work out each of these by using the memory keys on your calculator.
In each case, first make an estimate.

a $0.9 + 1.6 \times 2.1$ **b** $2 + 3 \times 1.8$ **c** $20 - 1.3 \times 5$
d $9 + 0.7 \div 1.4$ **e** $11.6 - 10 \div 2$ **f** $212 - 30.6 \div 3$

3 Work these out by using the bracket keys where appropriate.

a $5 \times (2 + 4)$ **b** $5 \times 2 + 4$ **c** $0.2 \times (3.6 + 4.4)$
d $8 \div (1.7 - 0.9)$ **e** $16.2 \times (4.3 + 1.2)$ **f** $174.8 \div (4.2 + 3.4)$

4 **a** Each day, John spends £2.65 and Paul spends £3.78 on sweets and comics. How much do they spend altogether in a week?
b Rukshana saves £12.45 per month. How much will she have saved after 3 years?
c Calculate the total cost of:
- 17 rulers at 32p each,
- 64 rubbers at 12p each,
- 14 pairs of compasses at £1.15 each and
- 27 refill pads at £1.99 each.

5 You will need to decide whether you should use a mental, written or calculator method to calculate.

a 12×11 **b** $84 \div 4$ **c** 23×25
d $85 \div 5$ **e** 37×23 **f** 0.72×100
g 3.8×5 **h** $14.2 \div 5$ **i** $36.4 \div 13$
j $0.9 + 5 \times 0.2$ **k** $4362 \div 100$ **l** $16.4 + 3.6 \times 2.8$

6 **The Approximation Game**
- Player 1 chooses 3 numbers from the grid.

42.3	14.9	38.4	16.4	7.3	29.2	9.2	23.1

- Player 1 places the 3 numbers into the equation: $\square \times \square + \square =$
- Player 2 gives an estimate of the answer.
  The difference between the estimate and the real answer is the score for Player 1.
- Player 2 now chooses 3 numbers and Player 1 gives an estimate of the answer.
- The winner is the first player to reach 100.

Summary

You should know how to ...

1 Know multiplication facts up to 10 × 10, and related division facts.

2 Know and use the order of operations including brackets.

3 Extend mental methods of calculation.

4 Identify and use appropriate operations to solve word problems involving numbers and quantities.

Check out

1 Work out each of these:

a 4×3
b $20 \div 5$
c 6×4
d 7×5
e $60 \div 10$
f 9×7
g $48 \div 6$
h 30×10
i 65×100
j $3200 \div 100$
k 4.5×10
j $52 \div 10$

2 Work out these by doing the operations in the correct order:

a $2 + 3 \times 5$
b $2 \times 8 - 3$
c $4 + 8 \div 2$
d $8 \div 4 - 2$
e $2 + (4 + 8) \div 3$
f $(3 + 2) \times (8 - 6)$

3 Work out these multiplications in your head:

a 23×11
b 45×9
c 36×12
d 27×19

4 For each of these questions, do a rough estimate first.

a Football stickers are sold in packs of 6.
Jane buys 38 packs.
How many stickers does she have?

b Candles are sold in packs of 25.
Ahmet buys 36 packs.
How many candles does he buy?

c Michelle works in an egg packing factory.
Eggs are packed in boxes of six.
If Michelle packs 492 eggs per hour, how many boxes does she fill in two hours?

d The students at Superspeed School run a 1600 metre relay race.
The winning team of four students run the race in 336 seconds.
How far did each student run?
What was the average time taken?

13 Functions and graphs

This unit will show you how to:

- Recognise and extend number sequences.
- Recognise multiples up to 10 ×10.
- Recognise prime numbers to at least 20.
- Know and apply simple tests of divisibility.
- Know squares of numbers to at least 10 × 10.
- Recognise the first few triangular numbers.
- Factorise numbers up to 100.
- Read and plot coordinates in the first quadrant.
- Solve mathematical problems or puzzles, recognise and explain patterns and relationships, generalise and predict.

Using formulae can help solve problems.

Before you start

You should know how to ...

1 Use the 2, 5, 10 times tables.

2 Recognise odd and even numbers.

3 Plot coordinates on a set of axes.

Check in

1 Copy and complete the following questions:

a	_ × 3 = 30	**b**	10 × 7 = _
c	_ × 5 = 35	**d**	15 ÷ 3 = _
e	2 × _ = 18	**f**	_ × 6 = 60

2 **a** Write whether each of these is odd or even.
12, 15, 100, 2657, 3001, 22 223

b Which of these would you be able to share equally with another person?

7 pens	12 pens	9 sweets
4 books	13 photographs	3 videos

3 What shape do you get if you plot the following coordinates on a grid labelled from 0 – 5 on the x and the y-axis? (Join the points as you go!)
(2, 1) (5, 1) (5, 2) (3, 2) (3, 5) (2, 5) (2, 1)

A3.1 Factors and primes

This spread will show you how to:

- Know and apply simple tests of divisibility.
- Factorise numbers up to 100.
- Recognise prime numbers up to 20.

KEYWORDS

Divide
Factor
Prime
Remainder

You can divide in your head when there is no remainder if you know your times tables.

- **A factor divides into a number without leaving a remainder:**
  $6 \div 3 = 2$, so 3 is a factor of 6.

There are some simple tests that will tell you if a number is a factor:

- 2 is a factor of numbers that end in 0, 2, 4, 6 or 8 – the even numbers.
- 3 is a factor of a number if you add the digits together and 3 is a factor of that.
- 4 is a factor of a number if 2 is a factor of it twice.
- 5 is a factor of numbers that end in 0 or 5 – the 5 times table.
- 10 is a factor of numbers that end in 0 – the 10 times table.

example

Is 2, 3, 4, 5 or 10 a factor of 651?

651 ends in 1 so 2, 4, 5 and 10 cannot be factors.
$6 + 5 + 1 = 12$ and 3 is a factor of 12 so 3 is a factor of 651.

- **You can list all the factor pairs of a number.**

example

List all the factor pairs of 15.

Start with 1: $15 = 1 \times 15$

15 ends in a 5 so 2 is not a factor.

Try 3: $1 + 5 = 6$ so 3 is a factor. $15 = 3 \times 5$

2 is not a factor so 4 is not a factor.

You already know 5 is a factor so you now have all the factor pairs.

If you follow the horseshoe you can list the factors of 15: 1, 3, 5, 15

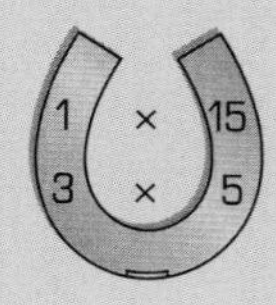

- **If there are only two different factors, the number is a prime number.**
  For example: $5 = 1 \times 5$, so 5 is prime.

Exercise A3.1

1 Is 2, 5 or 10 a factor of:

a 20 **b** 11 **c** 15 **d** 100 **e** 24
f 35 **g** 16 **h** 75 **i** 40 **j** 70

2 Is 2, 3, 4, 5 or 10 a factor of:

a 6 **b** 64 **c** 65 **d** 45 **e** 24
f 25 **g** 12 **h** 60 **i** 26 **j** 11

3 Find all the factor pairs of each of these numbers.

a 10 **b** 15 **c** 20 **d** 22 **e** 13
f 32 **g** 11 **h** 28 **i** 35 **j** 19

4 16 has 5 factors:

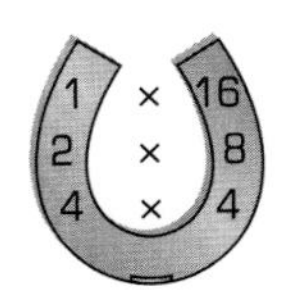

They are 1, 2, 4, 8 and 16.
How many factors do each of these numbers have?
List them.

a 10 **b** 15 **c** 20 **d** 22 **e** 13
f 32 **g** 11 **h** 28 **i** 35 **j** 19

5 A prime number has exactly two different factors.
List all the factors of each of these numbers and say whether they are prime or not.

a 9 **b** 14 **c** 17 **d** 21 **e** 23
f 11 **g** 27 **h** 2 **i** 3 **j** 4

6 Is 1 a prime number? Explain your answer.

7 **a** Use your answers to question 5 and 6 to list all the prime numbers between 1 and 20.

b Write down any even prime numbers.
What do you notice?
Explain your answer.

8 **Investigation**
Find a number between 0 and 100 with:

a exactly 3 factors
b exactly 5 factors
c exactly 7 factors

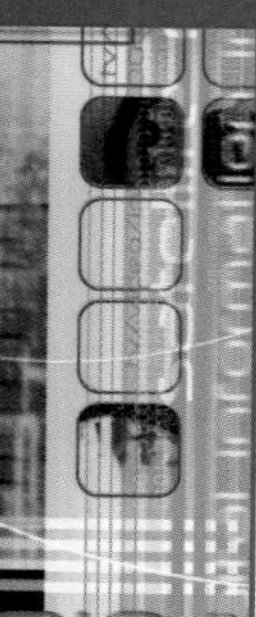

A3.2 Patterns in numbers

This spread will show you how to:

- Recognise squares of numbers.
- Factorise numbers up to 100.

KEYWORDS

Factor
Side
Length
Square number

You can draw any number as a rectangle.
The lengths of the sides of the rectangle are the factor pairs of the number.

example

Arrange 18 in as many different rectangles as you can.

You can have 1×18 – this is the same as 18×1

You can have 2×9 – this is the same as 9×2

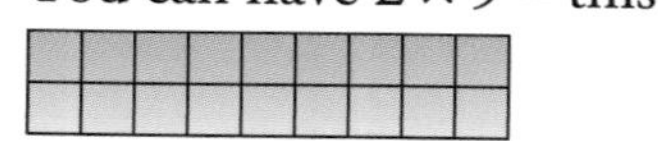

You can have 3×6 – this is the same as 6×3

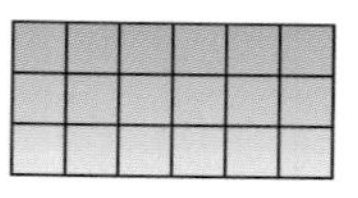

The factor pairs of 18 are:
1×18
2×9
3×6

- You can only draw one rectangle for a prime number.
  For example, 7 can be drawn as 1×7 only so it is prime.

A rectangle with all its sides equal is a square:

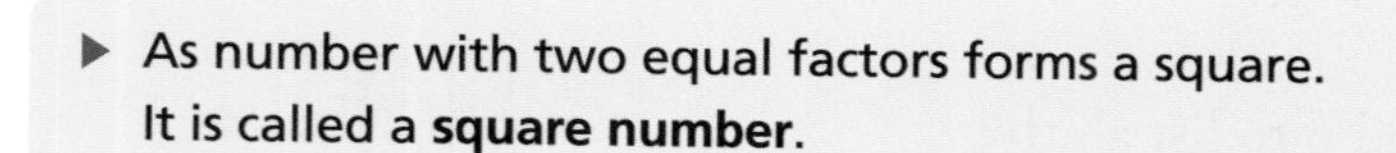

- As number with two equal factors forms a square.
  It is called a **square number**.

example

List the factor pairs of 16. Is it a square number? Give a reason for your answer.

The factor pairs of 16 are: 1×16
2×8
4×4

16 has two equal factors – 4×4 and so it is a square number.

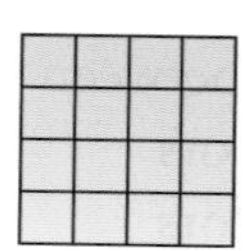

Exercise A3.2

1 The factor pairs of 32 are:

1×32 $\quad 2 \times 16$ $\quad 4 \times 8$

a How many factors does 32 have?
b Is 32 a square number?
Explain your answer.

2 Write down all the factor pairs of 24.
Is 24 a square number?
Explain your answer.

3 This rectangle measures 2 cm by 10 cm:

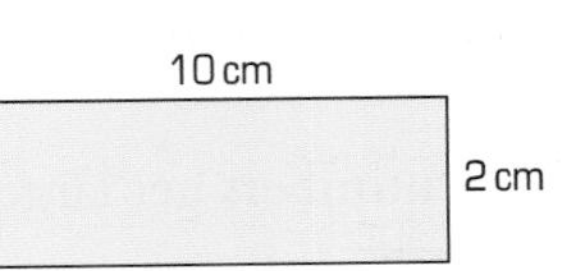

$2 \times 10 = 20$

The area of the rectangle is 20 cm^2.

a Draw as many different rectangles as you can with an area of 20 cm^2.
b Write down all the factor pairs of 20.
c Is 20 a square number? Explain your answer.

4 This rectangle measures 3 cm by 12 cm:

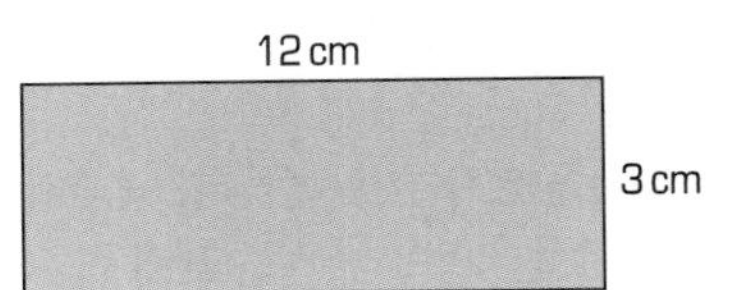

$3 \times 12 = 36$

The area of the rectangle is 36 cm^2.

a Draw as many different rectangles as you can with an area of 36 cm^2.
b Is 36 a square number? Explain your answer.

5 Draw a square with an **area** of 25 cm^2.
What is the length of each side?

6 Draw a square with an area of 49 cm^2.
What is the length of each side?

7 You find a square number when you multiply a number by itself.

For example: $1 \times 1 = 1$ and $2 \times 2 = 4$ so 1 and 4 are square numbers.

a Find all the square numbers up to 10×10.
b For each square number list all its factors.
c How many factors does each square number have?
Explain what you notice.

A3.3 Squares and triangles

This spread will show you how to:

- Recognise squares of numbers up to 10 × 10.
- Recognise triangular numbers.
- Recognise and extend number sequences.

KEYWORDS

Square number
Triangular number
Generate
Squared
Sequence

You can generate square numbers by multiplying.

The first square number is: $1 \times 1 = 2$
The second square number is: $2 \times 2 = 4$
The third square number is: $3 \times 3 = 9$

The sides of a square are the same length. To find the area you multiply the length by itself!

length
Area
length

Once the numbers get larger you can use a calculator.

Find the x^2 key on your calculator.
Check you can use it to show $11 \times 11 = 121$.
You say '11 squared is 121'.

▶ **The square numbers form a sequence.**
You can write the sequence using patterns or numbers.

Here are the first four square numbers:

1st

1

2nd

4

3rd

9

4th

16

You can draw 36 in many different ways.

As a rectangle:

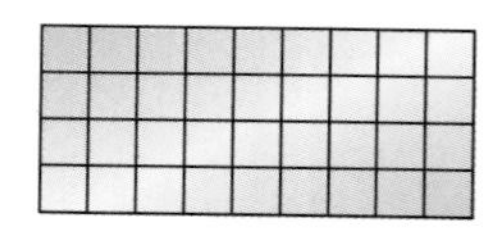

As a square:

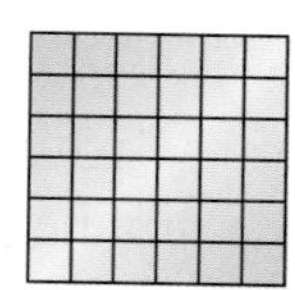

As a triangle:

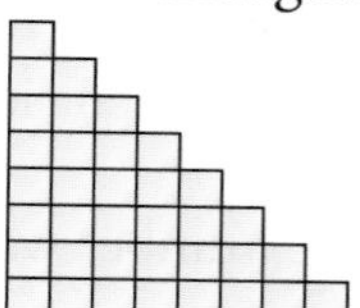

▶ A **triangular number** can be drawn as a right-angled isosceles triangle.

Here are the first four triangular numbers:

1st

1

2nd

3

3rd

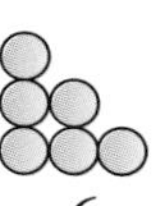

6

4th

10

Exercise A3.3

1 Use your calculator to work out the following.

a 7 squared **b** 12 squared **c** 6^2 **d** 13 squared **e** 29^2 **f** 18 squared
g 20^2 **h** 25 squared **i** 30^2 **j** 100 squared **k** 14^2 **l** 27 squared

2 Here are the first four square numbers:

1st:  1

2nd: 4

3rd: 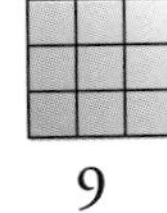 9

4th: 16

Copy and continue the sequence for the first 10 square numbers.

3 Here are the first four triangular numbers:

1st:  1

2nd: 3

3rd: 6

4th: 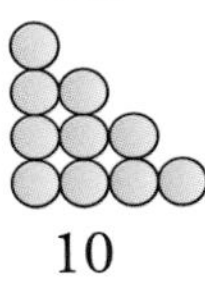 10

Copy and complete the sequence for the first 10 triangular numbers.

4 On a 10 × 10 multiplication grid, shade in the square numbers you found in question 2.
Describe any pattern you can see.

5 **a** On this multiplication grid:

The 1st square number: 1 × 1 is shaded red.

The extra squares for the 2nd square number: 2 × 2 are shaded blue.

The extra squares for the 3rd square number: 3 × 3 are shaded green.

1	2	3	4	5	6	7	8	9	10
2	4	6	8	10	12	14	16	18	20
3	6	9	12	15	18	21	24	27	30
4	8	12	16	20	24	28	32	36	40
5	10	15	20	25	30	35	40	45	50
6	12	18	24	30	36	42	48	54	60
7	14	21	28	35	42	49	56	63	70
8	16	24	32	40	48	56	64	72	80
9	18	27	36	45	54	63	72	81	90
10	20	30	40	50	60	70	80	90	100

Copy and complete the pattern up to the 10th square number.

b Copy and complete this table.

Square number	1st	2nd	3rd	4th	5th	6th
Number of **extra** squares	1	3				

A3.4 Functions and multiples

This spread will show you how to:

- Recognise multiples up to 10 × 10.
- Find simple common multiples.

KEYWORDS

Function, Multiple, Input, Output, Function machine

- In a function machine:
  - The **input** value is the value you put in to the machine.
  - The machine performs the **function**.
  - The **output** value is the result.

This is a function machine. The function is ×2:

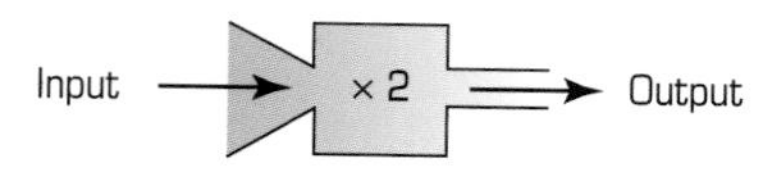

You input a number → The machine multiplies it by 2 → It outputs the result

If you input the counting numbers 1,2,3,4, ...

1
2
3
4
⋮

× 2

the output numbers are the 2 times table:

2
4
6
8
⋮

When you multiply by 2 you get a multiple of 2.
The numbers in the 2 times table are the multiples of 2.

- When you multiply by a whole number you get a **multiple** of the number.
  The multiple is a number from the times table.

This function machine generates multiples of 5:

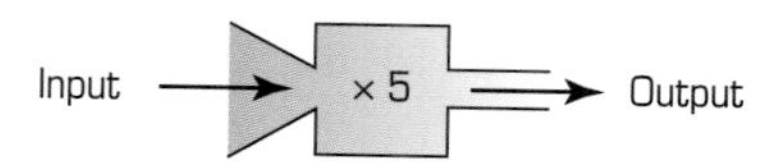

example

Write down three multiples of 5.

When you multiply by 5 you get a multiple of 5.
Multiply 5 by 1, 2 and 3:

$5 \times 1 = 5$
$5 \times 2 = 10$
$5 \times 3 = 15$

So 5, 10 and 15 are all multiples of 5.

To find a multiple you must use a whole number.

Exercise A3.4

1 Copy and complete these multiplication sums.

$1 \times 3 =$ $\quad 2 \times 3 =$ $\quad 3 \times 3 =$ $\quad 4 \times 3 =$

The answers are the first four multiples of 3.

2 Find the first 10 multiples of: **a** 3 **b** 4 **c** 6

3 This function machine generates multiples of 5:

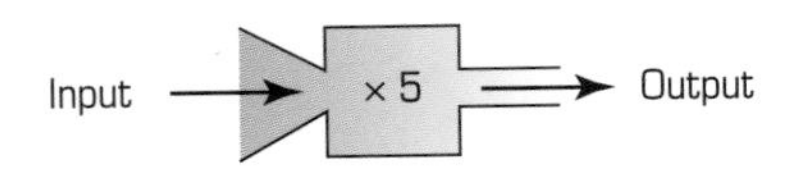

a Input 4 to find the 4th multiple of 5.
b Input 5 to find the 5th multiple of 5.
c Input 6 to find the 6th multiple of 5.
d Find the 8th multiple of 5.

4 The first 10 multiples of 2 are:
2, 4, 6, 8, 10, 12, 14, 16, 18, 20.
The first 10 multiples of 5 are:
5, 10, 15, 20, 25, 30, 35, 40, 45, 50.
The numbers common to both lists are 10 and 20.
Use your results from question 2 to answer these questions.
What numbers are common to the lists of:
a multiples of 3 and multiples of 4
b multiples of 4 and multiples of 6
c multiples of 3 and multiples of 4 and multiples of 6?

10 and 20 are common multiples of 2 and 5.
10 is the lowest common multiple of 2 and 5.

5 A dancefloor measures 12 metres by 12 metres.
Wooden tiles come in these sizes:
1×1 $\quad 2 \times 2$ $\quad 3 \times 3$ $\quad 4 \times 4$ $\quad 5 \times 5$ $\quad 6 \times 6$ $\quad 8 \times 8$
If you use 1×1 tiles you will need 144 tiles.

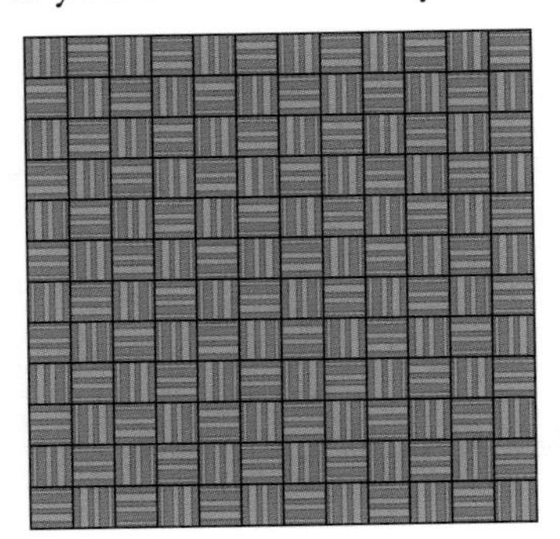

How many tiles will you need if you use:
a 2×2 tiles **b** 3×3 tiles **c** 4×4 tiles
d 5×5 tiles **e** 6×6 tiles **f** 8×8 tiles?

A3.5 Graphs of functions

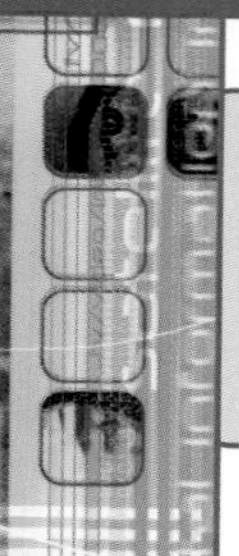

This spread will show you how to:

- Read and plot coordinates in the first quadrant.
- Recognise multiples up to 10 × 10.

KEYWORDS

Function machine
Coordinates
Multiple
Input
Output

This function machine generates multiples of 3:

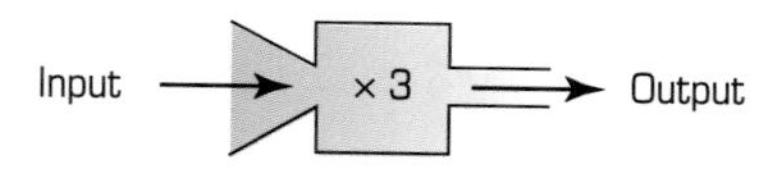

You only use whole number values for multiples.

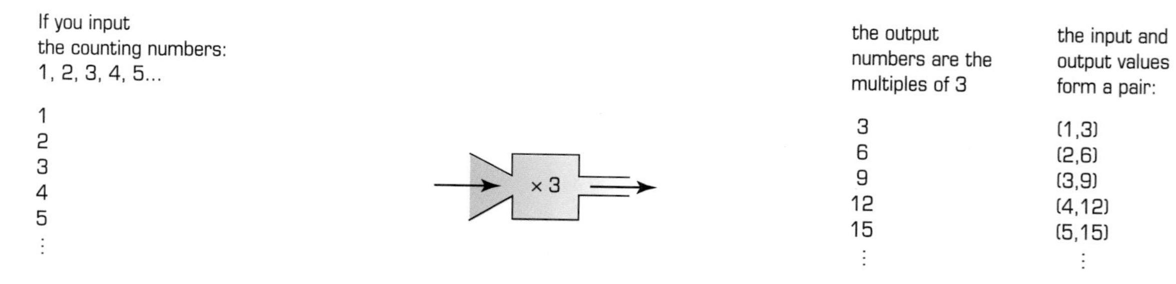

You can plot the input and output pairs of values on a coordinate grid.
The input is the x value and the output is the y value.

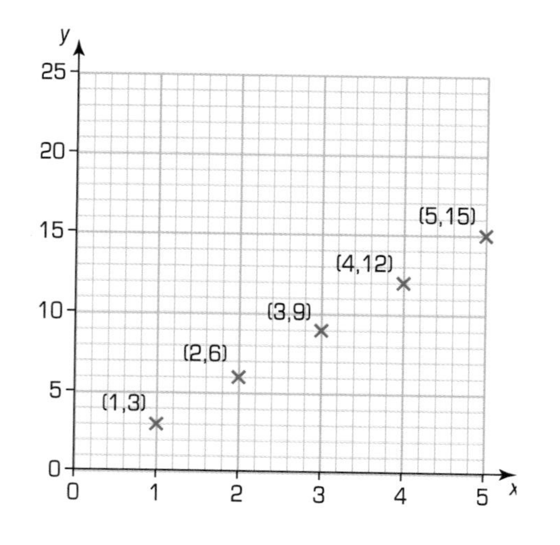

The coordinate pair (1, 3) means 1 across and 3 up

When the input or x value is 6, the output or y value is 18: (6, 18) is the next point.

▶ You can use the pattern of coordinates to predict an output for any given input value.

Exercise A3.5

1 Plot these sets of points on a coordinate grid.
Join each set together as you go.

Set A	(1, 8) (1, 10)
Set B	(4, 8) (3, 8) (3, 10)
Set C	(5, 8) (4, 9) (5, 10) (6, 9)
Set D	(6, 10) (7, 8) (8, 10)
Set E	(9, 8) (8, 8) (8, 10) (9, 10)
Set F	(8, 9) (9, 9)
Set G	(1, 4) (1, 7) (2, 6) (3, 7) (3, 4)
Set H	(4, 7) (5, 6) (6, 7)
Set I	(5, 6) (5, 4)

This should spell: I LOVE MY

2 Write a word to complete the sentence in question 1.
For example you might use: DOG, CAT, MUM, BIKE.
Write it on the coordinate grid.
Write down the coordinates to make your word.

3 This function machine generates multiples of 5:

Input	Function	Output	Coordinate pair
1	$\times 5$	5	(1,5)
2		10	(2,10)
3		15	(3,15)
4		20	(4,20)

Plot the pairs on a coordinate grid.
Copy the grid on page 126.

4 This machine generates multiples of 4:

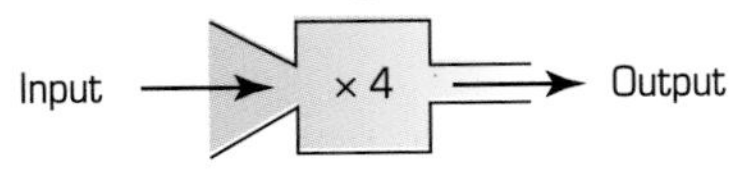

a Input the numbers 1, 2, 3, 4 and 5.
b Write down the coordinate pairs.
c Plot the pairs on a coordinate grid
Copy the grid on page 126.

5 Plot these coordinates on a grid:

(3, 4), (3, 8) and (7, 8).

The coordinates are three corners of a square.
Write down the coordinates of the point that completes the square.

A3.6 Using a table of values

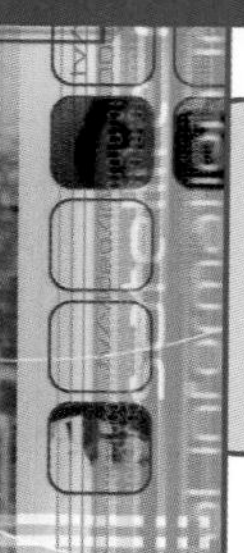

This spread will show you how to:

- Read and plot coordinates in the first quadrant.

KEYWORDS
Coordinate pair
Function machine
Straight line   Table

You can draw a graph for any function machine.
It can help to use a table for the input and output values.

example

Draw a graph of this function machine:

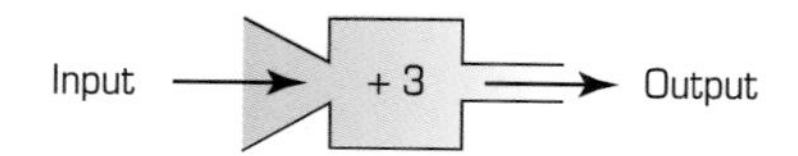

You need to generate some coordinate pairs.
Choose some input values and put them in a table.

Input	1	2	3	4	5
Output	4	5	6	7	8

It helps to keep the numbers small so they will fit on a grid.

Fill in the output values by adding 3 to each input.

Now write down the coordinate pairs:

(1, 4) (2, 5) (3, 6) (4, 7) (5, 8)

Plot the points:

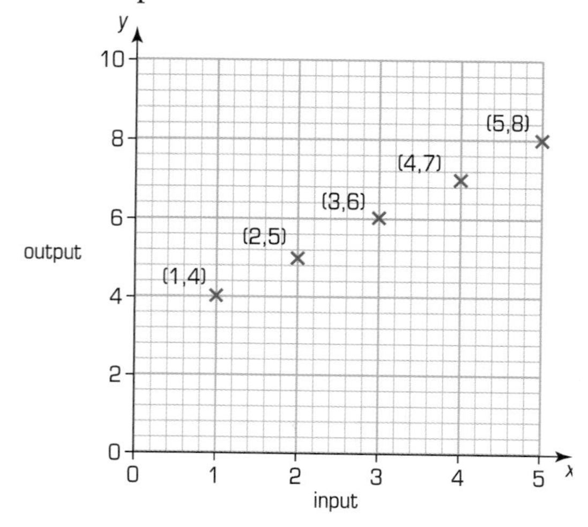

You can join the points to make a straight line.

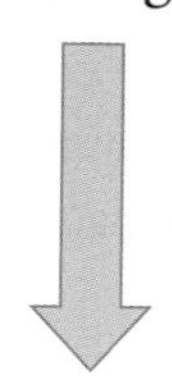

Now you can see that when the input is 0.5, the output is 3.5.

Exercise A3.6

1 **a** Copy and complete this function machine:

Input	Function	Output	Coordinate pair
1	+4	5	(1,5)
2			(2,)
3			(3,)
4			(4,)

b Draw the graph of the function machine.
Join the points together. You should get a straight line.

2 This table of values is generated by the function machine:

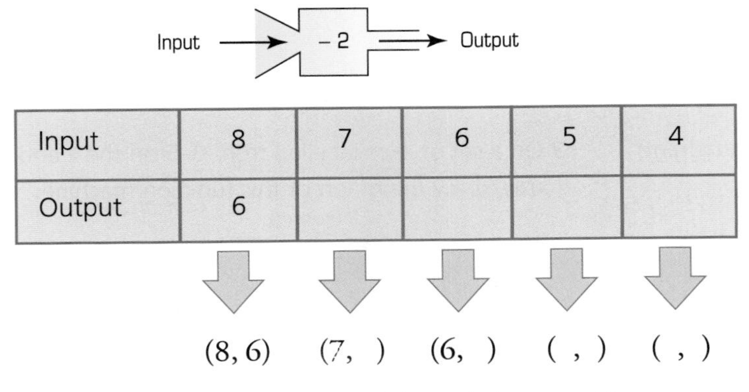

Input	8	7	6	5	4
Output	6				

(8, 6) (7,) (6,) (,) (,)

a Copy and complete the table and the coordinate pairs.
b Plot the pairs on a coordinate grid.
Join the points to make a straight line.

3 **a** Copy and complete this table of values for the function: -1.

Input	5	4	3	2
Output	4	3		

b Write down the coordinate pairs and plot them on a graph.
Join the points to make a straight line.
c What is the output when the input is 3.5?

4 **a** Choose some input values for the function: +2.
b Use the values to generate some coordinate pairs.
c Plot the pairs on a grid and join them to make a straight line.
d What input gives an output of 3.5?

Summary

You should know how to ...

1 Use simple tests of divisibility.

2 Factorise numbers up to 100 and recognise squares of numbers.

3 Read and plot coordinates in the first quadrant.

4 Solve mathematical problems or puzzles, recognise and explain patterns and relationships, generalise and predict.

Check out

1 Use simple tests of divisibility to decide which of the following numbers are multiples of 2, 3, 4, 5 or 10:

50 36 41 49 100 231 20 25

2 Find all the factor pairs of the square numbers in the above list.

3 On a set of axes labelled from 0–5 on the *x* and *y* axis, draw the graph of this function machine:

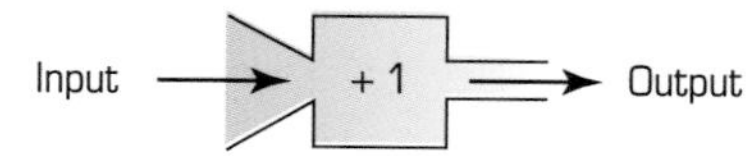

4 Here is a sequence of patterns made from matchsticks:

a Copy and complete this table:

Pattern number	1	2	3
Number of matchsticks	4		

b How many matchsticks will be needed for the 4th pattern?
Check your answer by drawing.

c How many matchsticks will be needed for the 20th pattern?

13 Triangles and quadrilaterals

This unit will show you how to:

- Recognise and estimate acute and obtuse angles.
- Use a protractor to measure and draw acute and obtuse angles to the nearest degree.
- Identify parallel and perpendicular lines.
- Calculate the sum of angles at a point, on a straight line and in a triangle.
- Check that the sum of angles in a triangle is 180°.
- Recognise properties of quadrilaterals.
- Visualise 3-D shapes from 2-D drawings.
- Solve mathematical problems or puzzles, recognise and explain patterns and relationships, generalise and predict.

Architects make accurate drawings of buildings.

Before you start

You should know how to ...

1 Recognise and name types of triangles.

Check in

1 Name each of these types of triangles:

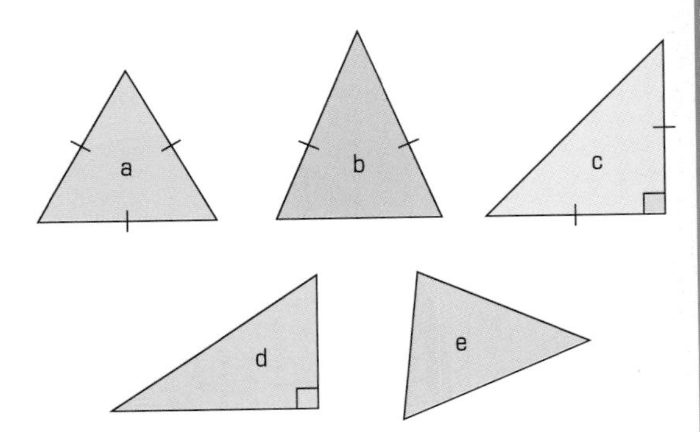

Measuring angles

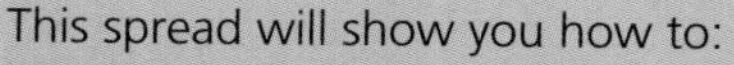

This spread will show you how to:

- Recognise and estimate acute and obtuse angles.
- Use a protractor to measure angles to the nearest degree.

KEYWORDS

Angle, Measure, Degree°, Protractor, Estimate

An angle is a measure of turn. You can measure the turn in degrees, ° for short.

Amount of turn	$\frac{1}{4}$ turn	$\frac{1}{2}$ turn	$\frac{3}{4}$ turn	full turn
	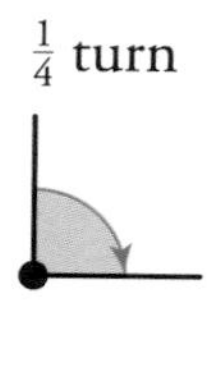			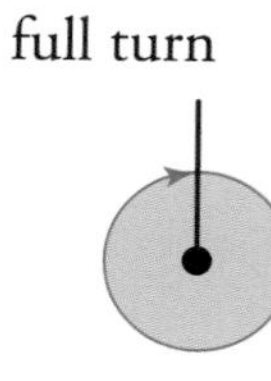
Angle in degrees	90°	180°	270°	360°

You can describe an angle depending on its size:

an **acute** angle is less than 90°

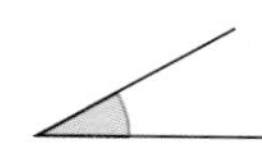

a **right** angle is exactly 90°

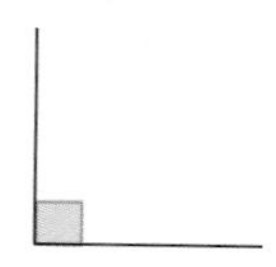

an **obtuse** angle is between 90° and 180°

a **reflex** angle is more than 180°

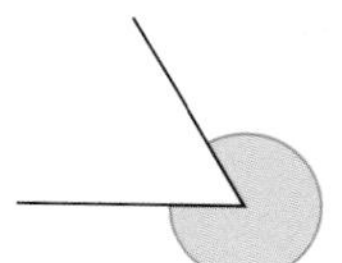

You measure an angle in degrees using a protractor.
You should estimate before measuring.

This angle is acute – it is less than 90°.
It is just over halfway so it is about 50°.

To use a protractor you must extend the arms of the angle:

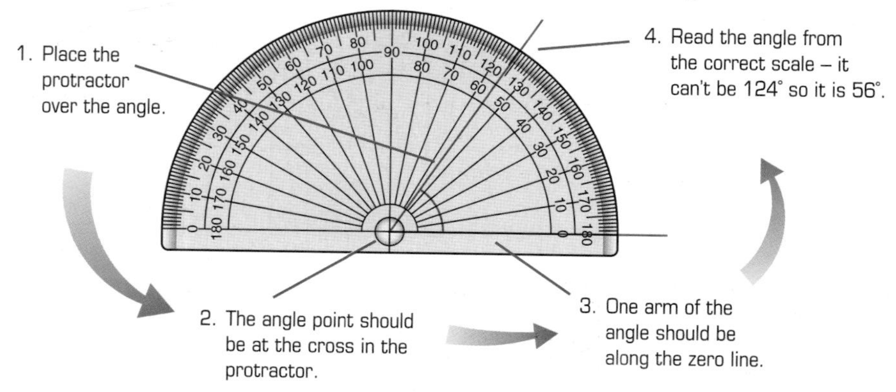

Exercise S3.1

Measure these angles.

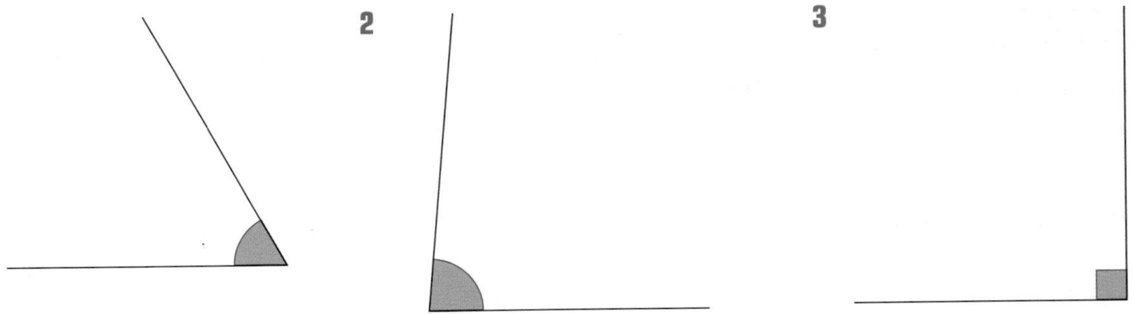

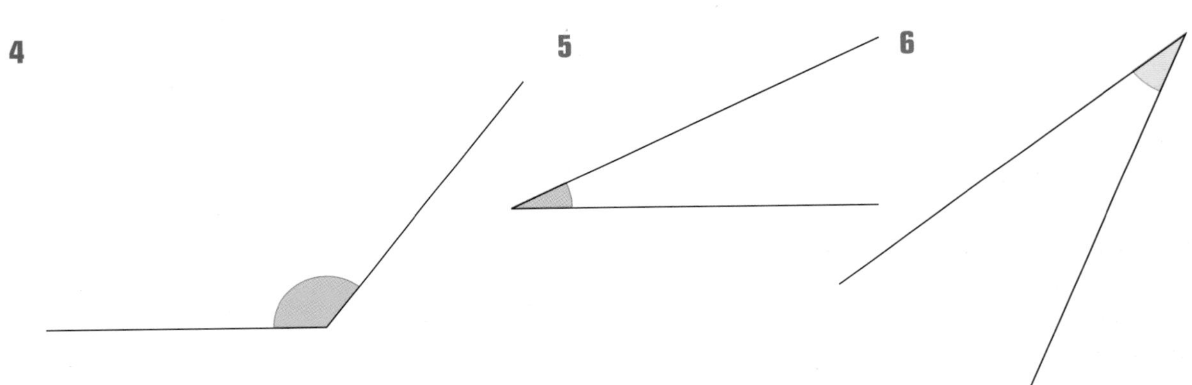

7 Use the diagram of the protractor to answer the question.

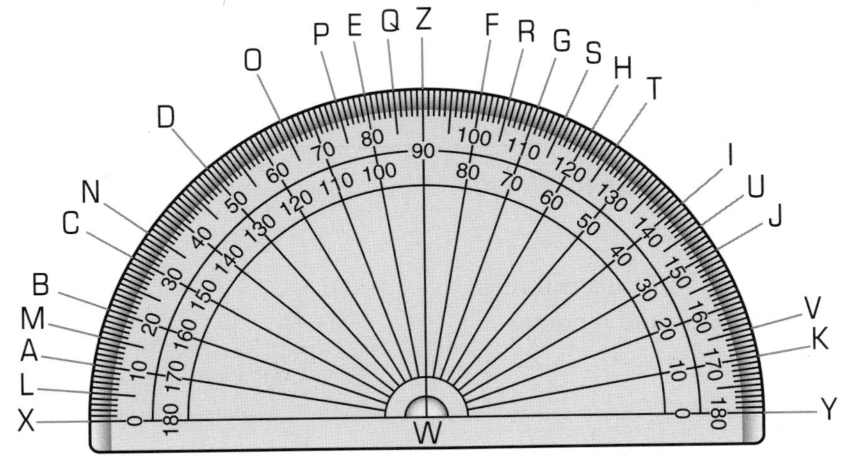

a Write down the angles shown by the letters A to K on the clockwise scale.

b Write down the angles shown by the letters L to V on the anticlockwise scale.

c On which scale does the letter X show 0°?

d On which scale does the letter Y show 180°?

S3.2 Finding angles

This spread will show you how to:

- Calculate angles at a point and on a straight line.

KEYWORDS

Degree° Protractor
Angles on a straight line
Angles at a point

There are 180° in a half turn:

A protractor measures 180°

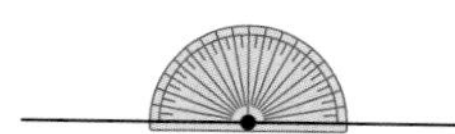

A half turn makes a straight line.

► **There are 180° on a straight line.**

This can help you solve problems involving angles.

example

Siân is tiling. She has a piece that is cut at 62°.
What angle tile does she need to make a straight line?

There are 180° on a straight line.

She has 62° already. 62° + 118° = 180°.
She needs a tile cut at 118° to make a straight line.

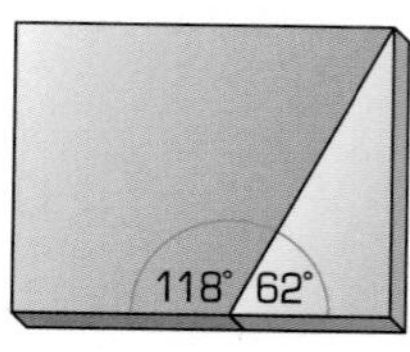

A full turn takes you back to where you started.

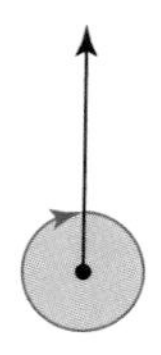

There are 360° in a full turn:
You turn around a point.

► **There are 360° at a point.**

The blue sector has an angle of 234°.

360° – 234° = 126°

The red sector has an angle of 126°.

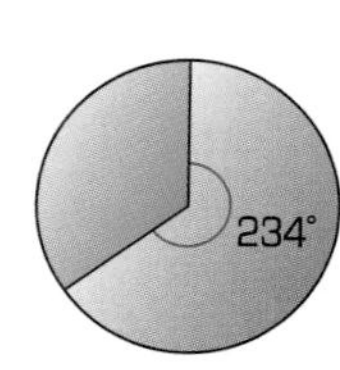

Exercise S3.2

Copy each diagram and find the unknown angles.

1

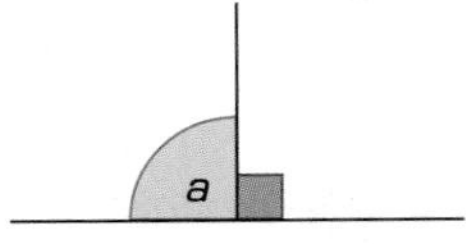

2

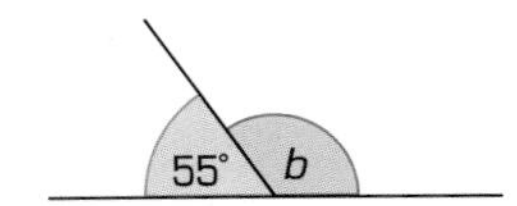

3

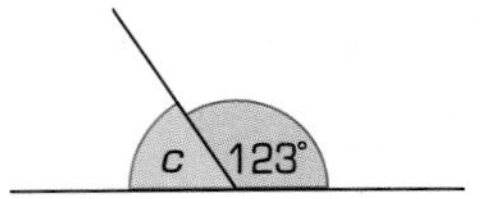

4

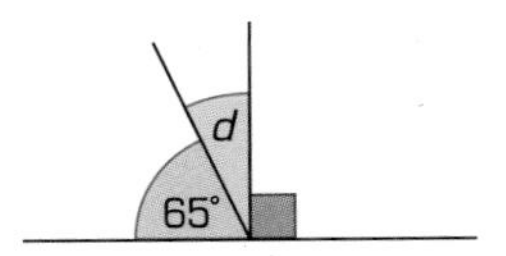

5

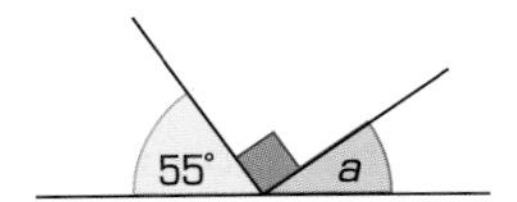

6

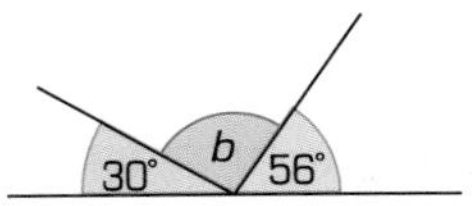

7

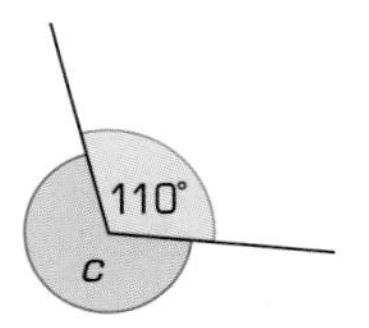

8

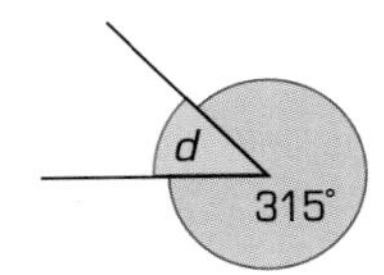

9

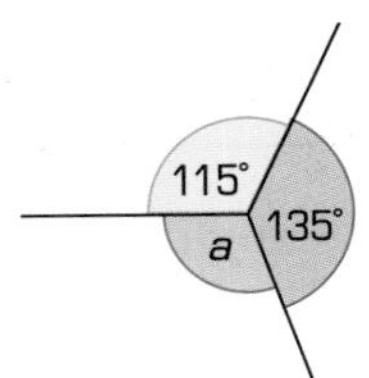

10

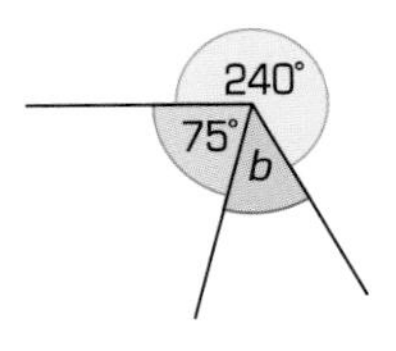

11

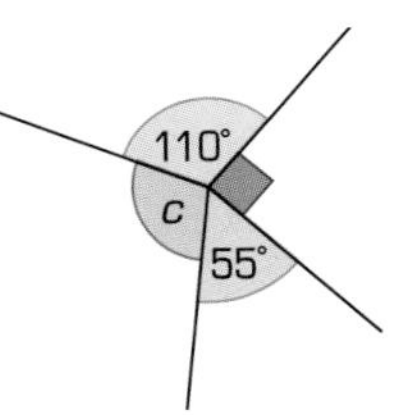

12

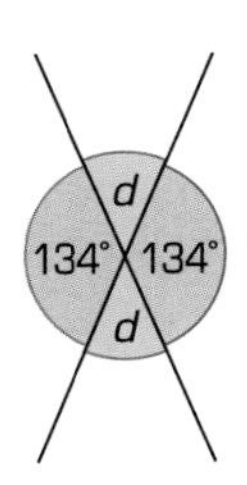

13

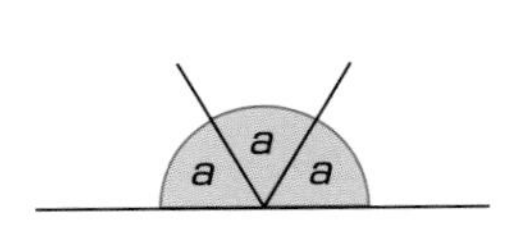

14

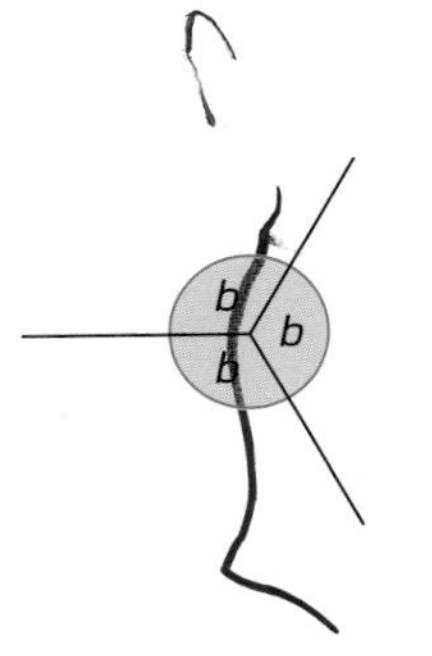

15

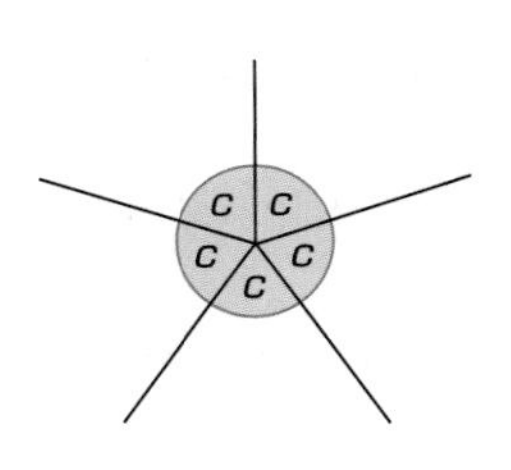

Drawing angles

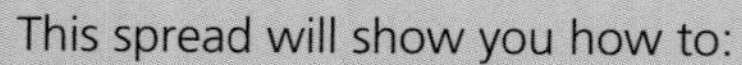

This spread will show you how to:

- Use a protractor to measure and draw acute and obtuse angles to the nearest degree.
- Recognise types of triangles.

KEYWORDS

Angle, Protractor, Degree°, Triangle, Angles on a straight line

To draw an angle of 43° using a protractor:

- Start with a straight line with a dot at one end.
- Place the protractor over the line so that the zero mark is at the dot:

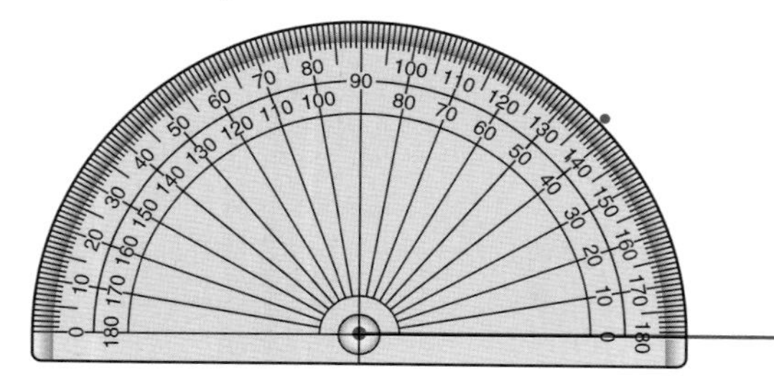

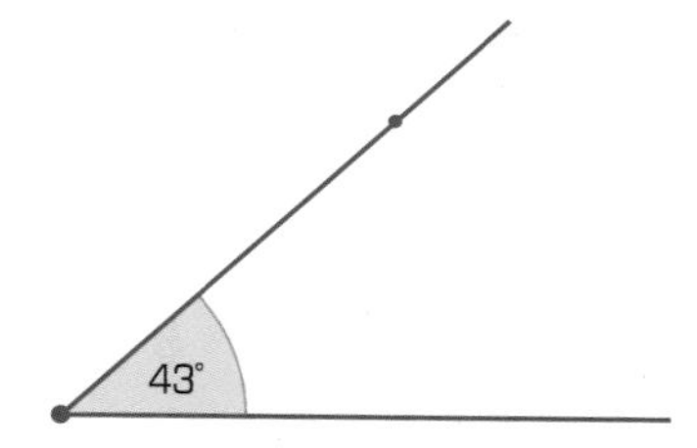

- Mark the angle – use the correct scale!
- Join the dots to complete the angle.

To draw an angle of 137° using a protractor:

- Start with a straight line with a dot at one end.

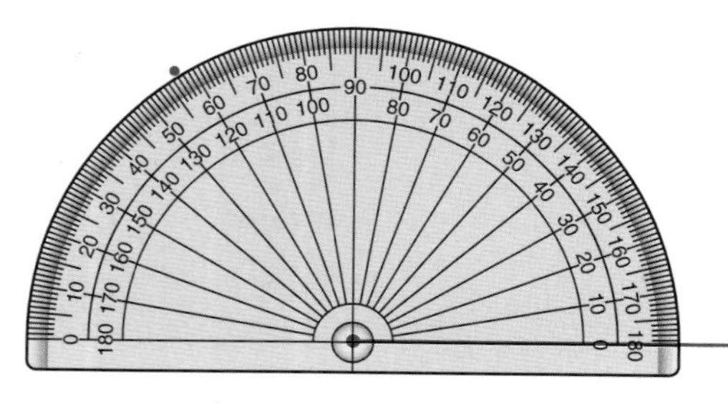

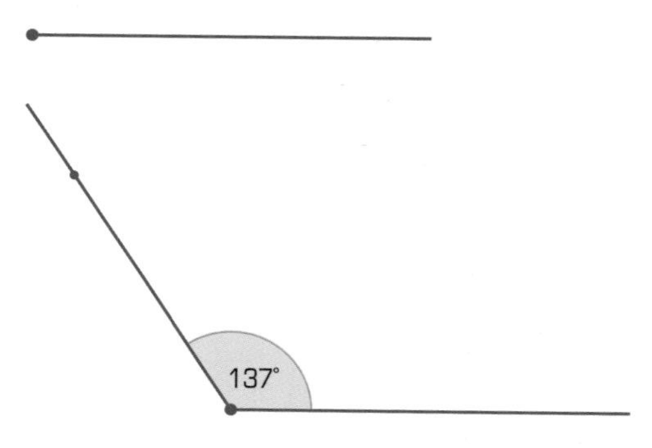

Mark the angle.

Join the dots to complete the angle.

- You should know these names for special triangles:

Right-angled

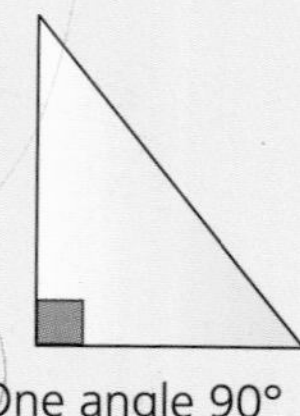

One angle 90°

Equilateral

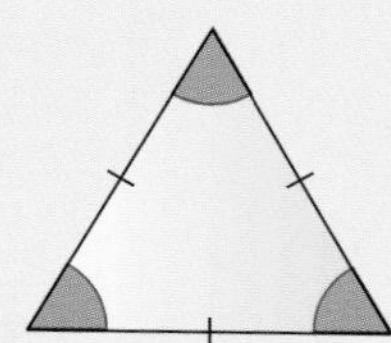

All angles 60°
All sides equal

Isosceles

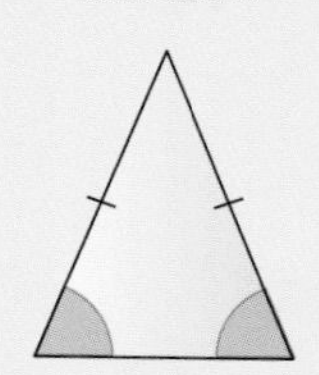

Two angles equal
two sides equal

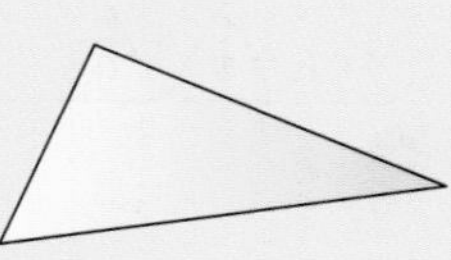

No angles equal
No sides equal

Exercise S3.3

1 Name each of these angles – choose from:

Acute Obtuse Right

Draw them accurately using a protractor.

a 45°	**b** 60°	**c** 120°	**d** 90°
e 56°	**f** 29°	**g** 132°	**h** 176°

2 Draw a line 8 cm long.
At one end, construct an angle of 56°.
At the other end, construct an angle of 32°.
Do your lines meet?

3 What are the names of each of these triangles?

a

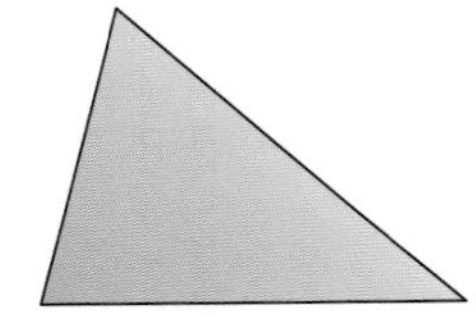

b

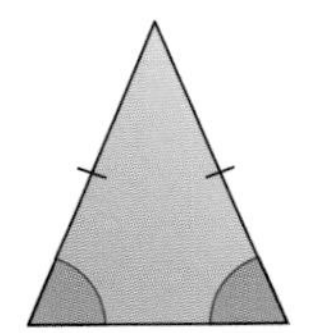

c

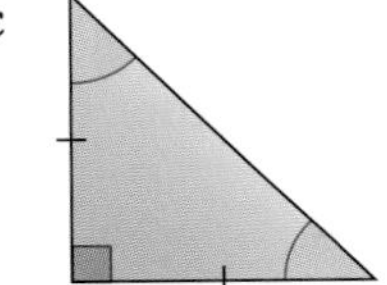

d

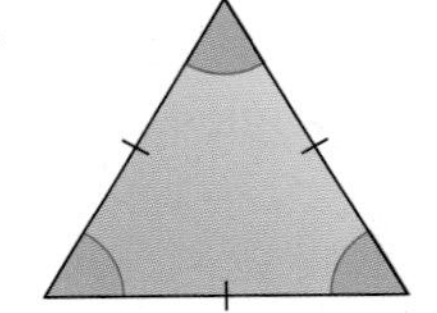

4 Find the angles or side lengths.

a Isosceles triangle

angle C
length AB

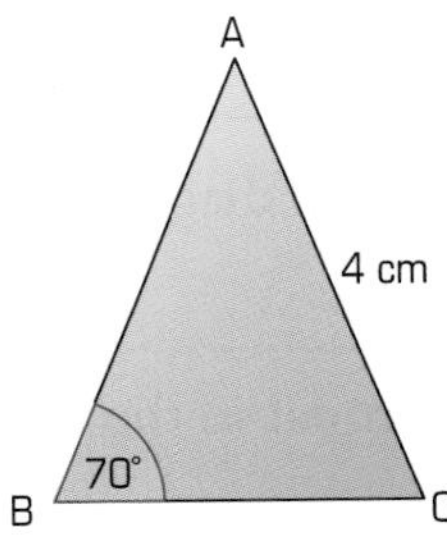

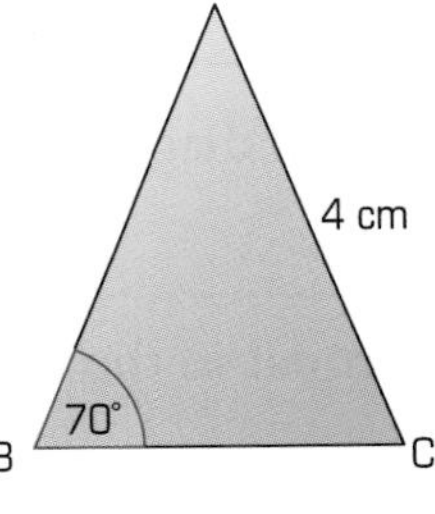

b Right-angled isosceles triangle

angle B
angle A
length BC

A
12 mm
45°
B
C

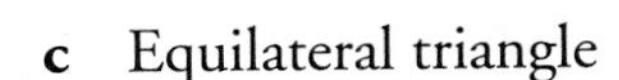

c Equilateral triangle

angle B
angle C
length AB
length AC

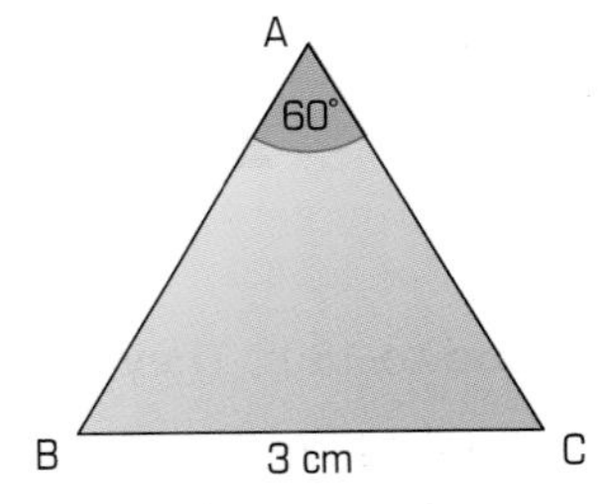

S3.4 Angles in triangles

This spread will show you how to:

- Check that the sum of angles in a triangle is 180°.
- Calculate angles at a point and on a straight line, and in a triangle.

KEYWORDS

Angle
Line
Triangle
Angles on a straight line

▶ **There are 180° on a straight line.**

These two straight lines cross:
Each line forms an acute and an obtuse angle.

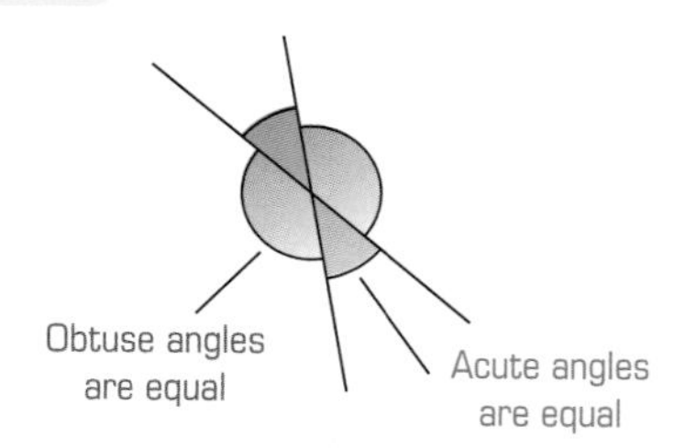

Acute + obtuse = 180° as they are on a straight line.

The angles that are opposite each other form pairs:

These pairs are called **vertically opposite** angles.

▶ **Vertically opposite angles are equal.**

example

Find all the missing angles on this diagram:

Vertically opposite angles are equal so the other acute angle is 43°.

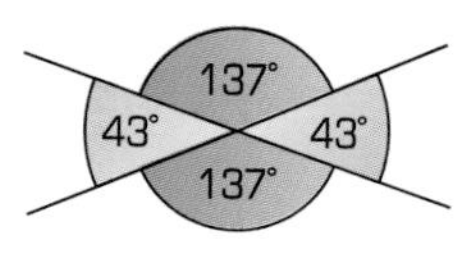

Acute + obtuse = 180°
43° + 137° = 180°

The obtuse angles are both 137°.

You can draw any triangle ... tear off the corners ... and put them together.

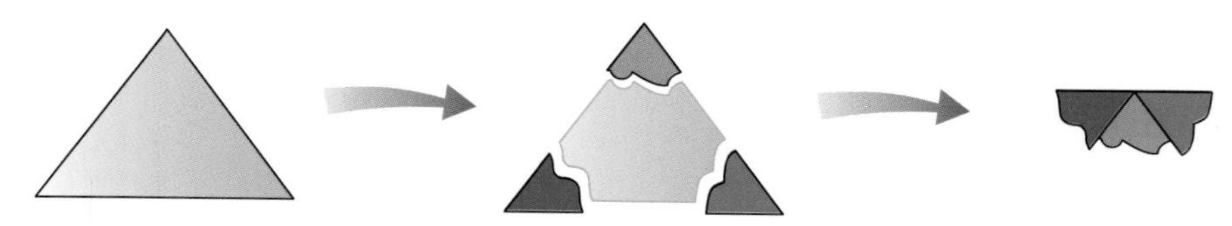

▶ **The angles in a triangle add to 180°. They make a straight line.**

Exercise S3.4

Copy each diagram and find the unknown angles.

1

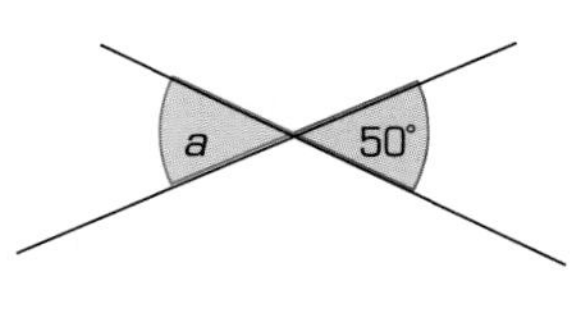

2

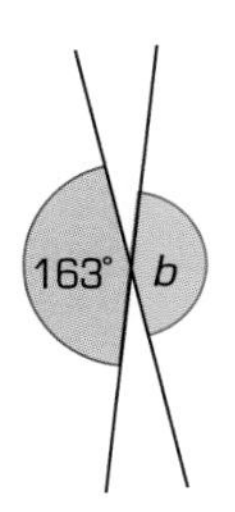

3

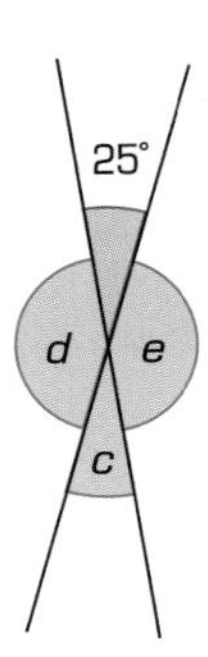

4

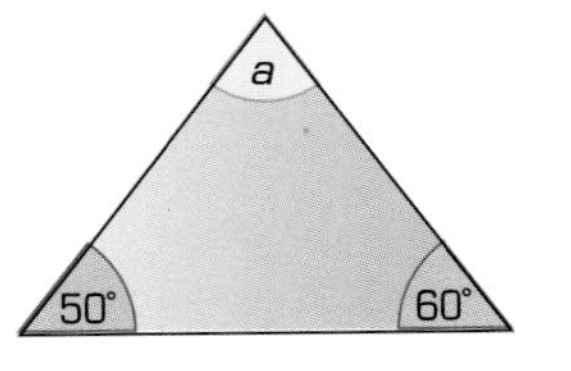

5

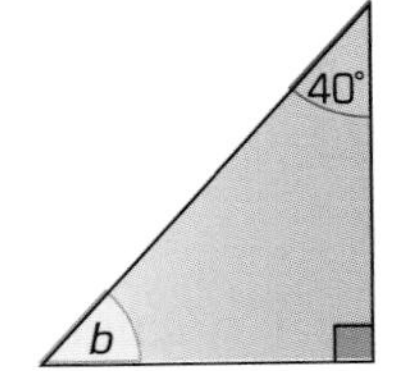

6

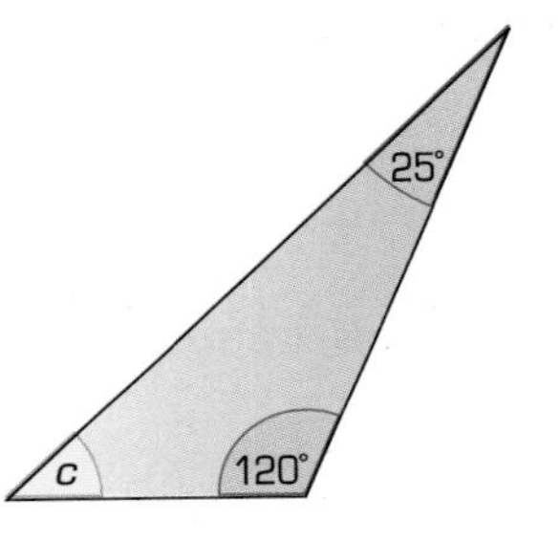

7

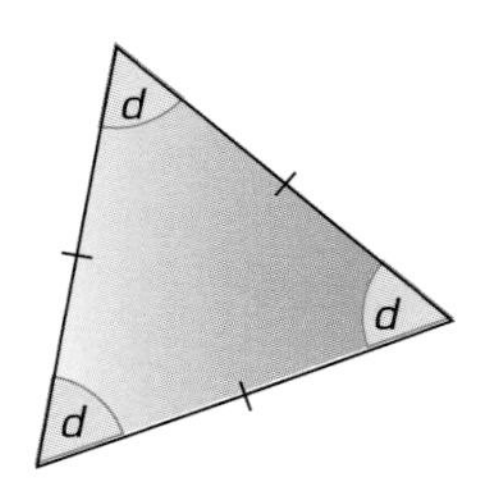

8

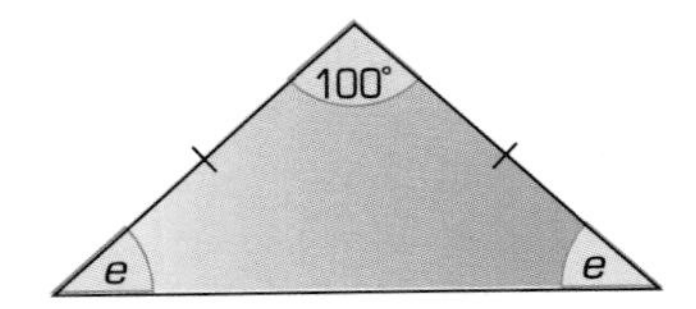

9

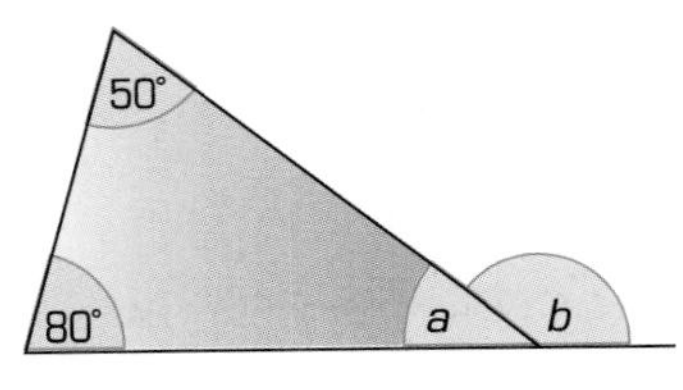

10

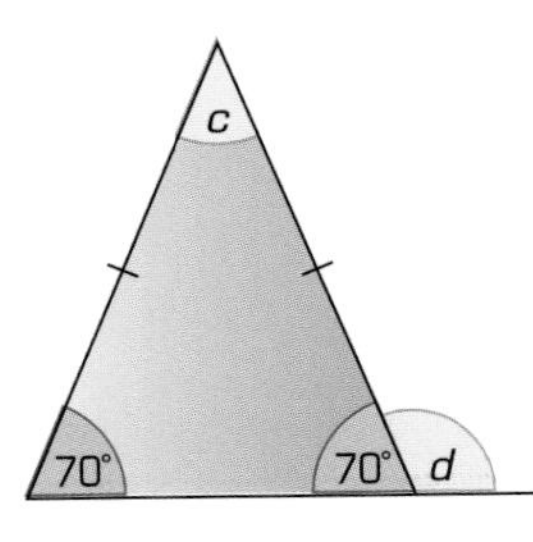

11

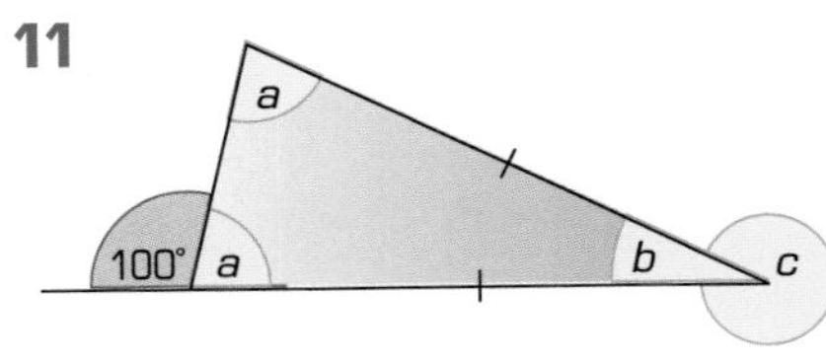

12

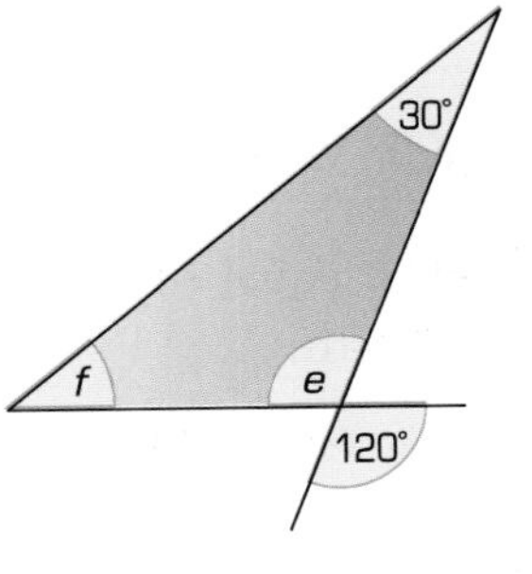

13

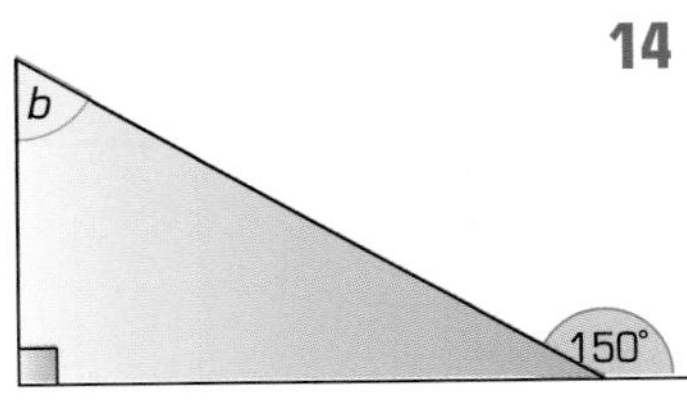

14

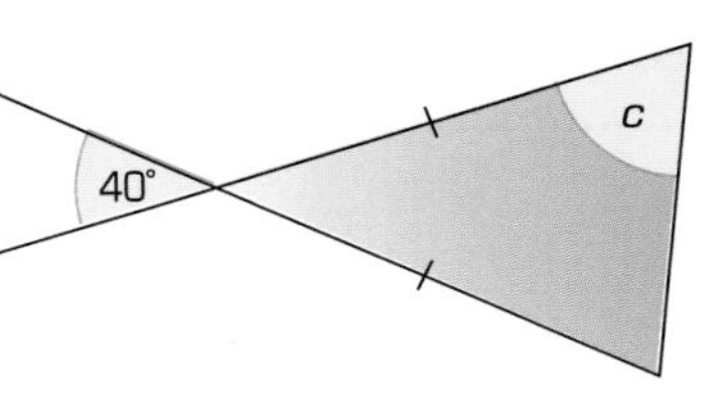

15

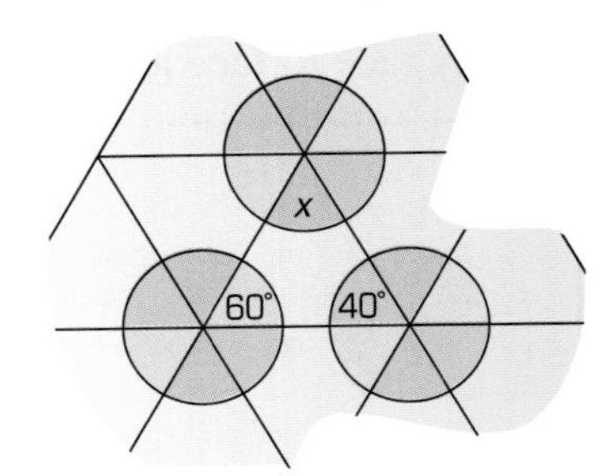

S3.5 2-D drawings of 3-D shapes

This spread will show you how to:

- Visualise 3-D shapes from 2-D drawings.
- Classify solids according to their properties.

KEYWORDS
Solid Face
Edge Vertex
Three-dimensional (3D)

A cube has 3 dimensions – length, width and height.

It has: 6 faces

12 edges

8 vertices

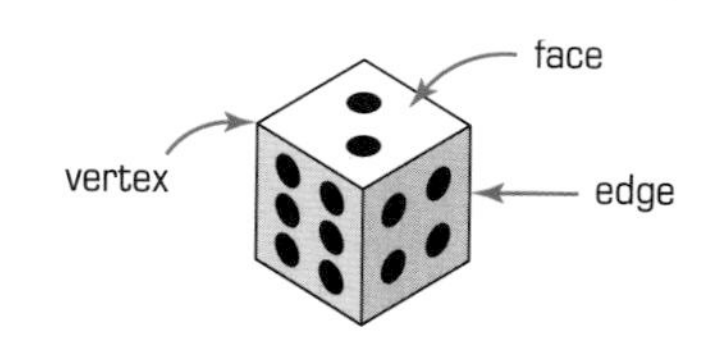

All the faces are squares and all the edges are equal in length.

To draw a cube:

Draw a square

Draw another square behind it

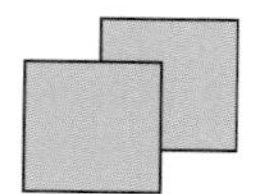

Join up the vertices

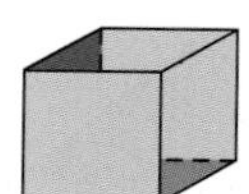

You can make more accurate drawings using isometric paper.
You draw uprights upright.

You show equal edges of the cube by lines of equal length.

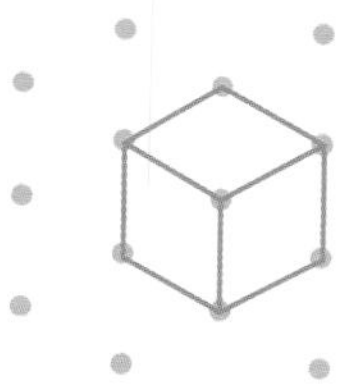

example

Draw the two possible 3D shapes that can be made using 3 cubes.
Only face to face joins are allowed.

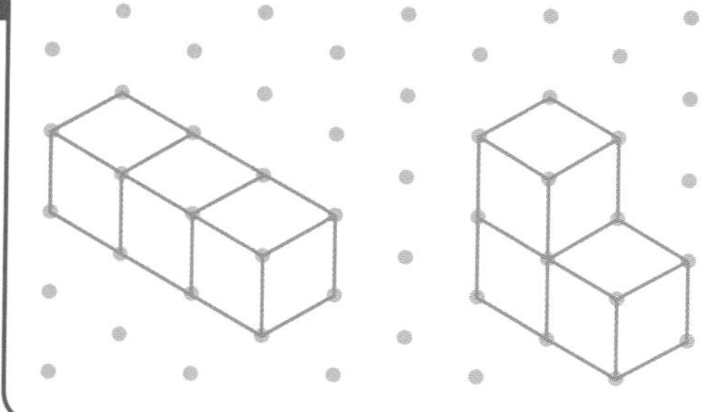

Exercise S3.5

1 Write down how many faces, edges and vertices each of these shapes has.

a

b

c

2 **a** How many cubes do you need to make each of these shapes?

A

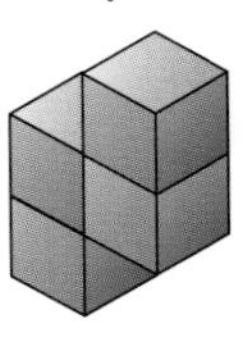

B

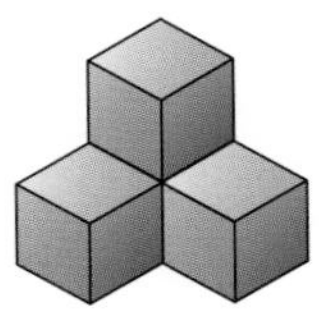

C

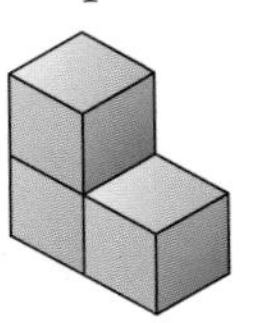

D

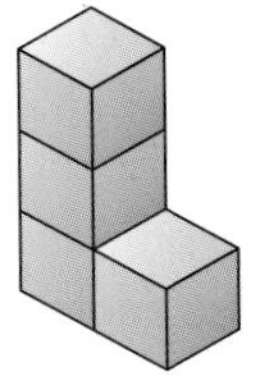

E

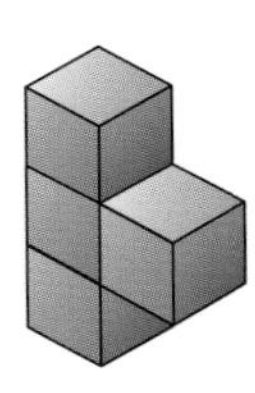

F

G

b Draw each shape on isometric paper.

c Make each of the shapes with multilink cubes.
Fit them together to make this cube:

3 Find the least number of cubes needed to cover and join the blue faces.
Justify your answer.

Summary

You should know how to ...

1 Recognise properties of rectangles.

2 Use a protractor to measure and draw acute and obtuse angles to the nearest degree.

3 Calculate angles on a straight line and in a triangle.

4 Visualise 3-D shapes from 2-D drawings.

5 Solve mathematical problems or puzzles, recognise and explain patterns and relationships, generalise and predict.

Check out

1 Measure the lengths of the diagonals of this rectangle.

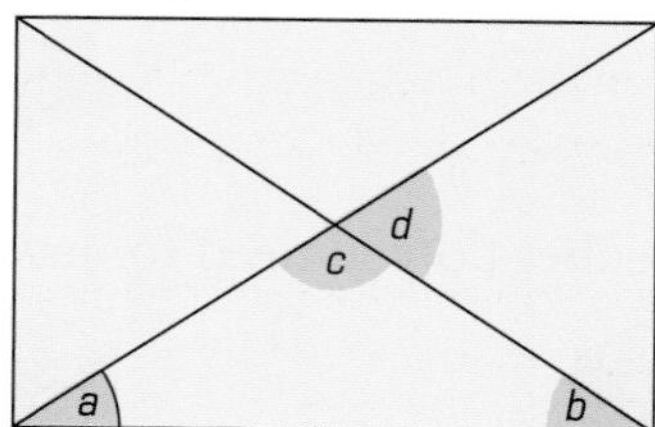

2 Use the diagram in question 1.

a Measure angle *a*

b Measure angle *b*

3 Use the diagram in question 1.

a Calculate angle *c*

b Calculate angle *d*

c Add up your answers for *a, b* and *c*

4 How many cubes are needed to make this shape?

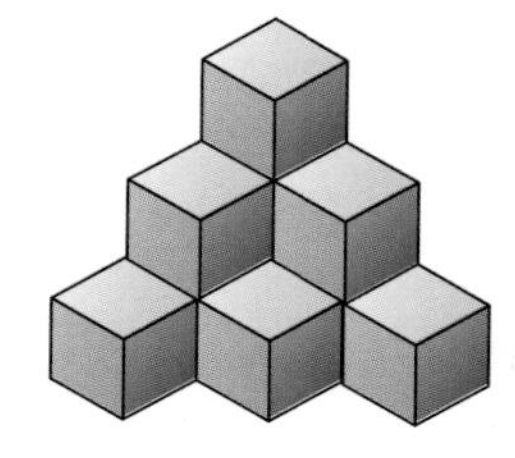

5 How many rectangles are there in this diagram?

Percentages, ratio and proportion

This unit will show you how to:

- Reduce a fraction to its simplest form by cancelling.
- Use a fraction as an 'operator' to find fractions.
- Order fractions and position them on a number line.
- Recognise the equivalence between decimal and fraction forms.
- Begin to convert a fraction to a decimal using division.
- Express simple fractions, as percentages.
- Understand percentage as the number of parts in every 100.
- Find simple percentages of quantities.
- Solve simple problems about ratio and proportion using informal strategies.
- Identify and use appropriate operations to solve word problems.

You can describe proportions in different ways.

Before you start

You should know how to ...

1 Understand percentage as 'parts per 100'.

2 Understand fractions and equivalence between different looking fractions.

3 Find fractions of amounts.

Check in

1 Look at these pictures.
Write down the % that is shaded.

a 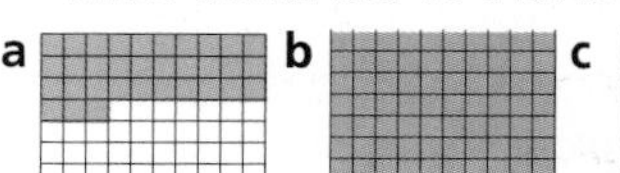 **b** **c** 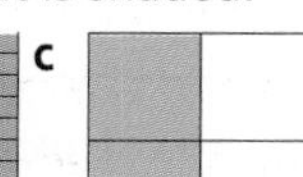 **d** 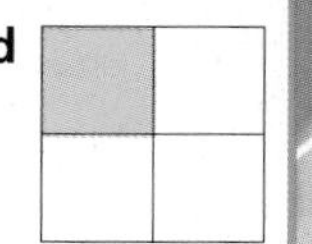

2 Here are 10 fractions. They make five pairs of equivalent fractions. Match the pairs together

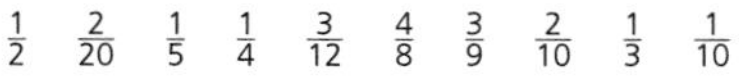

$\frac{1}{2}$ $\frac{2}{20}$ $\frac{1}{5}$ $\frac{1}{4}$ $\frac{3}{12}$ $\frac{4}{8}$ $\frac{3}{9}$ $\frac{2}{10}$ $\frac{1}{3}$ $\frac{1}{10}$

3 Work these out:

a $\frac{1}{2}$ of 28 cm **b** $\frac{1}{10}$ of 60 minutes
c $\frac{1}{2}$ of 90 cm **d** $\frac{1}{10}$ of £150
e $\frac{1}{4}$ of £80 **f** $\frac{3}{4}$ of 60 minutes

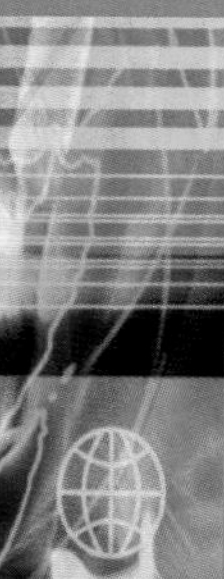

N4.1 Fraction, decimal and percentage equivalents

This spread will show you how to:
- Express simple fractions as percentages.
- Recognise the equivalence between fraction and decimal forms.
- Reduce a fraction to its simplest form by cancelling.

KEYWORDS
Cancel, Equivalent, Convert, Fraction, Decimal, Percentage

Percentages, fractions and decimals are different ways of writing a part of a whole.

- **You can write a percentage as a fraction.**
  48% means '48 parts out of 100' which is $\frac{48}{100}$.

example

Convert 25% to a fraction in its simplest form.

$25\% = \frac{25}{100}$

Now simplify your fraction by cancelling:

$\frac{25}{100} \xrightarrow{\div 5} = \frac{5}{20} \xrightarrow{\div 5} = \frac{1}{4}$

- **You can convert between percentages and decimals.**

It helps if you write them as fractions first.

example

Convert:

a 62% to a decimal

b 0.34 to a percentage

a $62\% = \frac{60}{100} + \frac{2}{100}$
$= \frac{6}{10} + \frac{2}{100}$
$= 0.6 + 0.02 = 0.62$

b $0.34 = \frac{3}{10} + \frac{4}{100}$
$= \frac{30}{100} + \frac{4}{100}$
$= \frac{34}{100} = 34\%$

You need to know some **equivalences** off by heart.

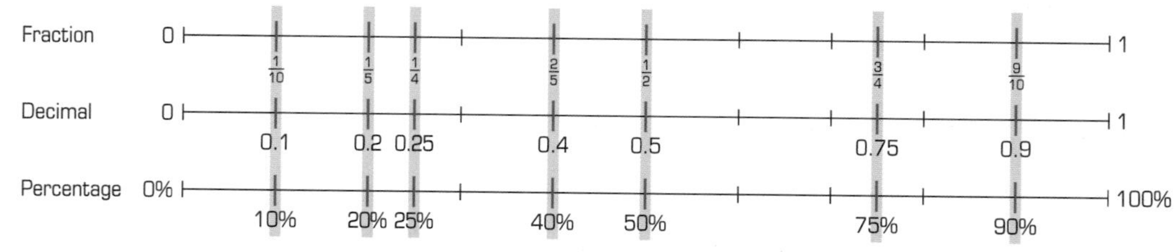

You can use this number line to convert between fractions, decimals and percentages.

$\frac{3}{5} = 3 \times \frac{1}{5}$
$= 3 \times 0.2$
$= 0.6$

$70\% = 7 \times 10\%$
$= 7 \times \frac{1}{10}$
$= \frac{7}{10}$

Exercise N4.1

1 Convert these percentages into decimals.
a 63% **b** 48% **c** 22% **d** 130% **e** 70% **f** 7% **g** 113% **h** 12%

2 Convert these decimals into percentages.
a 0.28 **b** 0.73 **c** 1.25 **d** 0.8 **e** 0.04 **f** 0.4 **g** 0.43 **h** 0.01

3 Copy and complete this table.

Percentage	Fraction	Decimal
60%	—	0.6
30%	$\frac{3}{10}$	—
—	—	0.1

Percentage	Fraction	Decimal
—	$\frac{7}{100}$	—
80%	—	—
—	—	0.35

4 Work out which is the biggest number in each question by changing the fraction into a percentage.
a 63% $\frac{1}{2}$ **b** $\frac{1}{4}$ 20% **c** 72% $\frac{3}{4}$ **d** 40% $\frac{3}{10}$ **e** $\frac{1}{4}$ 35%

5 Write these percentages as fractions in their simplest form.
a 60% **b** 42% **c** 51% **d** 72% **e** 135% **f** 18% **g** 4% **h** 30%

6 Write these numbers in order, starting with the smallest.
a 0.2 $\frac{1}{4}$ 22% **b** 0.7 73% $\frac{3}{4}$ **c** 38% 0.4 $\frac{4}{100}$ **d** $\frac{33}{100}$ 0.3 31%

7 Convert these fractions into percentages.
a $\frac{21}{50}$ **b** $\frac{37}{50}$ **c** $\frac{24}{25}$ **d** $\frac{8}{25}$ **e** $\frac{15}{20}$ **f** $\frac{7}{20}$ **g** $\frac{9}{75}$ **h** $\frac{15}{80}$

8 $\frac{1}{8} = 0.125$ and $\frac{3}{8} = 0.375$
a Use a calculator to work out $1 \div 8$.
Write down what you notice.
b Use a calculator to work out $3 \div 8$.
Write down what you notice.
c Use your calculator to work out:
$5 \div 8$, $1 \div 4$ and $7 \div 8$.
d Write down the decimal equivalents of:
$\frac{5}{8}$, $\frac{1}{4}$ and $\frac{7}{8}$.
e Find the decimal equivalents of $\frac{1}{3}$, $\frac{1}{9}$ and $\frac{1}{11}$.

9 **Challenge**
In this magic square each row, column and diagonal adds up to 3.
Unfortunately someone has muddled up the numbers.
Can you rearrange the numbers to make the square magic again?

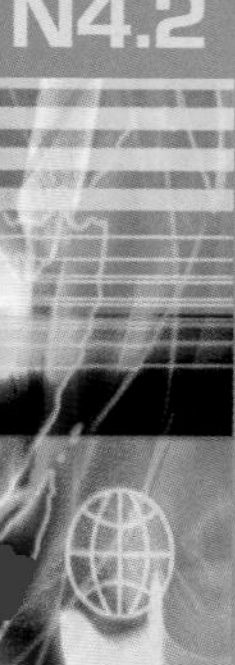

N4.2 Finding simple percentages

This spread will show you how to:

- Use a fraction as an operator to find fractions of quantities.
- Find simple percentages of quantities.

KEYWORDS

Calculate	Fraction
Convert	Percentage
Equivalent	Sale

Shops use percentages when they have a sale:

You can already find a fraction of an amount.

example

Find:

a $\frac{1}{2}$ of 42 **b** $\frac{3}{4}$ of 60

a $\frac{1}{2}$ of $42 = 42 \div 2$
$= 21$
$\frac{1}{2}$ of $42 = 21$

b $\frac{1}{4}$ of $60 = 60 \div 4 = 15$
$\frac{3}{4}$ of $60 = 3 \times \frac{1}{4}$ of $60 = 3 \times 15$
$= 45$
$\frac{3}{4}$ of $60 = 45$

Finding percentages of amounts is very similar to finding fractions.

▶ **You can find a percentage of an amount by converting it to a fraction first.**

example

Find:

a 50% of 82 **b** 10% of 30 cm

a Change 50% into a fraction:
$50\% = \frac{1}{2}$
50% of $82 = \frac{1}{2}$ of 82
$= 82 \div 2 = 41$
50% of $82 = 41$

b Change 10% into a fraction:
$10\% = \frac{1}{10}$
10% of 30 cm $= \frac{1}{10}$ of 30 cm
$= 30 \div 10 = 3$ cm
10% of 30 cm $= 3$ cm

Exercise N4.2

1 Work out these percentages.

a 50% of £20 **b** 50% of £28 **c** 50% of £80
d 50% of 64 cm **e** 50% of 32 km **f** 50% of 52 mm

2 Work out these percentages.

a 25% of £40 **b** 25% of £20 **c** 25% of £80
d 25% of 44 mm **e** 25% of 32 cm **f** 25% of 56 m

3 Now work out these percentages.

a 75% of £40 **b** 75% of £80 **c** 75% of £44
d 75% of 16 km **e** 75% of 36 cm **f** 75% of 52 m

4 Work out each percentage in these questions.
Write down which one gives the biggest answer.

a 50% of 30, 25% of 48 **b** 50% of 70, 25% of 120
c 25% of 24, 75% of 16 **d** 50% of 38, 75% of 28
e 75% of 60, 25% of 200 **f** 25% of 40, 10% of 80

5 Here are some lengths of rope. Find 10% of each length.

a 50 m **b** 30 cm **c** 800 cm
d 45 m **e** 95 cm **f** 18.5 m

6 Copy and complete these questions.

a 10% of £80 = £____ **b** 50% of £66 = £____
c 50% of £____ = £32 **d** 25% of £____ = £14
e ____% of £64 = £16 **f** ____% of £84 = £63

7 Three of the four percentages in each grid give the same answer.
Work out each answer and write down the odd one out.

a

25% of 40	10% of 100
50% of 20	75% of 12

b

10% of 200	75% of 16
25% of 48	50% of 24

c

75% of 28	25% of 84
10% of 120	50% of 42

d

10% of 90	25% of 32
50% of 18	75% of 12

8 Write down three percentage questions that give these answers. The first one is done for you.

a 4
50% of 8, 25% of 16, 10% of 40.

b 10 **c** 7 **d** 9

Finding harder percentages

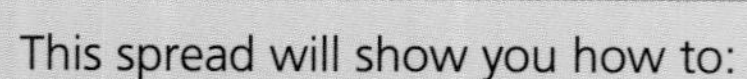

This spread will show you how to:

- Find simple percentages of quantities.
- Relate fractions to division.

KEYWORDS

Amount, Increase, Decrease, Percentage, Fraction

You already know how to find 10% of an amount.

example

Find 10% of £60.

10% is the same as $\frac{1}{10}$.
$\frac{1}{10}$ of 60 = 60 ÷ 10 = 6
So 10% of £60 is £6.

▶ **You can use 10% to find more difficult percentages.**
For example, 5% = 10% ÷ 2 and 35% = 3 × 10% + 5%

example

Find:

a 20% of 40

b 15% of 60 kg

a First find 10% of 40:
$\frac{1}{10}$ of 40 = 40 ÷ 10 = 4
20% = 10% + 10%
So 20% of 40 is 4 + 4 = 8

b First find 10% of 60:
$\frac{1}{10}$ of 60 = 60 ÷ 10 = 6
15% = 10% + 5%
5% = $\frac{1}{2}$ of 10%: 6 ÷ 2 = 3
So 15% of 60 kg is = 6 kg + 3 kg = 9 kg

Often you need to work out percentage increases and decreases.

example

Lotta bought a flat for £50 000.
In one year its value increased by 15%.
How much did her flat rise in value?

15% is 10% + 5%

10% of 50 000 = $\frac{1}{10}$ of 50 000
= 5000

5% of 50 000 = $\frac{1}{2}$ of 10% of 50 000
= $\frac{1}{2}$ of 5000
= 2500

So 15% of £50 000 = £5000 + £2500
= £7500

The flat rose in value by £7500.

Exercise N4.3

1 Find 10% of these amounts.

a £20 **b** £30 **c** £60
d £120 **e** £65 **f** £43

2 Work out the answers to these.

a 20% of £30 **b** 20% of £40
c 5% of £20 **d** 5% of £60
e 30% of £30 **f** 30% of £20
g 40% of £40 **h** 40% of £30
i 15% of £40 **j** 15% of £60

3 Calculate these percentages.

a 40% of £10 **b** 5% of £3
c 5% of £3.20 **d** 20% of £8.40
e 20% of £2.10 **f** 15% of £2.20
g 15% of £6.60 **h** 35% of £6.60

4 Sarah earns £280 a week.
She gets a rise of 10%.
Find 10% of £280 to find out how much more she will earn a week.

5 Will earns £360 a week.
He gets a rise of 5%.
Find 5% of £360 to find out how much more he will earn.

6 Jane bought a house for £30 000.
In one year its value increased by 15%.
Work out 15% of £30 000 to find out how much the value of the house went up.

7 Work out each of these amounts and put them in order, starting with the smallest.

a 20% of 60   15% of 50 $\frac{1}{4}$ of 40   35% of 40
b 45% of 160   35% of 180   15% of 700 $\frac{1}{4}$ of 300

8 You know that 5% is half of 10%. You should also know that $2\frac{1}{2}$% is half of 5%. Use this fact to work out:

a $2\frac{1}{2}$% of £40 **b** $12\frac{1}{2}$% of £40
c $7\frac{1}{2}$% of £60 **d** $12\frac{1}{2}$% of £80
e $7\frac{1}{2}$% of £120 **f** $17\frac{1}{2}$% of £160

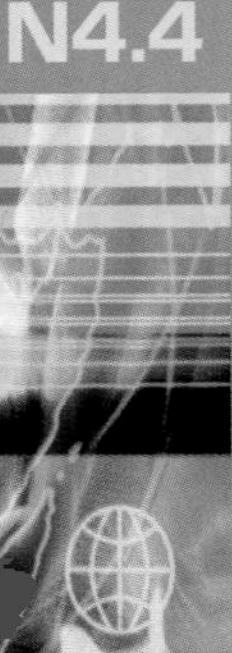

N4.4 Proportion

This spread will show you how to:
- Solve simple problems involving proportion.
- Recognise relationships between fractions.
- Understand percentage as the number of parts in every 100.

KEYWORDS
Fraction
Proportion
Percentage
Simplest form

A proportion compares the size of a part with the size of the whole.

Look at this pattern of tiles:

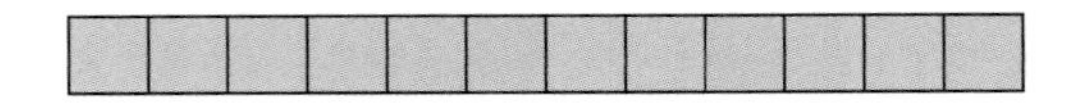

Two out of every three tiles are blue.
The **proportion** of blue tiles is $\frac{2}{3}$.

- You can write a proportion as a fraction, a decimal or a percentage.

example

In a class of 25 students, 15 are girls and 10 are boys.
What is the proportion of girls in the class?

Express your answer as:

a a fraction in its simplest form
b a decimal
c a percentage

---

a The proportion of girls is 15 out of 25, or $\frac{15}{25}$.
You can cancel the fraction:

$$\frac{15}{25} \overset{\div 5}{=} \frac{3}{5}$$

(÷5 applied to numerator and denominator)

As a fraction, the proportion of girls is $\frac{3}{5}$.

b You know that $\frac{1}{5}$ is the same as 0.2.
So $\frac{3}{5}$ is $0.2 \times 3 = 0.6$

As a decimal, the proportion of girls is 0.6.

c You know that $\frac{1}{5}$ is the same as 20%.
So $\frac{3}{5}$ is $20\% \times 3 = 60\%$

As a percentage, the proportion of girls is 60%.

$\frac{3}{5} = 0.6 = 60\%$
They all show the same proportion.

Exercise N4.4

1 Look at this pattern of tiles:

a What proportion of the tiles is yellow?

b What proportion of the tiles is purple?

2 Look at this pattern of tiles:

a What proportion of the tiles is yellow?

b What proportion of the tiles is black?

3 Look at this pattern of tiles:

a What proportion of the tiles is green?

b What proportion of the tiles is black?

c Write the proportions in **a** and **b** in their simplest form.

4 Look at this pattern:

a How many black tiles are there?

b How many orange tiles are there?

c What proportion of the tiles is black?

d What proportion of the tiles is orange?

e Write your answers to **c** and **d** in their simplest form.

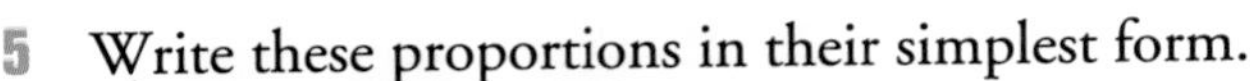

5 Write these proportions in their simplest form.

a $\frac{3}{6}$ **b** $\frac{4}{10}$ **c** $\frac{8}{16}$

d $\frac{5}{25}$ **e** $\frac{4}{16}$ **f** $\frac{3}{15}$

g $\frac{6}{9}$ **h** $\frac{8}{12}$ **i** $\frac{12}{18}$

6 Look at this grid:

a What proportion of the tiles is:

i green

ii yellow

iii purple?

b Write your answers to **a** in their simplest form.

7 Calculate the proportion of black cats in each of the following cat families.
Express each proportion in its simplest form.

a 3 white; 2 black; 5 tortoiseshell

b 6 white; 4 black; 4 tabbies

c 4 white; 7 black; 1 tabby; 3 tortoiseshell

d 11 white; 3 black; 6 tabbies

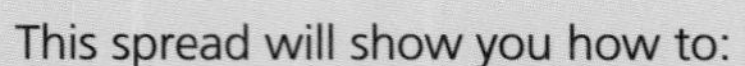

N4.5 Introducing ratio

This spread will show you how to:

- Solve simple problems involving ratio and proportion.

KEYWORDS
Compare
Proportion
Fraction
Ratio

You can look at this pattern of tiles in two ways:

1 in every 4 tiles is red.
The **proportion** of red tiles is $\frac{1}{4}$.

For every 3 blue tiles, there is 1 red tile.
The **ratio** of blue to red tiles is 3 : 1.

▶ **You use a ratio to compare the sizes of parts of a whole.**

Ratios can be cancelled down like fractions.

example

In a class of 30 there are 18 girls.
What is the ratio of girls to boys?

There are 18 girls, so there must be 30 − 18 = 12 boys.
The ratio of girls to boys is 18 : 12 (÷6, ÷6) → 3 : 2

The ratio of girls to boys in the class is 3 : 2.
This means 'for every 3 girls there are 2 boys'.

You can use tile diagrams to help with ratio and proportion problems.

example

There are 36 vehicles in a car park. For every 3 cars there is 1 motorcycle.
How many cars and how many motorcycles are there?

You could use a tile diagram:
3 cars and 1 motorcycle make 4 tiles.

Keep going until you get to 36 tiles like this:

C	C	C	M
C	C	C	M
C	C	C	M
C	C	C	M
C	C	C	M
C	C	C	M
C	C	C	M
C	C	C	M
C	C	C	M

There are 9 blocks of tiles.
$9 \times 4 = 36$.

Then count the tiles.
$9 \times 3 = 27$ and $9 \times 1 = 9$.

There are 27 cars and 9 motorcycles.

Exercise N4.5

Use the tile diagram shown in each question to help answer questions 1–5.

1 There are 20 pupils in a class.
There are 2 boys for every 3 girls.
Copy and continue the tile diagram to work out how many boys and how many girls there are in the class.

B	B	G	G	G

2 In a different class there are 24 pupils.
There is 1 girl for every 2 boys.
Copy and continue the tile diagram and work out how many boys and how many girls there are in the class.

B	B	G

3 1 kg of potatoes costs 35p.
How much will 5 kg of potatoes cost?

1 kg	35p

4 1 kg of broccoli costs £1.25.
How much does 6 kg of broccoli cost?

1 kg	£1.25

5 A bag of apples costs 63p.
How many bags could you buy if you had £5.12?

You can use tile diagrams for questions 6–10.

6 Jack has 15 sweets. He gives his brother Ewan 2 sweets for every 3 sweets that he takes.
How many sweets do they each get?

7 A painter mixes 1 tin of blue paint with 3 tins of white paint.
He needs 12 tins altogether.
How many tins of each colour will he need?

8 A gardener puts 2 spades of manure into his wheelbarrow for each 3 spades of soil.
The wheelbarrow holds 60 spadefuls in total.
How many are manure and how many are soil?

9 Kathy and Sharon put their money together and buy a raffle ticket for £1.
Kathy put in 30p and Sharon put in 70p.
They win the 1st prize of £25.
They share the prize in the ratio 30 : 70.
How much do they each get?

10 In a box of chocolates, the ratio of soft centres to hard centres is 2 : 5.
There are 42 chocolates in the box.

a How many have soft centres?

b How many have hard centres?

N4 Summary

You should know how to ...

1 Understand percentage as the number of parts in every 100; express simple fractions as percentages.

2 Reduce a fraction to its simplest form by cancelling.

3 Calculate simple percentages of amounts.

4 Solve simple problems using ideas of ratio and proportion.

5 Identify and use appropriate operations to solve word problems.

Check out

1 Change these percentages to fractions:

a 60% **b** 32% **c** 7% **d** 25% **e** 75%

Change these fractions to percentages:

f $\frac{1}{2}$ **g** $\frac{1}{10}$ **h** $\frac{3}{10}$ **i** $\frac{1}{5}$ **j** $\frac{2}{5}$

2 Write your answers to questions 1a–1e in their simplest form.

3 Work these out:

a 50% of £26 **b** 50% of £70

c 25% of £48 **d** 75% of £48

e 10% of 60 cm **f** 10% of 30 minutes

g 25% of 120 cm **h** 75% of 80 minutes

4

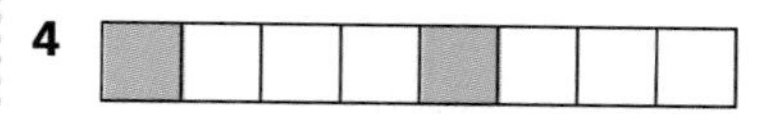

a What proportion of the tiles is shaded?

b What proportion of the tiles is unshaded?

c What is the ratio of shaded tiles to unshaded tiles?

Make sure your answers to a, b and c are in their simplest form.

5 **a** There are 30 pupils in a class.
There are 3 boys to every 2 girls.
Use the tile diagram to help you find how many boys and girls are in the class.

B	B	B	G	G

b 60 people go on a coach trip. The ratio of men to women is 3 : 7. Use the tile diagram to help you find out how many men there are and how many women.

M	M	M	F	F	F	F	F	F	F

c 35 people go on a coach outing.
The ratio of men to women is 3 : 4.
How many men are there?

14 Linear equations

This unit will show you how to:

- Understand the effect of and relationships between the four operations, and the principles of the arithmetic laws.
- Use brackets.
- Use the relationship between addition and subtraction.
- Develop from explaining a generalised relationship in words to expressing it in a formula using letters as symbols.
- Choose and use appropriate number operations to solve problems, and appropriate ways of calculating.

You can solve problems using algebra.

Before you start

You should know how to …

1 Work out simple calculations in your head.

Check in

1 Write down the answers to:

a 3×7 **b** 9×6
c 4×9 **d** 8×7
e $18 \div 3$ **f** $27 \div 9$
g $72 \div 8$ **h** $50 \div 5$
i $\frac{15}{3}$ **j** $\frac{24}{4}$
k $19 + 17$ **l** $28 + 27$
m $35 - 17$ **n** $52 - 28$
o 3.5×2 **p** $3.5 - 2$
q $^{-}3 + 5$ **r** $\frac{1}{2} + \frac{1}{4}$

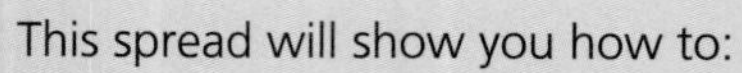

A4.1 Using algebraic expressions

This spread will show you how to:

- Develop from explaining a generalised relationship in words to expressing it in a formula using letters as symbols.

KEYWORDS

Expression, Variable, Value, Unknown

▶ A **variable** is a value that can change.

You use a letter to stand for a variable when ...

... the value is variable or ... when the value is unknown.

The number of biscuits in a jar changes as you eat them!

There are too many grains of rice in this bag to count!

You can say there are b biscuits and g grains of rice.
You have fixed the number using a letter.

There are b biscuits in a jar.
Siân eats 3 biscuits.

There are 3 less biscuits.
There are now $b - 3$ biscuits.

There are g grains of rice in a bag.
Jeff buys 2 bags.

There are 2 times the number of grains.
There are now $2g$ grains of rice.

$b - 3$ and $2g$ are algebraic expressions. b and g stand for variables.

▶ An **expression** is a sentence that uses letters to stand for variables.

Exercise A4.1

1 Write an algebraic expression for each of these sentences.

a There are x fleas on a dog.
12 jump off onto a cat.
How many fleas are left on the dog?

b There are f fish in a pond.
20 fish are added to the pond.
How many fish are there now?

c There are 22 players on a pitch.
Some get sent off.
How many are left?

d Add 5 to an unknown number, x.

e Subtract 3 from an unknown number, y.

f Divide an unknown number, z, by 3.

g Add two unknown numbers, r and s, together.

2 Write an algebraic expression for each of these sentences.

a There are lots of dogs at a show.
2 more dogs arrive.
How many dogs are there at the show now?

b There are lots of flowers in Ruby's garden.
She picks 8 flowers.
How many flowers are there in the garden now?

c Some students go to Blackpool on a coach.
2 students get left behind.
How many students are on the coach home?

d There are lots of videos in a rental shop.
They lend 25 videos out on Saturday.
How many videos are left?

e There are lots of books in the library.
20 people return a book and 5 people borrow a book.
How many books are there now?

f There are lots of cars in the car park.
Each car has 4 wheels.
How many wheels are there altogether?

g Jules has lots of sweets.
She shares them equally with a friend.
How many sweets do they each have?

3 On this square, all the edges are x cm long.
The distance around the edge of the square is the perimeter.
Write an expression for the perimeter of the square.

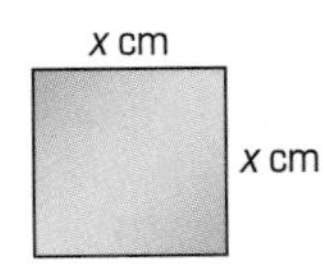

A4.2 Algebraic operations

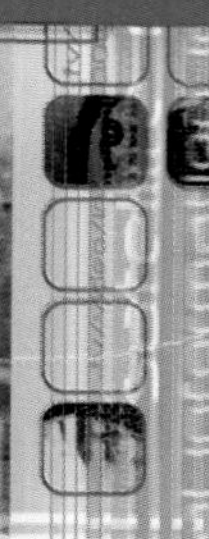

This spread will show you how to:
- Use the relationship between addition and subtraction.
- Understand and use the relationships between the four operations.

KEYWORDS
Operation Value
Unknown

In algebra you use letters to stand for unknown numbers or values.
The letters follow the same rules as numbers.

▶ **You can:**
add numbers together in any order and **multiply numbers together in any order.**

$1 + 2 = 2 + 1$	$2 \times 3 = 3 \times 2$
$8 + 5 = 5 + 8$	$8 \times 5 = 5 \times 8$
$a + b = b + a$	$a \times b = b \times a$

▶ **In algebra you write:**
the unknown first when you add and **the number first when you multiply:**

$1 + p$ is the same as $p + 1$ $1 + p = p + 1$ you usually write $p + 1$	and	$m \times 3$ is the same as $3 \times m$ or $3m$ $m \times 3 = 3 \times m$ you usually write $3m$
$8 + q$ is the same as $q + 8$ $8 + q = q + 8$ you usually write $q + 8$	and	$n \times 8$ is the same as $8 \times n$ or $8n$ $n \times 8 = 8 \times n$ you usually write $8n$

You should also know that:

▶ **Subtracting is the opposite of adding:**

$8 + 5 = 13$	or	$5 + 8 = 13$	so	$13 - 8 = 5$	and	$13 - 5 = 8$
$7 + 6 = 13$	or	$6 + 7 = 13$	so	$13 - 7 = 6$	and	$13 - 6 = 7$
$x + y = 13$	or	$y + x = 13$	so	$13 - x = y$	and	$13 - y = x$

▶ **Dividing is the opposite of multiplying:**

$3 \times 4 = 12$	or	$4 \times 3 = 12$	so	$12 \div 3 = 4$	and	$12 \div 4 = 3$
$2 \times 6 = 12$	or	$6 \times 2 = 12$	so	$12 \div 2 = 6$	and	$12 \div 6 = 2$
$a \times b = 12$	or	$b \times a = 12$	so	$12 \div a = b$	and	$12 \div b = a$

Exercise A4.2

1 Copy and complete these addition sums.

a $3 + 4 = 4 + \square$
b $6 + 9 = 9 + \square$
c $8 + 3 = 3 + \square$
d $7 + 1 = 1 + \square$
e $4 + 9 = \square + \square$
f $3 + 5 = \square + \square$
g $x + 2 = 2 + \square$
h $y + 4 = 4 + \square$
i $a + b = b + \square$
j $c + \square = d + \square$

2 Copy and complete these multiplication sums.

a $3 \times 4 = 4 \times \square$
b $6 \times 9 = 9 \times \square$
c $8 \times 3 = 3 \times \square$
d $7 \times 1 = 1 \times \square$
e $4 \times 9 = \square \times \square$
f $3 \times 5 = \square \times \square$
g $x \times 2 = 2 \times \square$
h $y \times 4 = 4 \times \square = 4y$
i $a \times b = b \times \square = ab$
j $c \times \square = d \times \square = \square$

3 Copy and complete these sums.

a $3 + 8 = 11$ or $8 + 3 = 11$ so $11 - 3 = \square$ and $11 - 8 = \square$
b $4 + 5 = 9$ or $5 + \square = \square$ so $9 - 4 = \square$ and $9 - \square = \square$
c $2 + 6 = \square$ or $6 + \square = \square$ so $8 - 2 = \square$ and $8 - \square = \square$
d $x + y = 7$ or $y + x = 7$ so $7 - x = \square$ and $7 - y = \square$
e $p + q = 8$ or $q + \square = \square$ so $8 - p = \square$ and $8 - \square = \square$

4 Copy and complete these sums.

a $3 \times 8 = 24$ or $8 \times 3 = 24$ so $24 \div 3 = \square$ and $24 \div 8 = \square$
b $4 \times 5 = 20$ or $5 \times \square = \square$ so $20 \div 4 = \square$ and $20 \div \square = \square$
c $2 \times 6 = 12$ or $6 \times \square = \square$ so $12 \div 2 = \square$ and $12 \div \square = \square$
d $x \times y = 7$ or $y \times x = 7$ so $7 \div x = \square$ and $7 \div y = \square$
e $p \times q = 8$ or $q \times \square = \square$ so $8 \div p = \square$ and $8 \div \square = \square$

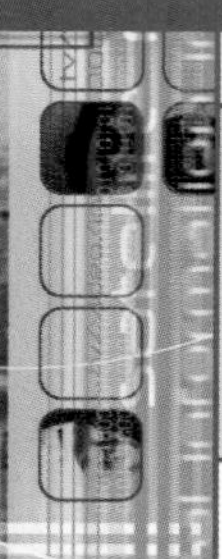

A4.3 Using brackets

This spread will show you how to:

- Understand and use the relationships between the four operations and the principles of the arithmetic laws.
- Use brackets.

KEYWORDS
Algebra
Operation
Brackets

Class 7B are trying to work out $2 + 3 \times 4 - 1$

They have four different answers:

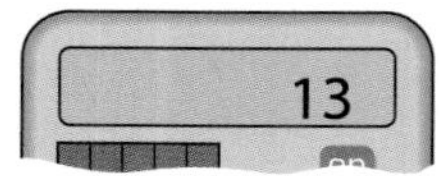

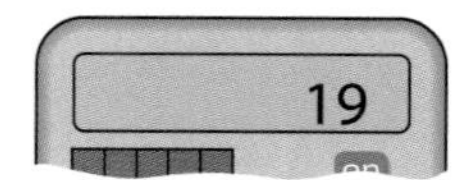

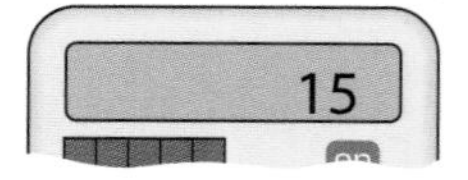

If everyone performs the operations in the same order they will all get the same answer.

In the sum $2 + 3 \times 4 - 1$
there are three operations: $+ \quad \times \quad -$

The conventional order of operations is:

multiply or divide first
add or subtract next

Multiply first: $2 + 12 - 1$

Then add: $14 - 1 = 13$
or subtract: $2 + 11 = 13$

The correct answer is 13.

To perform operations in a different order you use brackets.

To add first:

$$\begin{aligned} &(2 + 3) \times 4 - 1 \\ &= 5 \times 4 - 1 \\ &= 20 - 1 \\ &= 19 \end{aligned}$$

To subtract first:

$$\begin{aligned} &2 + 3 \times (4 - 1) \\ &= 2 + 3 \times 3 \\ &= 2 + 9 \\ &= 11 \end{aligned}$$

To add and subtract first:

$$\begin{aligned} &(2 + 3) \times (4 - 1) \\ &= 5 \times 3 \\ &= 15 \end{aligned}$$

▶ **You use brackets to change the conventional order of operations.**

The rule also applies in algebra.

$3(x + 1)$ is different to $3x + 1$

When $x = 5$:

$$\begin{aligned} &3(x + 1) \\ &= 3(5 + 1) \\ &= 3 \times 6 \\ &= 18 \end{aligned}$$

$$\begin{aligned} &3x + 1 \\ &= 3 \times 5 + 1 \\ &= 15 + 1 \\ &= 16 \end{aligned}$$

Exercise A4.3

1 Copy and complete the working out for these sums.

a

$3 + 4 \times 6 - 2$

$= 3 + 24 - 2$

$= \square - 2$

$= $ 

b

$2 + 9 \div 3 - 1$

$= 2 + \square - 1$

$= \square - 1$

$= $ 

c

$12 - 8 \div 2 + 5$

$= \square - \square + 5$

$= \square + 5$

$= $ 

2 Work out these sums.
Remember: multiply or divide first, then add or subtract.

a $7 + 2 \times 3 - 4$
b $5 + 8 \div 2 - 7$
c $8 \div 2 + 1 - 3$
d $9 - 2 + 4 \times 3$

3 $3 + 6 \div 3 - 1 = 4$
Use brackets to show how to make these answers.

a $3 + 6 \div 3 - 1 = 2$
b $3 + 6 \div 3 - 1 = 6$
c $3 + 6 \div 3 - 1 = 4.5$

4 Use $x = 2$ to find the value of:

a $2(x + 3)$
b $2x + 3$

Are $2(x + 3)$ and $2x + 3$ the same or different?
Explain your answer.

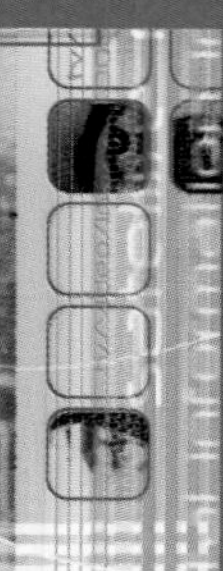

A4.4 Solving equations

This spread will show you how to:

- Develop from explaining a generalised relationship in words to expressing it in a formula using letters as symbols.

KEYWORDS

Expression, Unknown, Variable, Equation, Values

- An **expression** is a sentence that uses letters to stand for variables.

d is the score shown on a dice. The value of d can vary.

$2d + 1$ $\quad$ $d - 3$ $\quad$ and $\quad$ $5d$ $\quad$ are expressions.

If you know the value of d you can find the value of the expression.

If $d = 4$:

$$2d + 1 = 2 \times 4 + 1 = 8 + 1 = 9$$

$$d - 3 = 4 - 3 = 1$$

$$5d = 5 \times 3 = 15$$

Murray substitutes the score on a dice into two expressions to see which is bigger.
He records his results in this table:

d	$3d$	$d + 5$
2	6	7
1	3	6
		10
		8

Murray's dog treads on the sheet.
He has to work out the scores from the values:

$d + 5 = 10$ so $10 - 5 = d$
So $d = 5$ and $3d = 3 \times 5 = 15$.

$d + 5 = 8$ so $8 - 5 = d$
So $d = 3$ and $3d = 3 \times 3 = 9$.

$d + 5$ is an equation.
It links the expression $d + 5$ with its value, 10.

Murray solves the equation to find the value of d.

- An **equation** links an expression with its value using an equals sign.
- You can **solve** an equation to find the value of an unknown.

Exercise A4.4

1 Substitute $d = 3$ into each of these expressions.

a $2d$

b $d + 2$

c $2d + 1$

d $3d - 2$

2 Work out the value of d that makes these results true.
The first one is done for you.

a $d + 5 = 6$
$1 + 5 = 6$
so $d = 1$

b $d + 5 = 7$

c $d + 5 = 9$

d $d + 5 = 11$

3 Find the value of d when:

a $3d = 6$

b $3d = 12$

c $3d = 18$

d $3d = 3$

4 Find the value of d that makes these equations true.

a $d + 1 = 7$

b $d + 4 = 5$

c $d + 8 = 11$

d $d + 6 = 8$

e $3 + d = 5$

f $4 + d = 7$

g $d - 1 = 3$

h $d - 2 = 4$

i $d - 3 = 2$

j $d - 5 = 0$

5 Find the value of d that makes these equations true.

a $3d = 9$

b $4d = 8$

c $2d = 12$

d $3d = 15$

e $7d = 14$

f $5d = 25$

g $2 = 2d$

h $12 = 4d$

i $8 = 2d$

j $24 = 8d$

6 Find the value of x that makes these equations true.

a $x + 3 = 9$

b $x + 7 = 24$

c $x - 11 = 12$

d $x - 9 = 7$

e $2x = 16$

f $3x = 30$

g $\frac{x}{2} = 5$

h $\frac{x}{5} = 3$

i $2x + 1 = 5$

j $3x - 1 = 14$

A4 Summary

You should know how to ...

1 Develop from explaining a generalised relationship in words to expressing it in a formula using letters as symbols.

2 Choose and use appropriate number operations to solve problems, and appropriate ways of calculating.

Check out

1 Write expressions for each of these problems:

a 5 more than a number

b 3 or less than a number

c a number more than 2

d a number less than 4

e half a number

f a number divided by 3

g 3 times a number

h 4 more than 3 times a number.

Copy and complete:

i $35 \times 7 = 35$ so $35 \div \_ = 7$

j $5 \times x = 35$ so $35 \div \_ = x$

Solve the following equations:

k $x + 5 = 16$

l $y - 3 = 14$

m $3x = 18$

n $\frac{y}{3} = 5$

o $\frac{12}{x} = 4$

2 **a** Make up a number story to represent: 12×37

b Make up an algebra story to represent each of these expressions:

- ▶ $2 + x$
- ▶ $n - 5$
- ▶ 3m
- ▶ $\frac{d}{4}$

4 Transformations

This unit will show you how to:

- Recognise reflective symmetry in polygons.
- Recognise where a shape will be after reflection.
- Recognise where a shape will be after a translation.
- Recognise where a shape will be after a rotation through 180° or 90°.
- Solve mathematical problems or puzzles, recognise and explain patterns and relationships, generalise and predict.

You can make beautiful patterns using symmetry.

Before you start

You should know how to ...

1 Read and plot coordinates in the first quadrant.

2 Recognise common 2-D shapes.

Check in

1

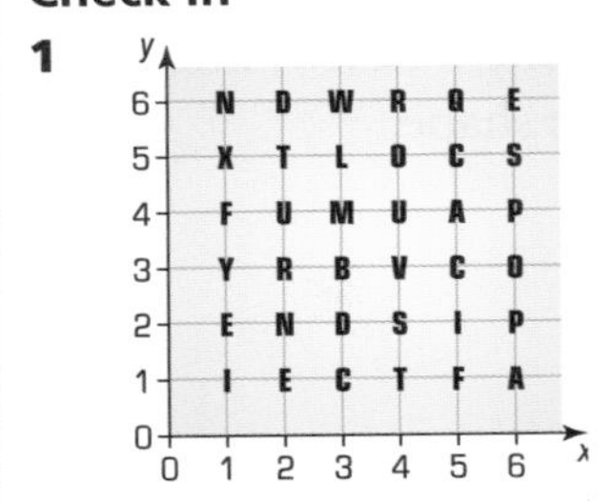

Find the letter at the given coordinates:

(2, 5) (4, 6) (5, 4) (2, 2) (4, 2) (1, 4) (4, 5) (2, 3)

T ___ ___ ___ ___ ___ ___ ___

(3, 4) (6, 1) (4, 1) (1, 1) (6, 3) (1, 6)

___ ___ ___ ___ ___ ___

2 Name these shapes:

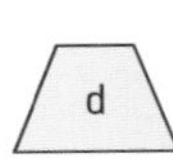

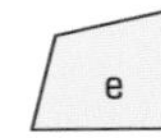

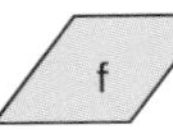

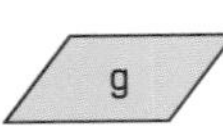

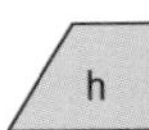

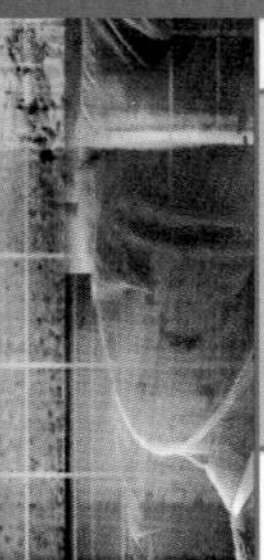

S4.1 Reflection symmetry

This spread will show you how to:

- Recognise reflection symmetry.

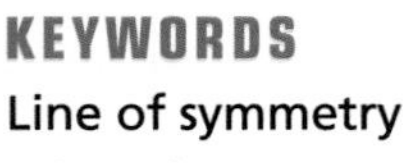

KEYWORDS

Line of symmetry
Mirror line
Reflection symmetry

This shape can be folded ... so one half fits exactly on the other:

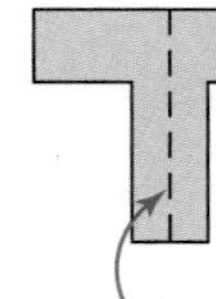

The shape has reflection symmetry.
The fold line is a line of symmetry.

- **A shape has reflection symmetry if you can fold it so that one half fits exactly on top of the other.**

You can use a mirror to find lines of symmetry.

Place your mirror along a mirror line and you will see the complete shape.

This shape does not have reflection symmetry:

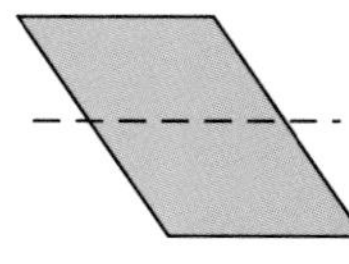

Using a mirror shows a different shape:

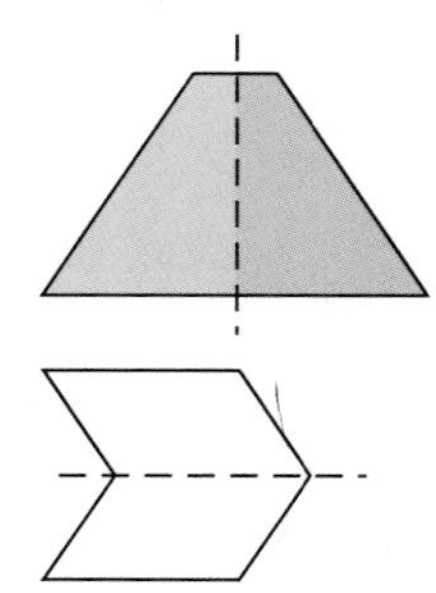

Exercise S4.1

Copy each shape and then draw its lines of symmetry.

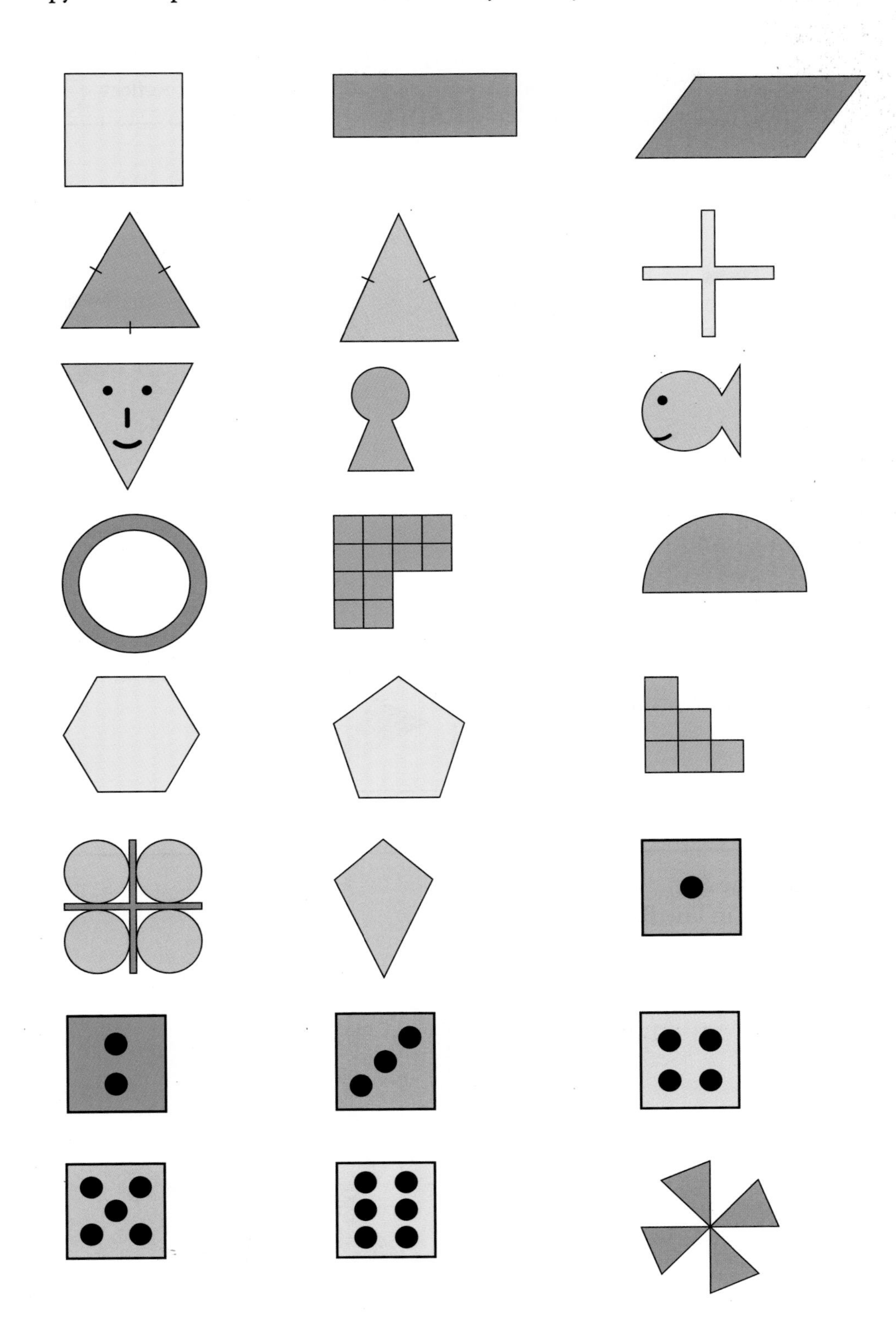

Reflecting shapes

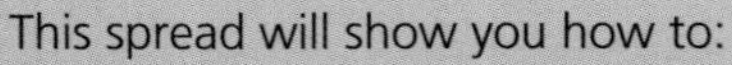

This spread will show you how to:

- Sketch the reflection of a simple shape in two mirror lines at right angles.

KEYWORDS

Image	Object
Mirror line	Reflect

In a mirror, your reflection is the same distance away.

If you stand close

... you look close

If you stand far

... you look far

You can reflect shapes in a mirror.
The shape is called the **object** and the reflection is called the **image**.

example

Reflect this shape in the mirror:

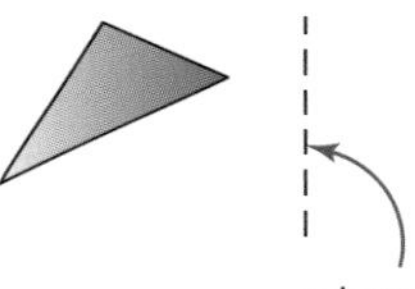

The image is the same distance from the mirror as the object.

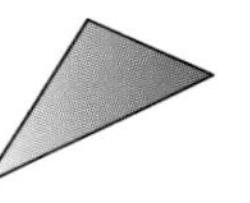

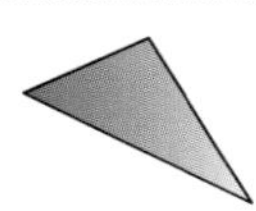

You can reflect shapes on a coordinate grid. It can help to use a mirror.

example

a Reflect this shape in line A

b Then reflect the image in line B.

c Then reflect the image in line A again.

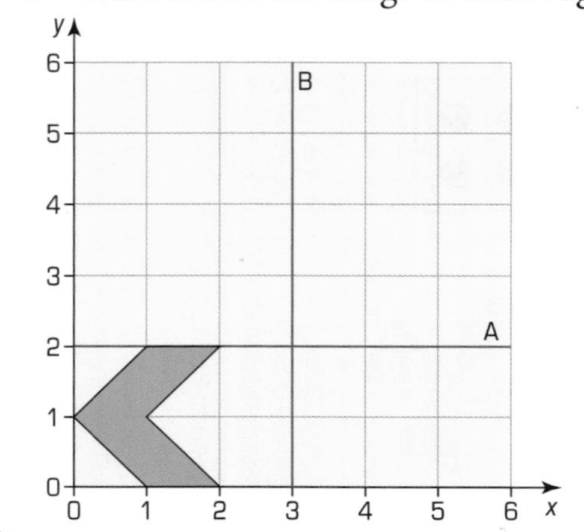

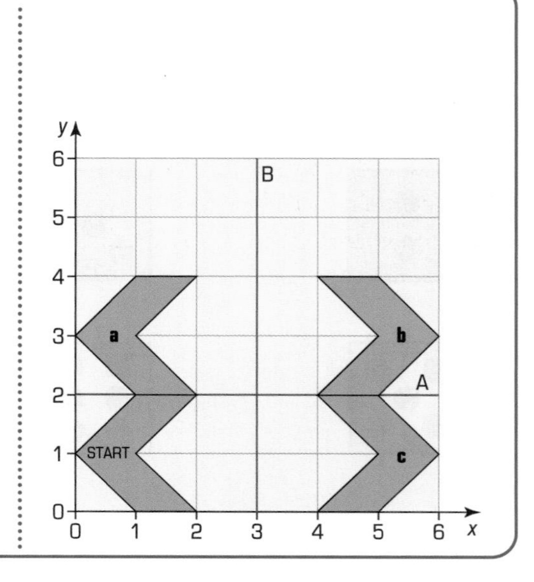

Exercise S4.2

Reflect each of these shapes in the mirror line.
What is the name of the new shape in questions 1, 2 and 3?

1

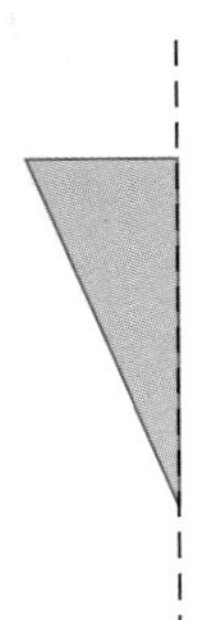

2

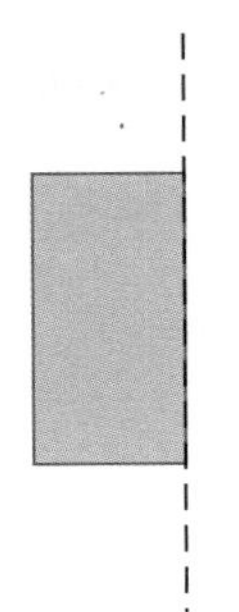

3

4

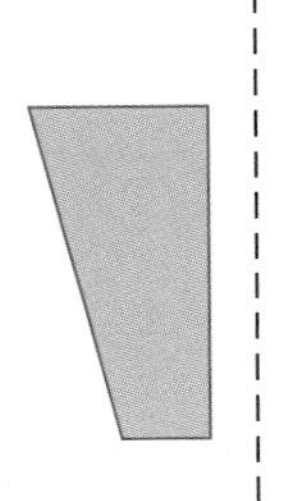

5

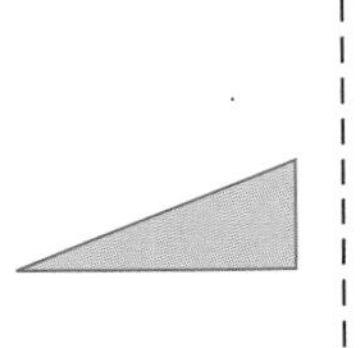

6

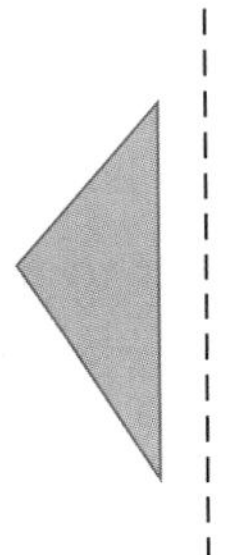

7

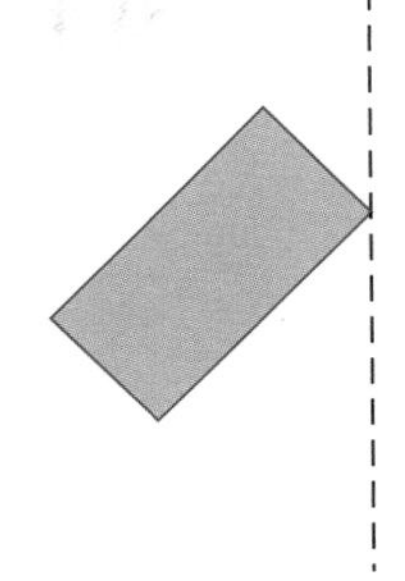

8

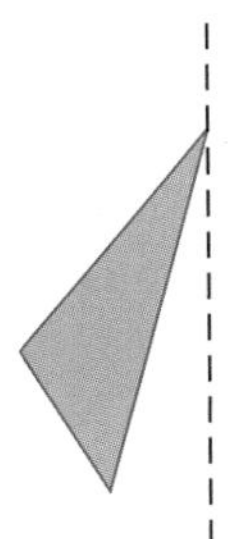

9

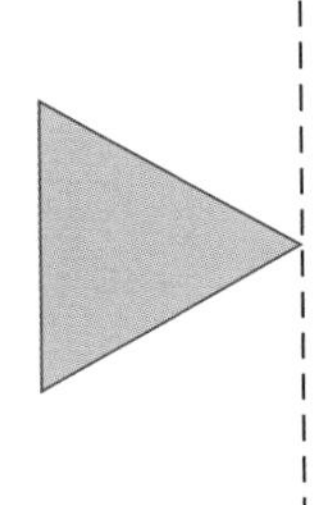

10 Copy this pattern onto squared paper.
Complete the pattern by drawing:

a the reflection in line A
b the reflection in line B
c the reflection of the reflection in line A.

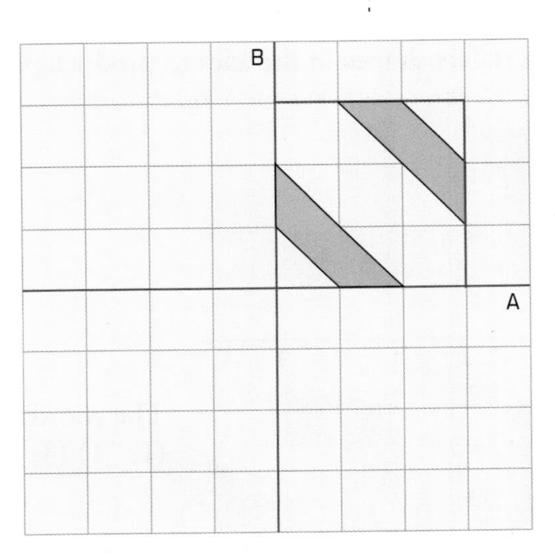

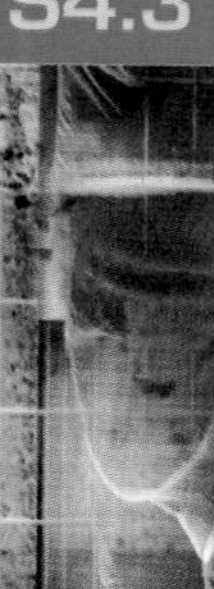

Reflecting in all four quadrants

This spread will show you how to:

- Read and plot coordinates in all four quadrants.
- Recognise where a shape will be after a reflection.

KEYWORDS

Horizontal	Vertical
Reflect	Vertices
Shape	Quadrant
Coordinates	Grid

You can count from $^-3$ to $^+3$ …
on a horizontal line:

$^-3$ $^-2$ $^-1$ 0 $^+1$ $^+2$ $^+3$

on a vertical line:

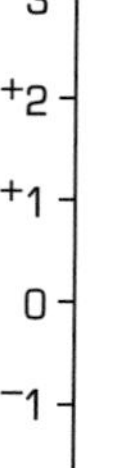

If you put the two lines together to meet at 0 you have a coordinate grid with four areas:

B has coordinates ($^-1$, 2)
1 back 2 up

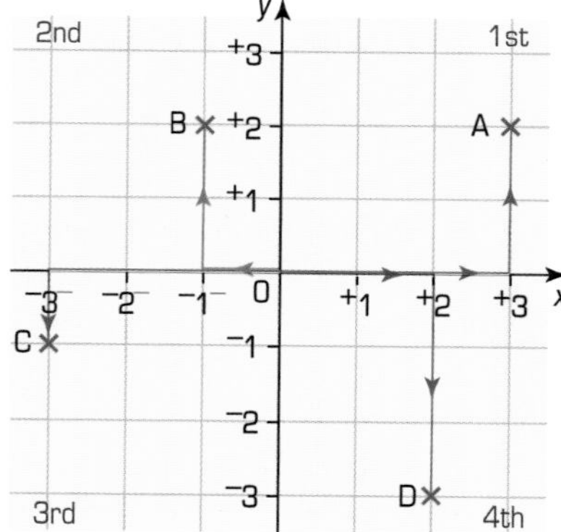

A has coordinates (3, 2)
3 across 2 up

C has coordinates ($^-3$, $^-1$)
3 back 1 down

D has coordinates (2, $^-3$)
2 across 3 down

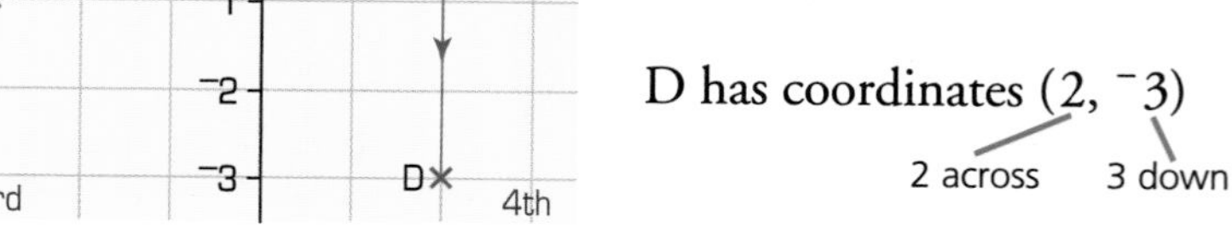

Now you can reflect shapes in the axes to produce patterns:

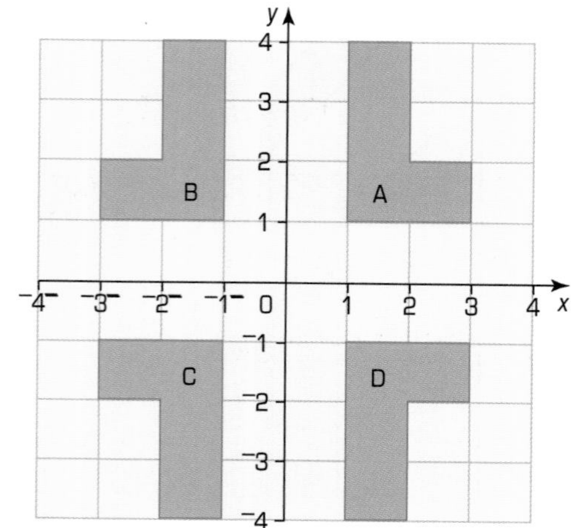

- Reflect A in the *y*-axis onto B.
- Reflect B in the *x*-axis onto C.
- Reflect C in the *y*-axis onto D.
- Reflect D in the *x*-axis onto A.

The coordinates of the vertices of D are:
(1, $^-1$) (3, $^-1$) (3, $^-2$) (2, $^-2$) (2, $^-4$) (1, $^-4$)

Exercise S4.3

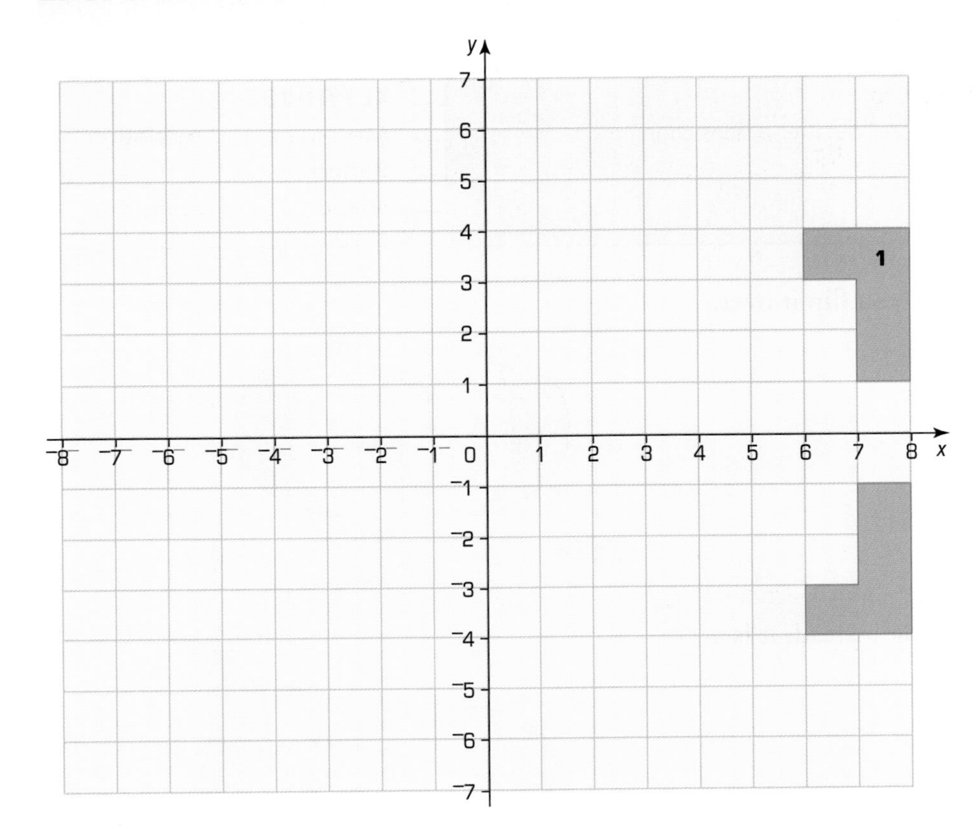

Each question gives the coordinates of the vertices of a shape.

- ▶ Copy the grid.
- ▶ Plot each set of points and join them to form a shape.
- ▶ Name the shape.
- ▶ Reflect it in the *x*-axis.
- ▶ Write down the coordinates of each point of the reflection.

The first one is started for you.

1 (8, 1) (8, 4) (6, 4) (6, 3) (7, 3) (7, 1) (8, 1) $\rightarrow$ (8, $^-1$) (8, $^-4$) ...

2 (1, 1) (3, 1) (1, 4) (1,1)

3 ($^-1$, 1) ($^-4$, 1) ($^-3$, 3) ($^-1$, 3) ($^-1$, 1)

4 (6, $^-2$) (5, $^-3$) (4, $^-2$) (5, $^-1$) (6, $^-2$)

5 ($^-4$, $^-2$) ($^-5$, $^-4$) ($^-4$, $^-6$) ($^-3$, $^-4$) ($^-4$, $^-2$)

6 (1, $^-5$) (2, $^-7$) ($^-2$, $^-7$) ($^-3$, $^-5$) (1, $^-5$)

7 (7, 5) (7, 7) (4, 7) (4, 5) (7, 5)

8 ($^-5$, 1) ($^-5$, 3) ($^-8$, 2) ($^-5$, 1)

9 (3, $^-2$) (3, $^-4$) (5, $^-4$) (3, $^-2$)

10 ($^-7$, 3) ($^-6$, 6) ($^-7$, 7) ($^-8$, 6) ($^-7$, 3)

S4.4 Translating shapes

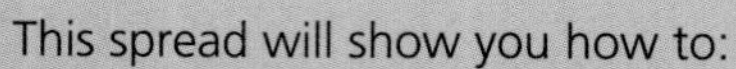

This spread will show you how to:

- Recognise where a shape will be after a translation.

KEYWORDS

Mirror line, Translation, Reflection, Shape

When you reflect a shape you flip it over:

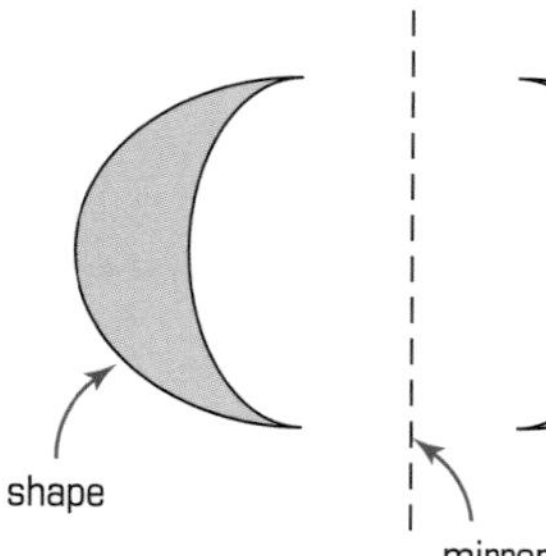

If you slide the shape along instead, it is a translation.

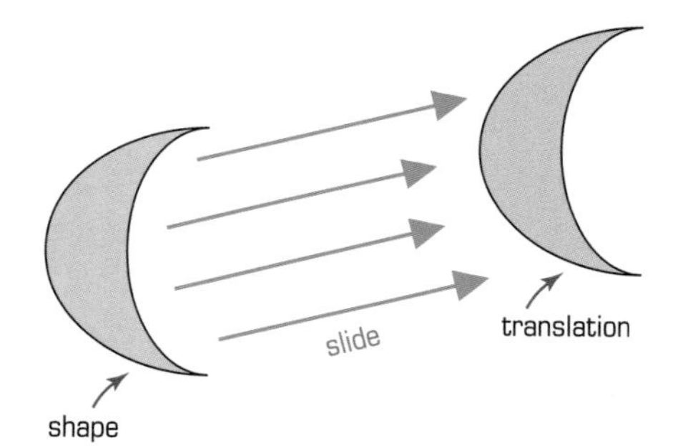

► A **translation** moves a shape along and then up or down.

example

Describe the translation that takes:

a A to B
b B to C
c A to C

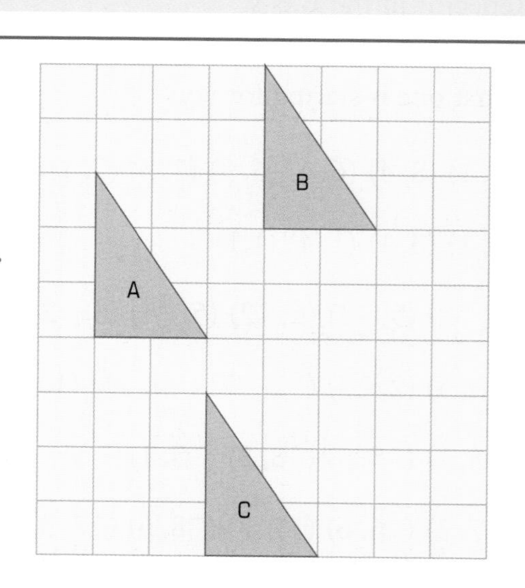

Count right or left first and then up or down.

a A to B: 3 right and 2 up
b B to C: 1 left and 6 down
c A to C: 2 right and 4 down

Exercise S4.4

1 Describe these translations.
The first one is done for you.

a A → B $\begin{pmatrix}\text{6 right}\\\text{1 up}\end{pmatrix}$
b D → B
c G → D
d E → F
e H → I
f I → H
g A → F
h F → D
i F → G
j A → D
k B → C

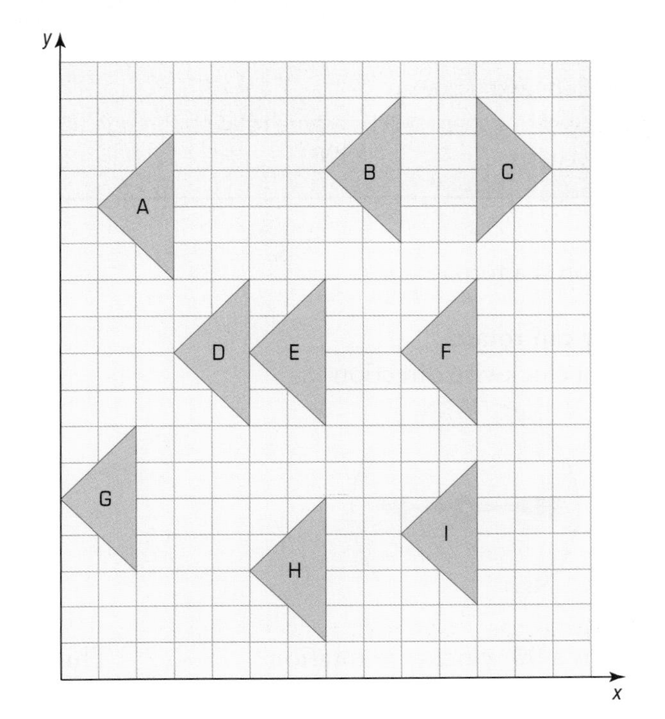

2 On squared paper draw: the x-axis from $x = 0$ to $x = 13$
the y-axis from $y = 0$ to $y = 13$

Plot each shape on the same grid.

- Name each shape.
- Translate each shape.
- Write down the coordinates of the translated shape.

A (3, 12) (6, 13) (6, 11) (3, 12) — Translate A $\begin{pmatrix}\text{1 to the left}\\\text{5 down}\end{pmatrix}$

B (0, 9) (0, 11) (6, 11) (6, 9) (0, 9) — Translate B $\begin{pmatrix}\text{5 to the right}\\\text{3 down}\end{pmatrix}$

C (0, 6) (1, 6) (1, 8) (0, 6) — Translate C $\begin{pmatrix}\text{4 to the right}\\\text{0 up}\end{pmatrix}$

D (0, 2) (1, 2) (1, 3) (2, 3) (2, 1) (0, 1) (0, 2) — Translate D $\begin{pmatrix}\text{4 to the right}\\\text{3 up}\end{pmatrix}$

E (3, 2) (5, 3) (3, 1) (3, 2) — Translate E $\begin{pmatrix}\text{8 to the right}\\\text{6 up}\end{pmatrix}$

F (5, 1) (5, 2) (6, 2) (6, 3) (7, 3) (7, 1) (5, 1) — Translate F $\begin{pmatrix}\text{4 to the right}\\\text{3 up}\end{pmatrix}$

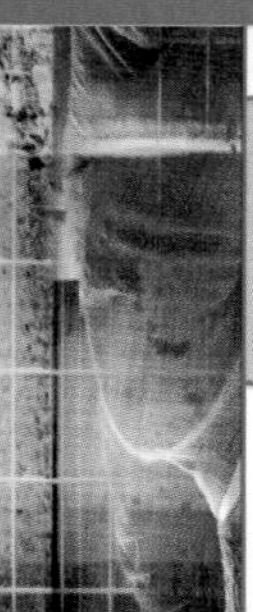

S4.5 Rotation

This spread will show you how to:

- Recognise where a shape will be after a rotation through 90° about one of its vertices.

KEYWORDS
Degree° Direction
Rotation

- A rotation is a turn.

This spinner can rotate:

in a clockwise direction ... or anticlockwise direction.

This is a 90° clockwise rotation

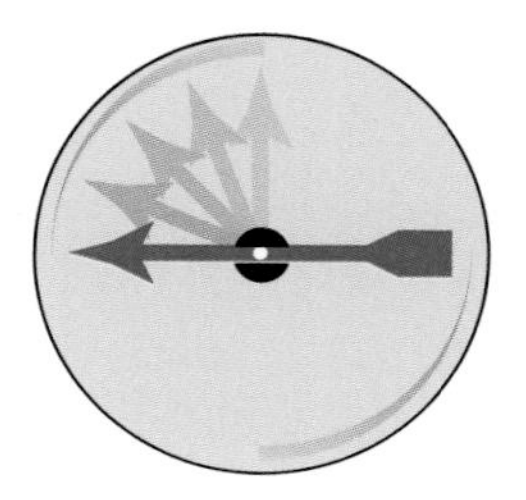

This is a 90° anticlockwise rotation

Rotating shapes often make a pattern.

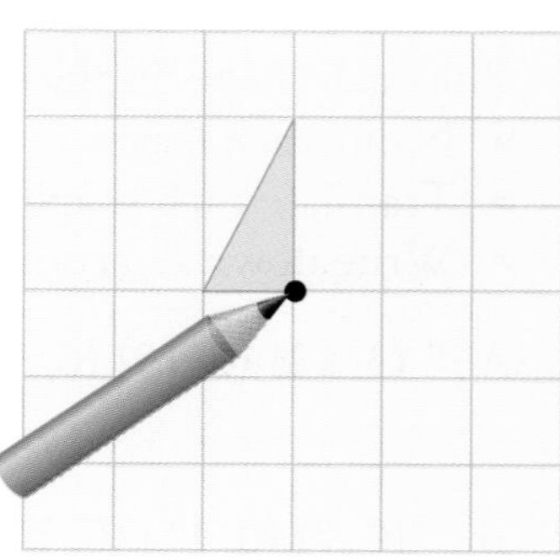

You can rotate a shape through 90°:

- Draw the shape on squared paper then cut it out.
- Place it on a grid and draw round it.
- Rotate the shape using a pencil point in the corner.
- Draw round it after each rotation.
- Continue until you get back to the start.

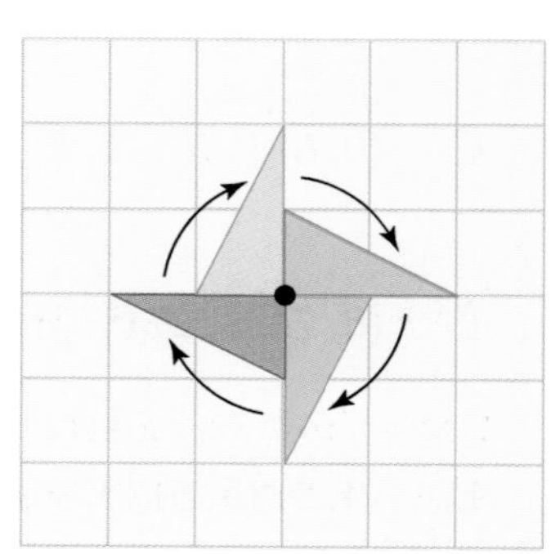

Four rotations of 90° make 360°.
$4 \times 90 = 360$.

Exercise S4.5

1 Give the angle and direction of each of these rotations.

a

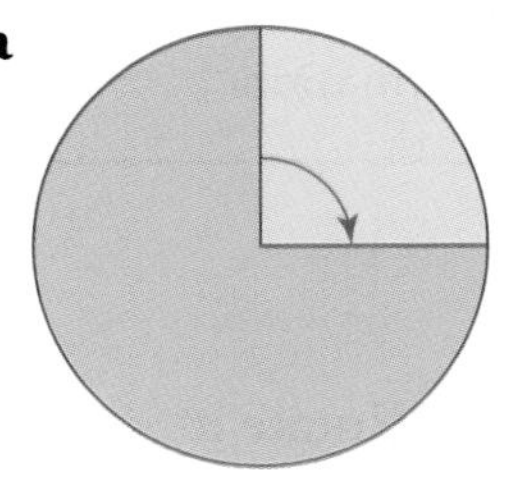

b

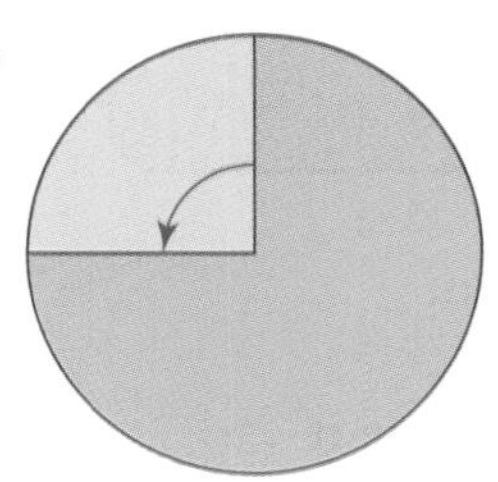

c

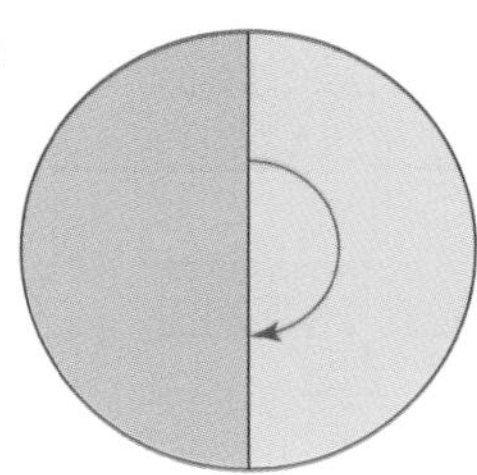

d

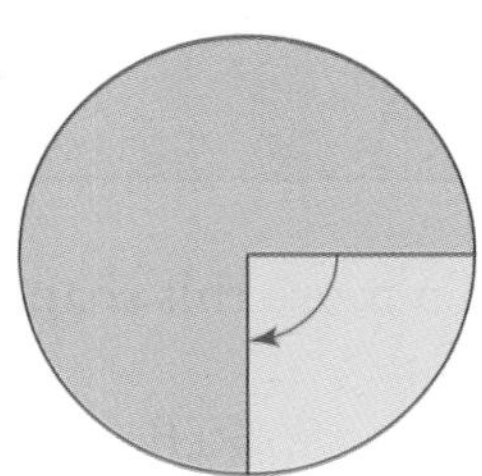

2 Copy and complete this table.

	Start	Finish	Angle and direction
a	12	3	
b	6	3	
c	1	4	
d	9		90° clockwise
e	8		180°
f	7		90° clockwise
g		6	90° anticlockwise
h		10	180°
i		4	180°

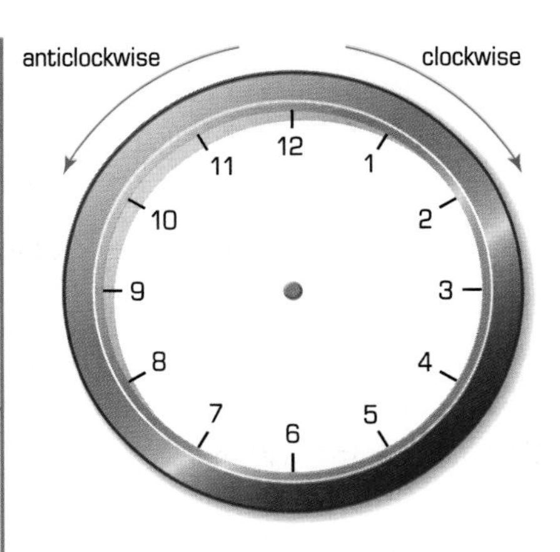

3 Copy and rotate these shapes about the dot through the angle given.

a 90° clockwise

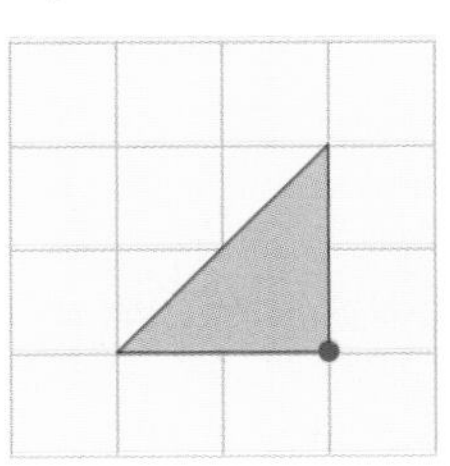

b 90° clockwise

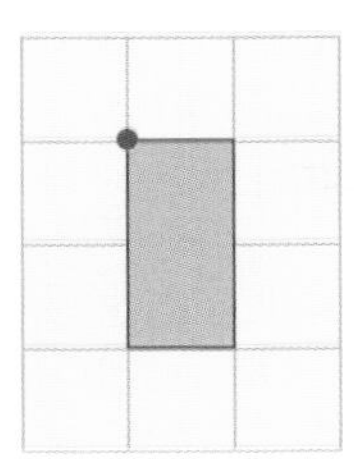

c 180°

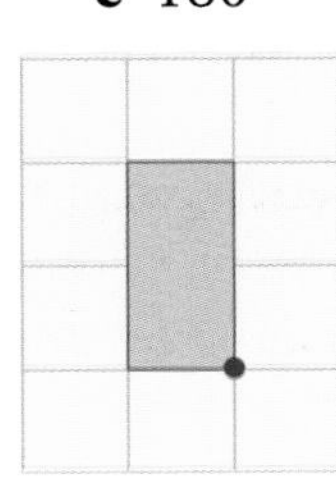

d 90° anticlockwise

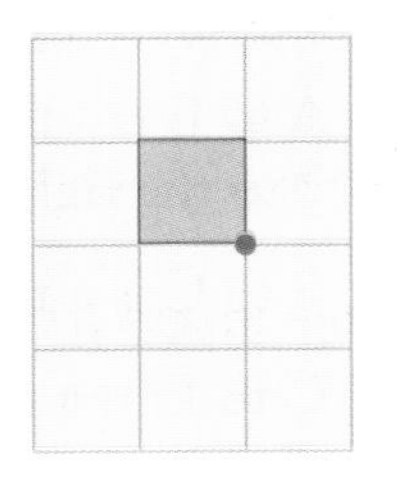

4 When a right-angled triangle is rotated, these are the results.

Draw similar rotations for:

a an isosceles right-angled triangle

b a square

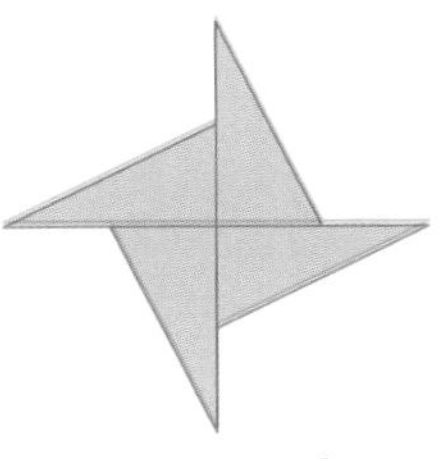

rotations of 90°

rotations of 45°

Transformations

This spread will show you how to:

- Recognise reflective symmetry in 2-D shapes, reflections and translations.

KEYWORDS

Reflect
Translate
Rotate
Transformation

To move this shape you can:

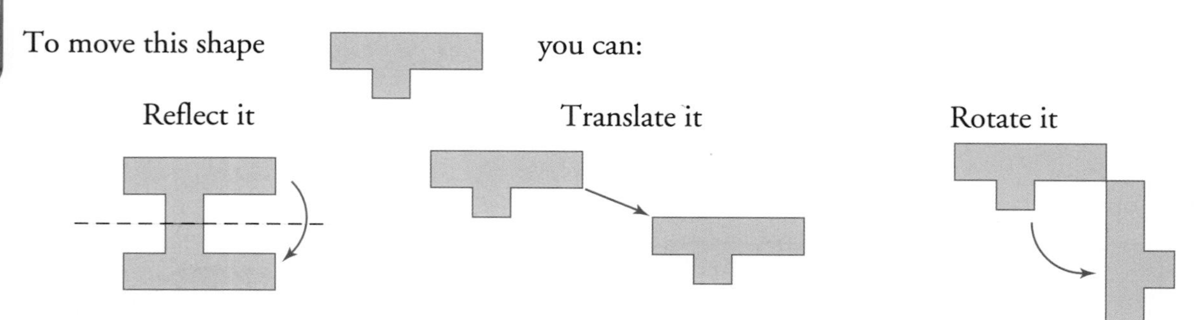

Flip it over

Slide it right or left then up or down

Turn it clockwise or anticlockwise

Reflections, translations and rotations are transformations. They change the position but not the size of a shape.

To describe a transformation you should give as much information as possible.

example

Describe the transformation that takes

a A to B **b** B to C **c** C to D
d D to E **e** A to E **f** B to D

a A to B is a translation: 2 to the right and 2 up

b B to C is a reflection

c C to D is a reflection

d D to E is a translation: 3 to the left and 0 up.

e A to E is a rotation of 180°

f B to D is a rotation of 180°

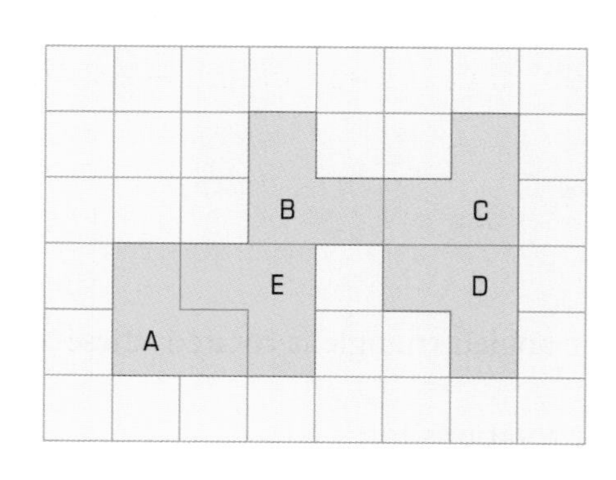

Exercise S4.6

1 Reflect each of these shapes in the mirror line.
What is the name of each new shape?

a

b

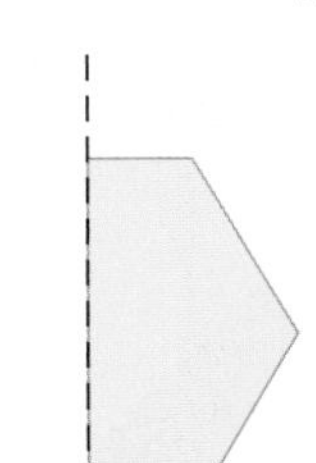

c

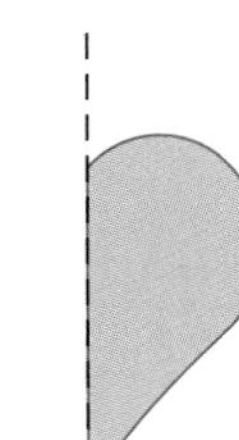

2 Copy and complete this table.

	Start	Finish	Angle and direction
a	N	E	
b	NE	NW	
c	SE	NW	
d	W		90° anticlockwise
e	SW		180°
f	S	SW	

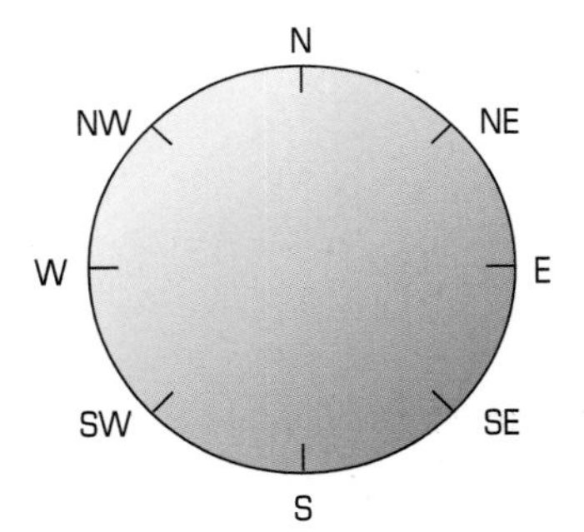

3 A shape is translated 2 units to the right, 3 units down and 4 units to the left.
Describe this translation with only two instructions.

4 Copy this grid and the triangle.
Rotate the triangle:
a through 90° clockwise about the point (0, 3)
b through 90° anticlockwise about the point (2, 0)

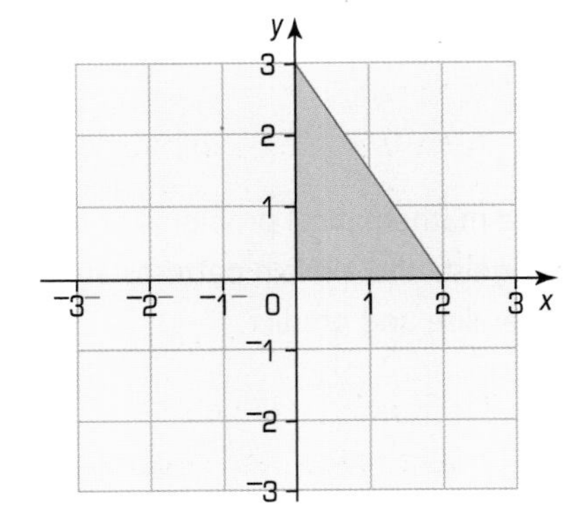

5 Translate the triangle in question 4 using the instructions in question 3.
What are the coordinates of the translated shape?

S4 Summary

You should know how to ...

1 Read and plot coordinates in all four quadrants.

Check out

1

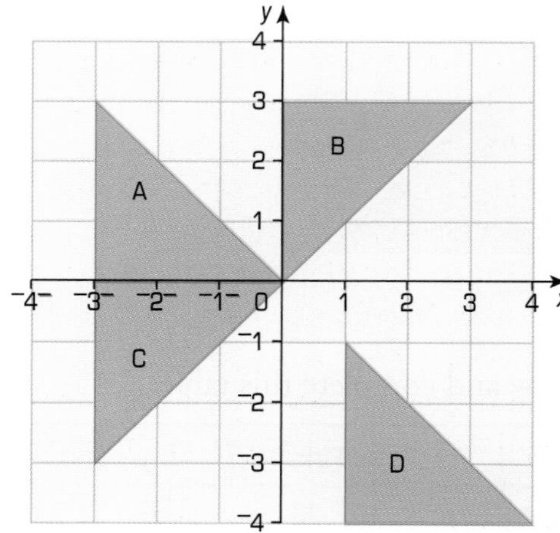

The coordinates of each vertex of triangle A are (0, 0) (⁻3, 0) and (⁻3, 3)

Write down the coordinates of each vertex of

a triangle B

b triangle C

c triangle D

2 Recognise reflections, translations and rotations through 90°.

2 Which transformation will move:

a A → D

b A → C

c A → B

d C → B

e D → A?

3 Solve mathematical problems or puzzles, recognise and explain patterns and relationships, generalise and predict.

3 This is half of a shape:

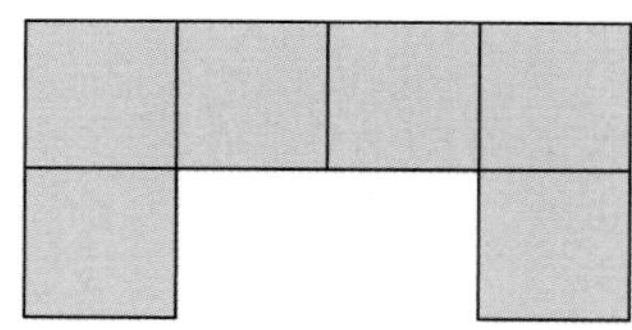

Sketch possible complete shapes, marking any lines of symmetry.

15 More number calculations

This unit will show you how to:

- Round positive whole numbers to the nearest 10, 100 or 1000 and decimals to the nearest whole number or one decimal place.
- Recognise multiples up to 10 ×10.
- Recognise prime numbers to at least 20.
- Factorise numbers up to 100.
- Reduce a fraction to its simplest form by cancelling common factors.
- Use a fraction as an operator.
- Compare and order simple fractions.
- Use decimal notation for tenths and hundredths.
- Find simple percentages of quantities.
- Recognise the equivalence of fraction and decimal forms.
- Understand the effect of and relationships between the four operations, and the principles of the arithmetic laws as they apply to multiplication.
- Approximate first. Use informal pencil and paper methods to support, record, or explain multiplications and divisions.
- Extend written methods of multiplication and division.
- Use the vocabulary of estimation and approximation.
- Consolidate and extend mental calculation strategies.
- Explain methods and reasoning.

You often need to round measured values.

Before you start

You should know how to …

1 Recall 10 × 10 time table facts.

2 Convert between fractions and percentages.

Check in

1 Work out:

a 5×5 **b** 6×6 **c** $30 \div 6$
d $28 \div 7$ **e** 9×8 **f** $81 \div 9$

2 Change these fractions to percentages:

a $\frac{1}{2}$ **b** $\frac{1}{4}$ **c** $\frac{3}{4}$ **d** $\frac{1}{10}$ **e** $\frac{7}{10}$ **f** $\frac{1}{5}$

Change these fractions to percentages:

g $\frac{31}{100}$ **h** $\frac{63}{100}$ **i** $\frac{7}{100}$ **j** $\frac{1}{2}$ **k** $\frac{1}{4}$ **l** $\frac{1}{10}$

N5.1 Rounding

This spread will show you how to:

- Round numbers up to 10 000 to the nearest 10, 100 and 1000.
- Round numbers with one or two decimal places to the nearest whole number.

KEYWORDS
Decimal
Nearest
Round

Often it makes sense to round numbers up or down.

Matt earned £203.56 last week

Matt earned about £200

There were 8942 people at the match

There were about 9000 spectators

▶ **You usually round numbers to the nearest 1000 or 10, or 10.**

A number line can help.
To round to the nearest 1000, use a number line marked in 1000s.

example

Round 3250 to the nearest 1000.

- ▶ 3250 is between 3000 and 4000.
- ▶ It is nearer to 3000.
- ▶ 3250 rounds **down** to 3000.

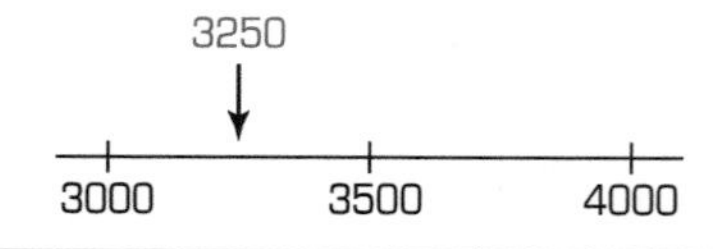

You can also round decimals to the nearest whole number.

example

Round 3.5 to the nearest whole number.

- ▶ 3.5 is between 3 and 4.
- ▶ It is exactly halfway.
- ▶ 3.5 rounds **up** to 4.

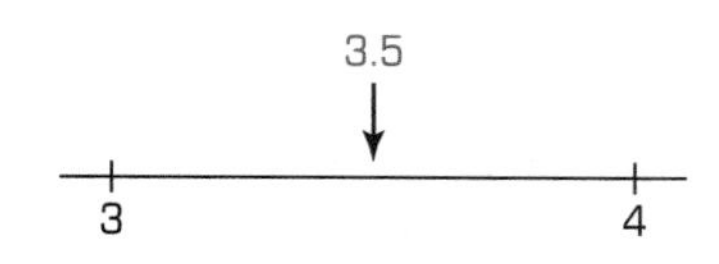

▶ **When a number is exactly halfway between two numbers on your number line, you round up.**

Exercise N5.1

1 Round these numbers to the nearest 10.

a 21 **b** 53 **c** 67 **d** 89 **e** 55 **f** 45
g 99 **h** 127 **i** 253 **j** 3 **k** 7 **l** 1563

2 Round these numbers to the nearest 100.

a 167 **b** 293 **c** 832 **d** 749 **e** 350 **f** 150
g 907 **h** 951 **i** 1371 **j** 82 **k** 31 **l** 2593

3 Round these numbers to the nearest 1000.

a 5200 **b** 5907 **c** 1350 **d** 2795 **e** 12 302 **f** 950
g 3500 **h** 7500 **i** 172 **j** 3962 **k** 9563 **l** 29 807

4 Round these numbers to the nearest whole number.

a 3.2 **b** 7.8 **c** 4.5 **d** 12.3 **e** 18.59 **f** 3.8
g 21.5 **h** 0.8 **i** 0.2 **j** 9.8 **k** 29.3 **l** 99.5

5 Round the heights of these people to the nearest 10 cm.

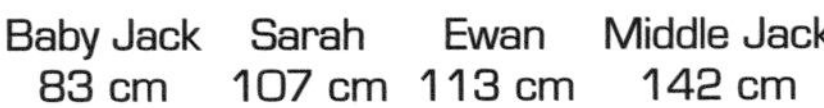

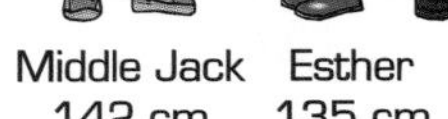

6 These are the results of a 400 m race.

1st	Jim	48.3 secs	2nd	Joe	48.8
3rd	Bill	49.2	4th	Mac	49.4
5th	Samir	49.9	6th	Nick	50.1
7th	David	53.5	8th	Tom	54.1

Round each of the times to the nearest second.

7 The crowd at a football match was 3752.
Round this number to the nearest 1000.

8 For each question use the numbers only once.
You do not have to use them all.

6 9 3 7 5

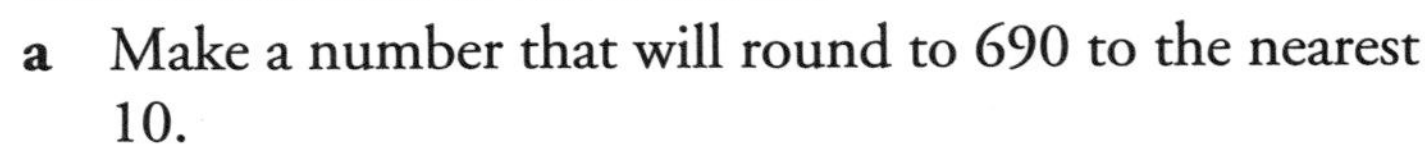

a Make a number that will round to 690 to the nearest 10.

b Make as many numbers as you can that will round to 570 to the nearest 10.

c Make as many numbers as you can that will round to 5700 to the nearest 100.

N5.2 Factors, multiples and primes

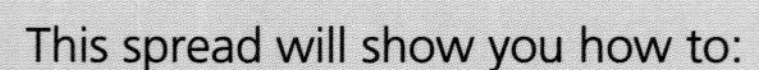

This spread will show you how to:

- Recognise multiples up to 10 × 10.
- Factorise numbers up to 100.
- Recognise prime numbers.

KEYWORDS
Factor
Multiple
Prime

You can list the **multiples** of a number by using its times table.

example

List the first four multiples of 7.

$1 \times 7 = 7$
$2 \times 7 = 14$
$3 \times 7 = 21$
$4 \times 7 = 28$

The first four multiples of 7 are 7, 14, 21 and 28.

You can break any whole number down into its **factors**.

$12 = 3 \times 4$, so 3 and 4 are factors of 12.

There are other factors of 12:

$12 = 1 \times 12$ $\quad\quad$ $12 = 2 \times 6$ $\quad\quad$ $12 = 3 \times 4$

The factors of 12 are 1, 2, 3, 4, 6 and 12.

- All whole numbers have at least two factors.

Some numbers only have two factors.

$5 = 5 \times 1$ The only factors of 5 are 5 and 1.
$29 = 29 \times 1$ The only factors of 29 are 29 and 1.

These are called **prime** numbers.

- A **prime number** only has two factors.
  These are 1 and the number itself.

Exercise N5.2

1 Write down the first five multiples and the 10th multiple of:

a 2 **b** 5 **c** 6 **d** 3 **e** 9 **f** 8

2 Write down all the factors of:

a 7 **b** 6 **c** 8 **d** 12 **e** 15
f 11 **g** 20 **h** 21 **i** 30

3 Think of the numbers on a dice.
Which numbers are :

a Multiples of 2
b Prime numbers
c Factors of 9
d Factors of 24?

4 Look at this list of numbers.
8, 9, 10, 11, 12, 13, 14, 15, 16, 17, 18, 19, 20
Which numbers are:

a Multiples of 4
b Multiples of 6
c Prime numbers
d Multiples of both 3 and 4
e Multiples of both 2 and 5
f Factors of both 6 and 15?

5 Look at this list of numbers.
19, 20, 21, 22, 23, 24, 25, 26, 27, 28, 29, 30
Which numbers are:

a Multiples of 3
b Multiples of 4
c Multiples of both 3 and 4
d Factors of 60
e Prime numbers?

6 Work out what these numbers are and place them in order starting with the smallest.

a The 5th multiple of 7
b The 4th multiple of 8
c The 9th multiple of 3
d The 2nd multiple of 10

7 **Investigation**
Find a number that has

a 4 factors **b** 5 factors **c** 6 factors **d** 10 factors

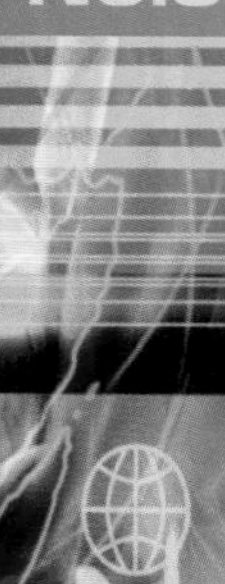

N5.3 Multiplying and dividing mentally

This spread will show you how to:

- Consolidate and extend strategies for mental multiplication and division by using factors.

KEYWORDS
Division
Partition
Factor

Here are two methods that can help you to multiply and divide in your head.

Factor method

This method involves splitting one of the numbers into its factors.
To work out 13×6:

- Split 6 into its factors: $6 = 3 \times 2$
- So $13 \times 6 = \underbrace{13 \times 3}_{39} \times 2$

$$\underbrace{39 \times 2}_{78}$$

So $13 \times 6 = 78$

You can also do divisions using the factor method.

example

Work out $72 \div 6$

$72 \div 6$ — Split 6 into its factors ($6 = 2 \times 3$)

$72 \div 2 = 36$

$36 \div 3 = 12$ So $72 \div 6 = 12$

Partition method

You can break one of the numbers down by adding or subtracting.

example

Work out

a 32×12 **b** $120 \div 8$

a 32×12 — Split 12 into 10 and 2 ($12 = 10 + 2$)

$32 \times 10 = 320$

$32 \times 2 = 64$

$320 + 64 = 384$

So $32 \times 12 = 384$

b

$$\begin{array}{rl} 120 & \\ -80 & \quad 10 \times 8 \\ \hline 40 & \\ -40 & \quad 5 \times 8 \\ \hline 0 & \end{array}$$

So $15 \times 8 = 120$
$120 \div 8 = 15$

Exercise N5.3

Do questions 1–3 using the factor method.

1 **a** 23×20 $(20 = 2 \times 10)$
b 31×30 $(30 = 3 \times 10)$
c 25×6 $(6 = 2 \times 3)$
d 32×8
e 43×20
f 45×30

2 **a** $84 \div 8$ ($\div 2$ then $\div 4$)
b $96 \div 6$ ($\div 3$ then $\div 2$)
c $252 \div 12$ ($\div 3$ then $\div 4$)
d $288 \div 9$
e $126 \div 6$
f $372 \div 12$

3 **a** 42×4 **b** $156 \div 6$ **c** 3.5×20
d 4.2×30 **e** $330 \div 15$ **f** 5.3×40

Do questions 4–6 using the partition method.

4 **a** 21×12 **b** 24×13 **c** 33×12
d 22×22 **e** 51×14 **f** 3.5×21

5 **a** $390 \div 13$ **b** $264 \div 12$ **c** $1331 \div 11$
d $433 \div 13$ **e** $467 \div 15$ **f** $450 \div 14$

6 **a** 55×13 **b** $417 \div 14$ **c** 4.5×31
d 27×22 **e** 121×14 **f** $609 \div 12$

In questions 7–10 choose the most appropriate method to UK.

7 Eight people go on a coach trip for the day.
They each pay £63.
How much do they pay in total?

8 Jim buys 20 pizzas at £2.25 each.
How much does he spend in total?

9 Harry buys 21 CDs at £12 each.
How much does he spend in total?

10 Gina buys 8 CDs at £10.99 each.
How much does she spend in total?

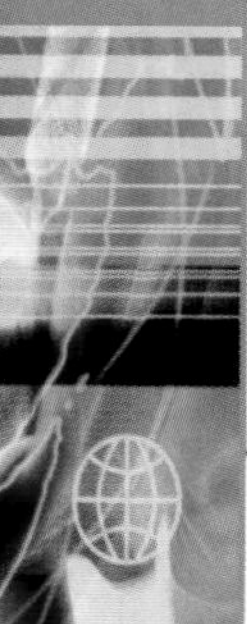

N5.4 Standard written calculations

This spread will show you how to:

- Approximate first.
- Develop and refine written methods for multiplication.

KEYWORDS

Decimal
Multiply
Digit
Estimate

You can already multiply numbers with two digits using the grid method.

To work out 21 × 23, first estimate: 20 × 20 = 400. Then partition the numbers in a grid. Adding the answers gives 483 which is close to 400.

21 × 23

×	20	1
20	400	20
3	60	3

$$\begin{array}{r} 400 \\ 20 \\ 60 \\ +\ \ 3 \\ \hline 483 \end{array}$$

so 21 × 23 = 483

You can still use the grid method with larger numbers.

example

Work out 323 × 24 using the grid method.

First estimate: 323 × 24 ≈ 300 × 25 = 7500

×	300	20	3
20	6000	400	60
4	1200	80	12

$$\begin{array}{r} 6000 \\ 1200 \\ 400 \\ 80 \\ 60 \\ +\ \ 12 \\ \hline 7752 \end{array}$$

323 × 24 = 7752

You can also use the grid method to multiply decimals.

example

Work out 3.28 × 6 using the grid method.

First estimate: 3.28 × 6 ≈ 3 × 6 = 18

×	3	0.2	0.08
6	18	1.2	0.48

0.08 0.08 0.08 0.08 0.08 0.08
0 0.08 0.16 0.24 0.32 0.40 0.48

$$\begin{array}{r} 18 \\ 1.2 \\ 0.48 \\ \hline 19.68 \end{array}$$

So 3.28 × 6 = 19.68

Exercise N5.4

1 Use the grid method to work out these questions.

a 21×4 **b** 35×4
c 42×5 **d** 83×4
e 72×5 **f** 81×6
g 22×32 **h** 25×23
i 32×34 **j** 56×38
k 37×26 **l** 95×78

2 Work out the answers to these questions.

a 217×31 **b** 315×24
c 422×25 **d** 153×35
e 325×42 **f** 527×31
g 325×38 **h** 577×27
i 627×35 **j** 895×56
k 343×32 **l** 415×26

3 A school trip costs £325.
25 pupils go on the trip.
How much do the pupils pay in total?

4 The head gardener of a park buys 375 trees each costing £27.
How much does he pay for the trees in total?

5 Work out the answers to these using the grid method.

a 32.5×7 **b** 25.7×8 **c** 2.75×6 **d** 8.25×3
e 2.91×9 **f** 8.35×7 **g** 2.32×7 **h** 52.1×8
i 43.5×4 **j** 4.13×8 **k** 35.7×7 **l** 29.2×6

6 A record shop owner buys 85 CDs each costing £5.75.
How much does he pay in total for the CDs?

7 One of these numbers is the answer to 347×27:

8249 899 9369 3642 134 799 1439

a Write down which number you think is the answer.
b Work out 347×27 by the grid method to check whether you are correct.

8 In each question, three of the four multiplications give the same answer.
Find the odd one out.

a

52×86	104×43
63×74	26×172

b

68×156	51×208
34×312	42×234

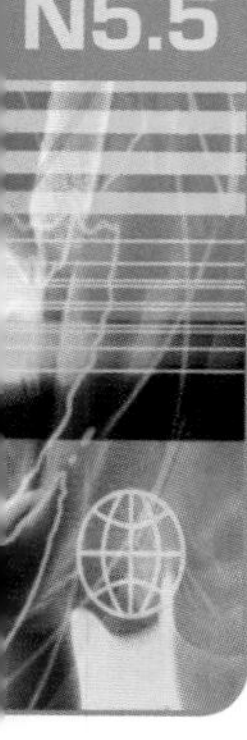

N5.5 Standard written division

This spread will show you how to:

- Understand the operation of division and its relationship to subtraction.
- Develop and refine written methods for division.

KEYWORDS

Divide
Subtraction
Multiply
Remainder

If you know how to multiply, you can often divide.

To work out $48 \div 6$: $8 \times 6 = 48$
So $48 \div 6 = 8$

- **Division is the opposite of multiplication.**

When you divide larger numbers, it helps to use a written method, like **repeated subtraction**.

example

Work out $264 \div 12$

$$
\begin{array}{r l}
12\overline{)264} & \\
-120 & \quad 10 \times 12 \\
\hline
144 & \\
-120 & \quad 10 \times 12 \\
\hline
24 & \\
-24 & \quad 2 \times 12 \\
\hline
0 & \\
\hline
\end{array}
$$

$22 \times 12 = 264$
So $264 \div 12 = 22$

There is often a remainder left over.

example

Work out $378 \div 13$

$$
\begin{array}{r l}
13\overline{)378} & \\
130 & \quad 10 \times 13 \\
\hline
248 & \\
-130 & \quad 10 \times 13 \\
\hline
118 & \\
-117 & \quad 9 \times 13 \\
\hline
1 & \\
\hline
\end{array}
$$

It helps if you write down the 13 times table: 13, 26, 39, 52, 65, 78, 91, 104, 117, 130.

$29 \times 13 + 1 = 378$
So $378 \div 13 = 29$ remainder 1

Exercise N5.5

1 Do these division questions. Write out the times table by the side of each question to help you.

a $170 \div 5$ **b** $204 \div 4$ **c** $384 \div 6$
d $258 \div 3$ **e** $504 \div 9$ **f** $304 \div 8$
g $609 \div 7$ **h** $405 \div 6$ **i** $648 \div 7$
j $457 \div 8$ **k** $336 \div 6$ **l** $774 \div 9$

2 Now do these division questions. Again, write out the times table by the side to help you.

a $336 \div 12$ **b** $473 \div 11$ **c** $675 \div 15$
d $592 \div 16$ **e** $416 \div 16$ **f** $728 \div 14$
g $444 \div 17$ **h** $743 \div 16$ **i** $953 \div 15$
j $623 \div 18$ **k** $728 \div 13$ **l** $703 \div 19$

3 A football player wins the 'Man of the Match' award of £375. He shares the money equally with the other 10 members of the team.

a How much does each player get?
b How much money is left over?
c Suppose the two substitutes are included as well. How much does each player get now?

4 James buys 32 graphic calculators at a total cost of £576. How much does each calculator cost?

5 Freja has £835 to spend on graphic calculators. Each calculator costs £18.

a How many calculators can she buy?
b How much money will she be left with?

6 18 friends spend £612 on a school reunion dinner. They split the cost equally between them. How much do they each pay?

7 In each of these questions, three of the four divisions give the same answer. Find the odd one out.

a

$504 \div 12$	$756 \div 18$
$688 \div 16$	$630 \div 15$

b

$441 \div 7$	$832 \div 13$
$945 \div 15$	$756 \div 12$

N5.6 Using equivalent fractions

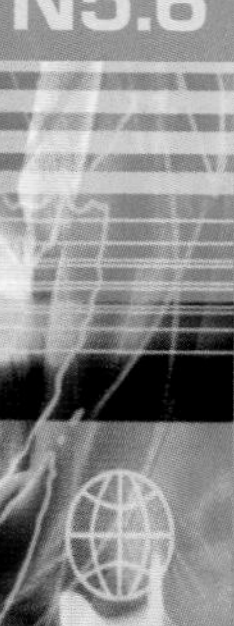

This spread will show you how to:

- Order and compare simple fractions.
- Reduce a fraction to its simplest form by cancelling.
- Recognise multiples up to 10 × 10.

KEYWORDS

Compare, Multiple, Denominator, Numerator, Equivalent, Fraction, Simplest form

Equivalent fractions describe the same proportion:

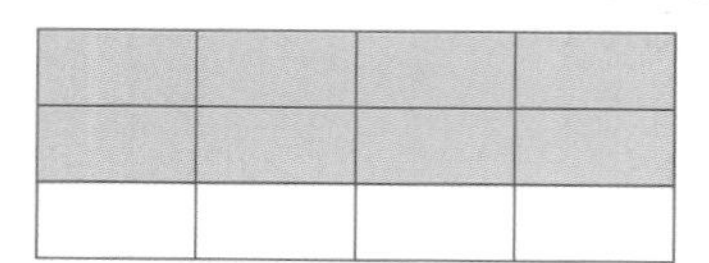

$\frac{2}{3}$ of the rectangle is shaded
$\frac{8}{12}$ of the rectangle is shaded
$\frac{2}{3}$ and $\frac{8}{12}$ are equivalent

- You can find equivalent fractions by multiplying or dividing the numerator and denominator by the same amount.

For example, $\frac{4}{10} = \frac{8}{20}$ (×2 top and bottom), and $\frac{4}{10} = \frac{2}{5}$ (÷2 top and bottom)

Note: $\frac{4}{10}$ can be simplified to make $\frac{2}{5}$.

- You can compare fractions with the same denominator.

example

Which is larger, $\frac{2}{3}$ or $\frac{5}{7}$?

Think of the 3 times table: 3, 6, 9, 12, 15, 18, 21, 24, ...

Think of the 7 times table: 7, 14, 21, 28, 35, ...

21 is a **common multiple** of 3 and 7.

Now find equivalent fractions with a denominator of 21:

$\frac{2}{3} = \frac{14}{21}$ (×7) $\qquad \frac{5}{7} = \frac{15}{21}$ (×3)

$\frac{15}{21}$ is larger than $\frac{14}{21}$.

So $\frac{5}{7}$ is larger than $\frac{2}{3}$.

- If you want to compare the size of two fractions:
  - Find equivalent fractions with the same denominator
  - Compare the numerators

Exercise N5.6

1 Copy and complete these equivalent fraction problems.

a $\frac{2}{3} \overset{\times 3}{=} \frac{\square}{9}$ (× 3)

b $\frac{3}{4} \overset{\times 2}{=} \frac{\square}{\square}$ (× 2)

c $\frac{1}{3} \overset{\times 4}{=} \frac{\square}{\square}$ (× 4)

d $\frac{3}{5} \overset{\times 3}{=} \frac{\square}{\square}$ (× 3)

2 Copy and complete these simplifying problems.

a $\frac{10}{12} \overset{\div 2}{=} \frac{\square}{6}$ (÷ 2)

b $\frac{15}{25} \overset{\div 5}{=} \frac{\square}{\square}$ (÷ 5)

c $\frac{8}{24} \overset{\div 8}{=} \frac{\square}{\square}$ (÷ 8)

d $\frac{20}{30} \overset{\div 10}{=} \frac{\square}{\square}$ (÷ 10)

3 Copy and complete this question.
Which is bigger:
$\frac{3}{5}$ or $\frac{5}{7}$?

$\frac{3}{5} \overset{\times 7}{=} \frac{\square}{\square}$ (× 7) $\frac{5}{7} \overset{\times 5}{=} \frac{\square}{\square}$ (× 5)

4 Which is bigger: $\frac{2}{3}$ or $\frac{7}{10}$?

5 Which is smaller: $\frac{3}{10}$ or $\frac{1}{3}$?

6 Which is bigger: $\frac{5}{11}$ or $\frac{3}{7}$?

7 Put these fractions in order, starting with the smallest first.
$\frac{3}{10}, \frac{2}{5}, \frac{7}{15}, \frac{13}{20}$
You need to find equivalent fractions to do this.

8 Put these fractions in order, starting with the smallest first.
$\frac{2}{3}, \frac{5}{6}, \frac{1}{2}, \frac{5}{7}$

9 Look at these fractions:
$\frac{2}{6}, \frac{4}{10}, \frac{3}{9}, \frac{5}{15}, \frac{20}{30}$
Find which two of them are not equivalent to $\frac{1}{3}$.

10 Look at these fractions:
$\frac{6}{10}, \frac{18}{25}, \frac{30}{50}, \frac{9}{15}, \frac{20}{35}$
Find which two of them are not equivalent to $\frac{3}{5}$.

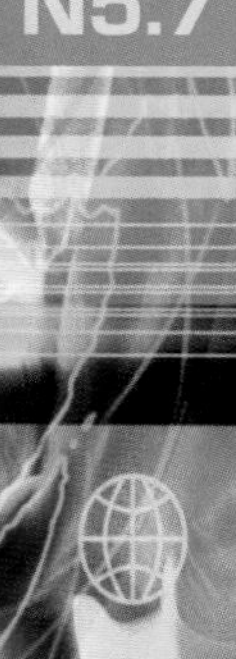

N5.7 Converting fractions, decimals and percentages

This spread will show you how to:

- Round numbers to two decimal places.
- Begin to convert a fraction to a decimal using division.
- Recognise the equivalence between fraction and decimal forms.

KEYWORDS

Convert | Fraction
Decimal | Percentage
Equivalent | Order

Doctor Quotient has a machine that converts fractions to decimals.

$\frac{1}{8}$ goes in

... and 0.125 comes out.

This is how the machine works:

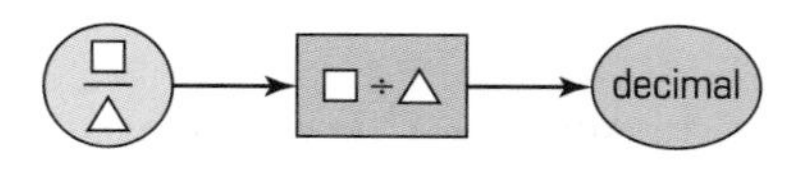

You can use a calculator to do the same job!

- **You convert a fraction to a decimal by dividing the numerator by the denominator.**

You can use this fact to put fractions in order of size.

example

Put these fractions in order, starting with the smallest: $\frac{3}{8}, \frac{4}{10}, \frac{5}{20}$.

Using a calculator:

$\frac{3}{8} = 3 \div 8 = 0.375$ $\quad\frac{4}{10} = 4 \div 10 = 0.4$ $\quad\frac{5}{20} = 5 \div 20 = 0.25$

- Put the decimals in order of size: 0.25, 0.375, 0.4
- Now put the fractions in order: $\frac{5}{20}, \frac{3}{8}, \frac{4}{10}$

Often you need to round off a calculator answer.

example

Convert $\frac{3}{7}$ to a decimal. Give your answer to 2 decimal places.

- Using a calculator, $\frac{3}{7} = 0.4285714$ ...
- Look at the **3rd** digit after the decimal point: 0.42**8**5714 ...
- It is 5 or above, so round the **previous** digit up: 0.43

You can change percentages to decimals as well.

- A percentage is a fraction out of 100.

example

Convert 37% to a decimal.

$37\% = \frac{37}{100} = 37 \div 100 = 0.37$ So 37% = 0.37

Exercise N5.7

1 Use a calculator to change these fractions into decimals.

a $\frac{1}{2}$ b $\frac{3}{6}$ c $\frac{5}{10}$

d $\frac{1}{4}$ e $\frac{2}{8}$ f $\frac{5}{20}$

g $\frac{1}{5}$ h $\frac{3}{15}$ i $\frac{10}{50}$

2 Use a calculator to change these fractions into decimals. (Only write down the first three decimal places.)

a $\frac{1}{3}$ b $\frac{3}{9}$ c $\frac{5}{15}$

d $\frac{1}{9}$ e $\frac{4}{9}$ f $\frac{7}{9}$

g $\frac{3}{7}$ h $\frac{4}{7}$ i $\frac{6}{7}$

3 Put these fractions in order, smallest first, by changing them into decimals.

a $\frac{5}{7}, \frac{13}{16}, \frac{18}{25}$ b $\frac{13}{23}, \frac{7}{13}, \frac{25}{48}$ c $\frac{13}{41}, \frac{21}{53}, \frac{18}{49}$

d $\frac{21}{32}, \frac{9}{13}, \frac{16}{23}$ e $\frac{15}{18}, \frac{7}{8}, \frac{12}{13}$ f $\frac{27}{83}, \frac{13}{51}, \frac{83}{127}$

4 Rewrite these percentages as fractions out of 100.

a 67% b 83% c 29%

d 8% e 35% f 42%

g 19% h 81% i 90%

5 Rewrite these fractions as percentages.

a $\frac{21}{100}$ b $\frac{63}{100}$ c $\frac{38}{100}$

d $\frac{49}{100}$ e $\frac{32}{100}$ f $\frac{99}{100}$

g $\frac{69}{100}$ h $\frac{3}{100}$ i $\frac{5}{100}$

6 Rewrite these fractions as percentages.

a $\frac{1}{2}$ b $\frac{1}{4}$ c $\frac{3}{4}$

d $\frac{1}{10}$ e $\frac{3}{10}$ f $\frac{7}{10}$

g $\frac{1}{5}$ h $\frac{3}{5}$ i $\frac{4}{5}$

7 Change these percentages into fractions then decimals.

a 18% b 53% c 59%

d 71% e 77% f 23%

g 28% h 39% i 43%

8 Put these percentages, fractions and decimals in order, starting with the smallest first.

a $\frac{2}{3}$, 65%, 0.62 b 32%, 0.4, $\frac{3}{8}$

c $\frac{2}{9}$, 20%, 0.02 d 62%, 0.7, $\frac{5}{8}$

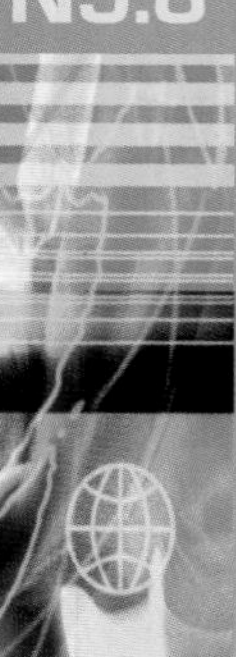

N5.8 Calculating parts of quantities

This spread will show you how to:

- Use a fraction as an operator.
- Find simple percentages of quantities.

KEYWORDS

Amount, Numerator, Denominator, Percentage, Fraction

Siân has 60 sweets. She wants to share them equally between her three friends, not forgetting herself!

She gives herself $\frac{1}{4}$ of the sweets:

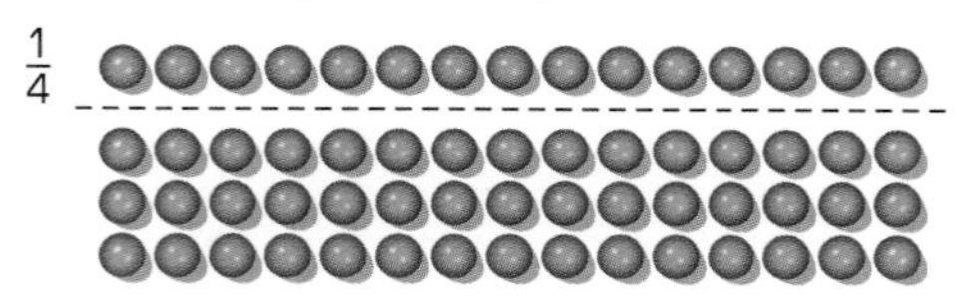

$\frac{1}{4}$ of $60 = 60 \div 4 = 15$

She gives her friends $\frac{3}{4}$ of the sweets:

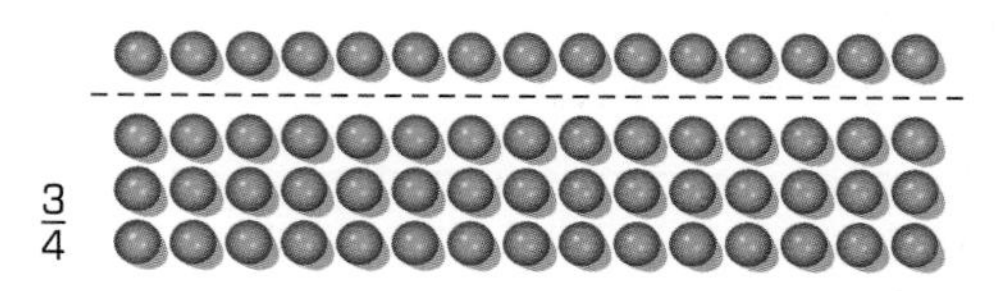

To find $\frac{3}{4}$ of 60:

- First find **one** quarter: $60 \div 4 = 15$
- Then **three** quarters: $3 \times 15 = 45$

- **You find a fraction of an amount by:**
  - **Dividing the amount by the denominator**
  - **Multiplying the answer by the numerator**

example

Find $\frac{4}{5}$ of £3.50.

- First find $\frac{1}{5}$: $3.50 \div 5 = 0.7$
- Then find $\frac{4}{5}$: $4 \times 0.7 = 2.8$

So $\frac{4}{5}$ of £3.50 is £2.80

You can find a percentage of an amount by first changing it into a fraction.

example

Find:

a 10% of 30 kg

b 35% of 80 cm

a $10\% = \frac{1}{10}$

$\frac{1}{10}$ of $30 = 30 \div 10 = 3$

So 10% of 30 kg is 3 kg.

b $35\% = 10\% + 10\% + 10\% + 5\%$

10% of $80 = \frac{1}{10}$ of $80 = 80 \div 10 = 8$

So 5% of $80 = 4$

35% of $80 = 8 + 8 + 8 + 4 = 28$

So 35% of 80 kg is 28 kg.

- **If you know 10% of an amount, you can work out other percentages.**
  For example, 11% = 10% + 1%

Exercise N5.8

1 Work out the following amounts.

a $\frac{1}{2}$ of 84 cm **b** $\frac{1}{2}$ of 28 g

c $\frac{1}{3}$ of 60 minutes **d** $\frac{2}{3}$ of 30 minutes

e $\frac{1}{4}$ of £8 **f** $\frac{3}{4}$ of £8

g $\frac{2}{5}$ of 30p **h** $\frac{7}{10}$ of 50p

2 Work out the following amounts.

a 10% of £40 **b** 10% of £80

c 10% of 60 cm **d** 10% of 20 minutes

e 10% of £4 **f** 10% of 65p

g 10% of 950 cm **h** 10% of 9 m

3 Now work out these amounts.

a 20% of 90p **b** 20% of 50p

c 20% of 20 minutes **d** 5% of 40 metres

e 5% of 60 minutes **f** 30% of 30 litres

g 30% of 50p **h** 40% of 25 minutes

4 Which is bigger:
$\frac{2}{3}$ of 63 metres or $\frac{2}{5}$ of 100 metres?

5 Which is bigger:
$\frac{5}{7}$ of 35 minutes or $\frac{3}{4}$ of 44 minutes?

6 Which is bigger:
$\frac{1}{3}$ of 60p or 20% of 80p?

7 Which is longer:
$\frac{2}{5}$ of 35 metres or 20% of 60 metres?

8 Which is heavier:
5% of 600 g or $\frac{3}{10}$ of 90 g?

9 Find three different ways of calculating the following.

a 25% of 60

b 35% of 60

c 15% of 60

10 Find three different ways of calculating 120% of 80.

Summary

You should know how to ...

1 Round to the nearest 10, 100, 1000 and whole number.

2 Multiply and divide three-digit by two-digit whole numbers.

3 Calculate simple fractions and percentages of amounts.

4 Relate fractions to division.

5 Explain methods and reasoning.

6 Convert a fraction to its simplest form by cancelling.

Check out

1 Round each of these numbers:

a to the nearest 10
b to the nearest 100
c to the nearest 1000

i 2732   ii 5765   iii 572   iv 329

2 Work out these:

a 121×12 b 253×21
c 207×15 d 517×23
e $672 \div 12$ f $816 \div 12$
g $676 \div 13$ h $532 \div 14$

3 Work these out:

a 50% of £64   b $\frac{1}{4}$ of £18
c 10% of 50 cm   d 20% of 50 cm
e 10% of 20 mins   f $\frac{3}{4}$ of 200 metres
g $\frac{7}{9}$ of 90°   h 5% of 1 hour
i 15% of 1 hour   j 30% of 70 cm

4 Work these problems out in your head:

a $\frac{1}{4}$ of 20   b $\frac{3}{4}$ of 20
c $\frac{1}{5}$ of 35   d $\frac{3}{5}$ of 35
e $\frac{1}{10}$ of £400   f $\frac{7}{10}$ of £400
g $\frac{2}{3}$ of 90 litres   h $\frac{3}{8}$ of 40 kg
i $\frac{4}{5}$ of 30 students   j $\frac{5}{6}$ of 30 students

5 Work out these problems. Explain the method you used giving reasons.

a 62×15
b 17.5% of £2000
c 37×19
d $\frac{3}{10}$ of £4.50

6 Simplify where possible:

a $\frac{3}{12}$ b $\frac{15}{21}$ c $\frac{9}{30}$ d $\frac{7}{35}$

Write as fractions in their simplest form:

e 60%   f 0.4   g 35%   h 0.15

3 Analysing statistics

This unit will show you how to:

- Solve a problem by representing, extracting and interpreting data in tables, charts, graphs and diagrams.
- Find the mode and range of a set of data.
- Begin to find the median and mean of a set of data.

You can collect data in a survey.

Before you start

You should know how to ...

1 Read and use scales.

2 Construct simple statistical diagrams, including bar charts.

Check in

1 What reading does each arrow show on these scales?

a

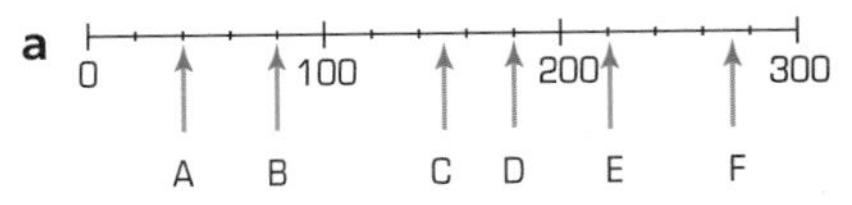

b

0 100 200

A B C D E

2 The numbers of ice creams sold each day one week (Monday – Friday) are:
22, 18, 45, 35, and 50.
Make a **bar chart** for this information.

D3.1 Planning the data collection

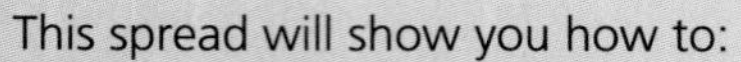

This spread will show you how to:

- Solve a problem by collecting and organising data into tables.

KEYWORDS
Data
Data collection sheet
Survey

Mrs Rogers is planning a school ski trip.
She wants to check the weather conditions before deciding where to go.

She collects some ...

primary data and secondary data

She surveys students preferred destination at an assembly.

She uses the Internet to find information about weather conditions at the resort.

- **Primary data** is data that you collect yourself.
- **Secondary data** is data that you look up.

The web site contains a lot of information about weather conditions.
Mrs Rogers decides to use the snowfall data for the last three years as complete data is available.
Here is her data collection sheet:

Snowfall		Month				
(cm)		Dec	Jan	Feb	Mar	Apr
Year	98–99	281	335	449	311	60
	99–00	251	226	98	182	90
	00–01	171	163	53	124	137

Source: The Whistler Blackcomb ski resort: www.whistler-blackcomb.com/weather/stats/index/asp

- To collect the right data for your enquiry:
  1. Decide which data is relevant
  2. Research possible sources of the data
  3. Plan and design a data collection sheet.

Exercise D3.1

1 Copy this table:

Primary Data	Secondary Data

Put each of these sources of data into the correct column:
- A reference book in a library
- A survey you do in your class
- A Web Site on the Internet
- An experiment you do in a science lesson
- A newspaper artical
- This week's football league table

2 Here are some questions that could be investigated.
- Do boys spend more time than girls playing computer games?
- What is your class's favourite television programme?
- Is there more rain in London or in Liverpool?

Explain where you could find information to help you answer each question.
Explain whether the information would be *primary* or *secondary* data.

3 Design a questionnaire or a data collection sheet for one of the three questions given in Question 2.

4 Design a data collection sheet to collect these sets of data:
- **a** Number of vehicles of different types (car, van...) in a car park.
- **b** Lengths of the world's five longest rivers, in kilometres and miles.
- **c** How often people in your class use the Internet to help with homework.
- **d** Attendance at the UK's top five tourist attractions over the past 10 years.

5 For each part of question 4:
- Explain whether you would need to collect *primary* data or *secondary* data.
- Describe a *source* for the data required.

D3.2 Constructing statistical diagrams

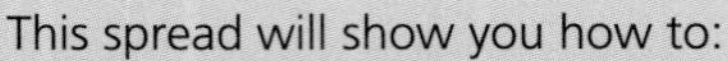

This spread will show you how to:

- Represent and interpret data in tables, charts, graphs and diagrams.

KEYWORDS

Bar chart
Interpret
Table
Line graph
Pie chart

Mrs Rogers will present the data to the parents.
This will help people decide whether their children can go on the trip.

She can use:

Pie charts

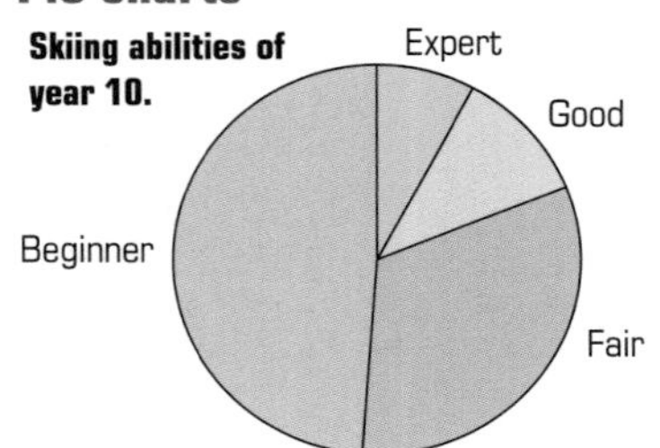

The pie chart shows that most people interested in the trip are beginners or fair skiers.
It does not show you how many people there are in each category.

If there were many more categories, it would be harder to interpret the pie chart:

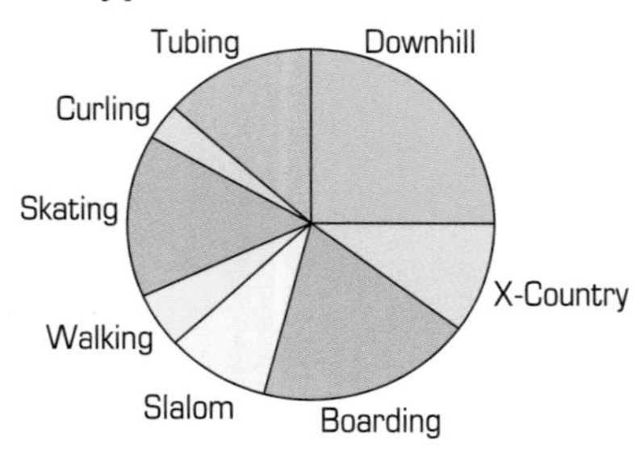

The smaller categories are more difficult to read.
The large number of categories makes it harder to see an overall picture.

Line graphs

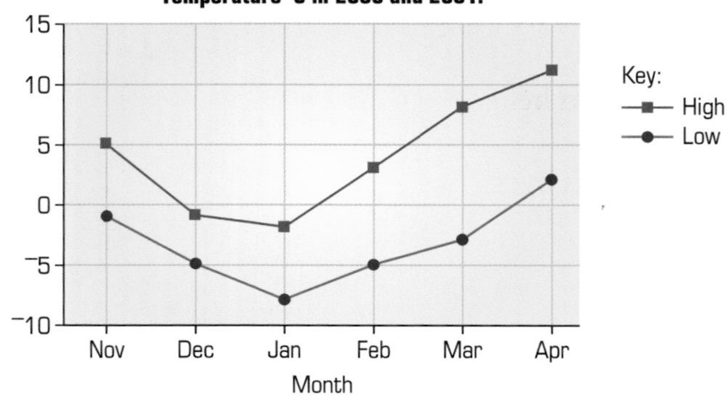

The calculated data for each month is plotted, and is accurate.
The lines between the marks indicate the trend.

- You use a line graph to show numerical data.

Exercise D3.2

1 The bar chart shows the number of times Sophie's computer crashed one week.

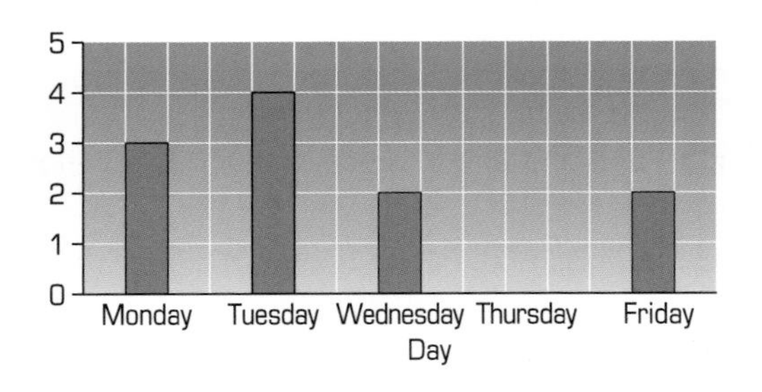

a How many times did Sophie's computer crash on Monday?

b On which day of the week did Sophie's computer not crash at all?

c How many times did her computer crash altogether?

2 The chart shows the number of people attending a club each month.
Copy and complete the table using the data from the chart.

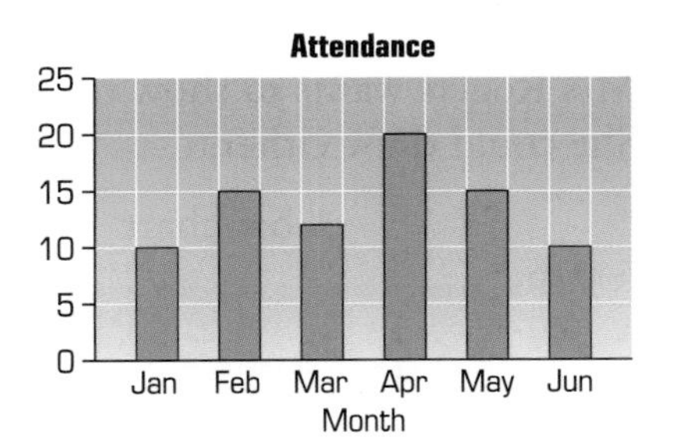

Month	Attendance
January	10
February	
March	
April	
May	
June	

3 The pie chart shows the colours of 14 cars.

a What was the most popular car colour?

b Copy and complete the table using the data from the pie chart.

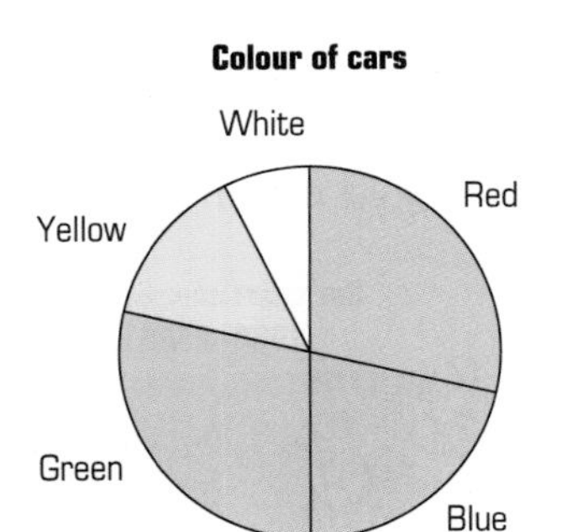

Colour	Frequency
Red	4
Blue	
Green	
Yellow	
White	

4 The line graph shows the temperature at midday each day one week.

a Which day was the hottest?

b Which day was the coldest?

c What was the range of temperatures during the week?

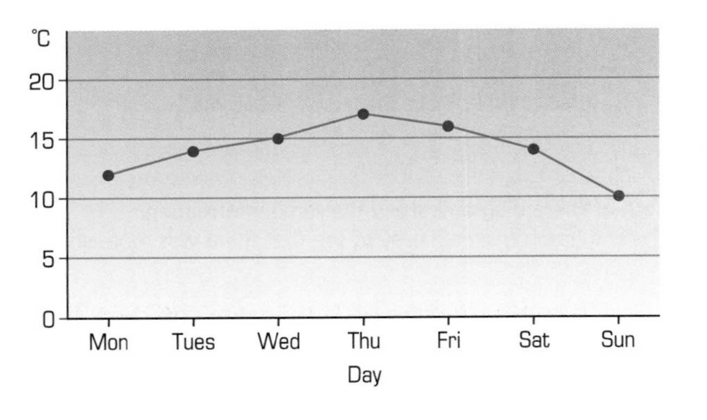

D3.3 Comparing data using diagrams

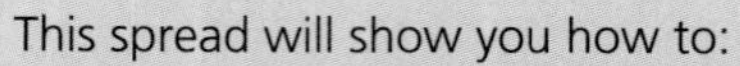

This spread will show you how to:

- Represent and interpret data in tables, charts, graphs and diagrams.

KEYWORDS

Bar chart
Line graph
Interpret

It is easier to interpret data when you draw a diagram.
Diagrams often allow you to see patterns and trends.

Mrs Rogers wants to show the monthly snowfall for the ski resort in the 2000–2001 season.
She could draw either...

A bar chart or a line graph.

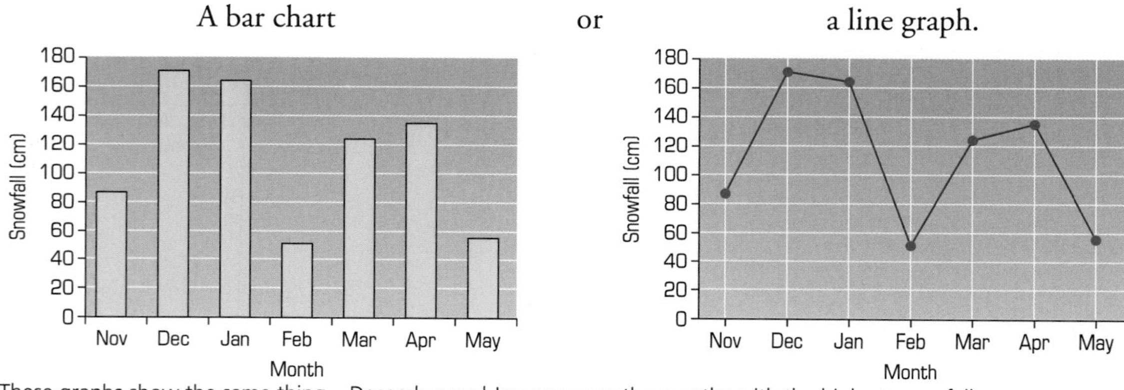

These graphs show the same thing – December and January were the months with the highest snowfall.

Mrs Rogers wants to compare the snowfall for the 2000–2001 season and the 1999–2000 season.
She could do this with a bar chart or a line graph.

The bar chart is more accurate as it shows only the measured data.

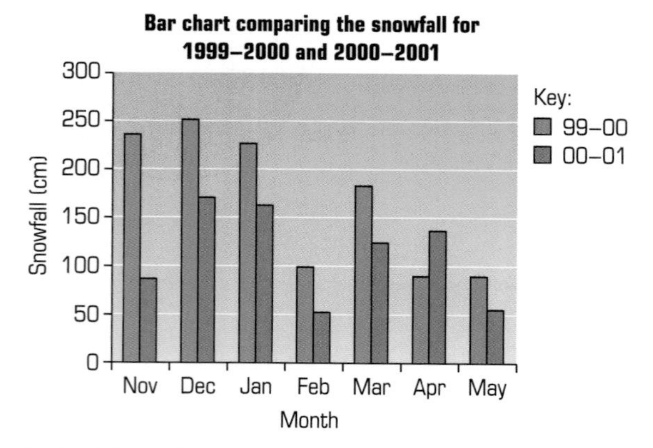

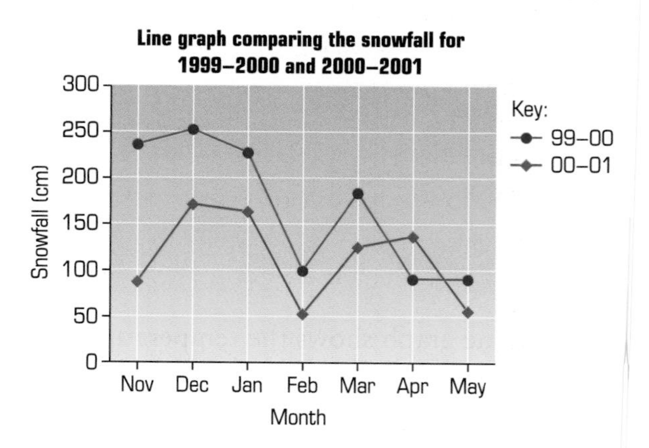

Both of these diagrams show the same information.
The line graph makes it easy to see that there was more snow in 1999–2000 than in 2000–2001.

▶ To compare data on a diagram the data should fit on the same scale.

Exercise D3.3

1 The graph shows the number of absences for two classes last week.

a How many people in 7A were absent on Tuesday?

b How many people were absent altogether on Monday?

c On which day was *nobody* absent in 7B?

d Jack was away all week. Which class is he in?

e What was the total number of absences, for both classes, for the whole week?

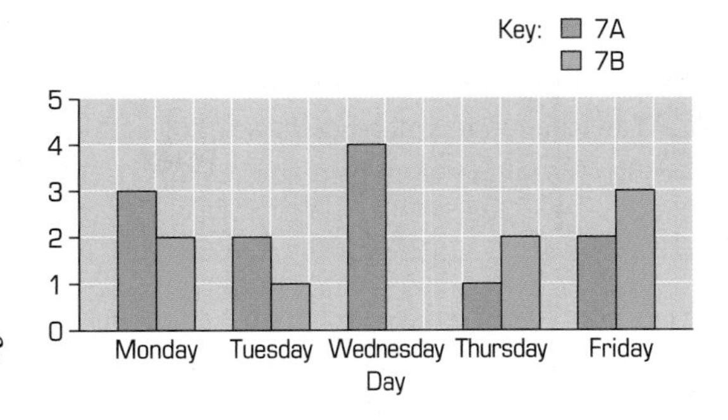

2 Students at Piltdown School are awarded Gold Certificates for outstanding work.
The table shows the number of certificates awarded to boys and girls last year.

	Certificates Awarded	
Term	**Boys**	**Girls**
Autum	5	6
Spring	8	7
Summer	12	14

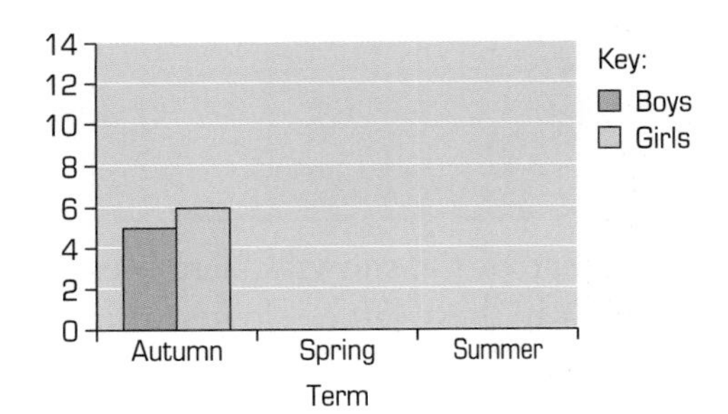

Copy and complete the bar chart to show these data.

3 A club has a pool table and a pinball machine.
The chart shows how many times they were played last year.

a Which game was most popular?
Explain your answer.

b The pool table was out of order for a while.
Which month did this breakdown happen in?

c The club decides to get the machines serviced next year.
They will be out of action for about a week.
What would be the best month to do this?

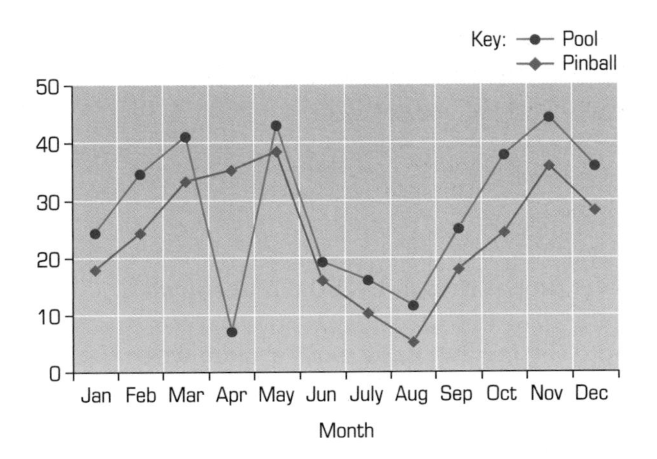

D3.4 Describing data using statistics

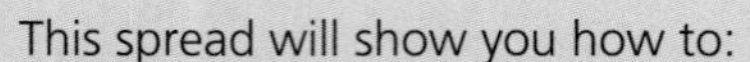

This spread will show you how to:

- Find the mode and range of a set of data.
- Begin to find the median and mean of a set of data.

KEYWORDS

Average, Mean, Median, Modal class, Mode, Range

You can summarise data using the average and the range.

- The average is a value that is typical of the data.
- The **range** = the highest value – the lowest value.

The ski trip will take place in February. Here is the February snowfall for previous seasons:

Season (Ending)	02	01	00	99	98	97	96	95	94	93	92	91
February Snowfall (cm)	235	53	98	449	263	60	177	133	426	14	170	182

The range of the snowfall data is easy to find.
Range = highest value – lowest value
= 449 – 14 = 435 cm.
There are three different averages:

- The **mode** is the value that occurs most often.

It is not sensible to find the **mode** of these data. You can organise the data into classes, and then find the *modal class*:

Snowfall (cm)	1–100	101–200	201–300	301–400	401–500
Frequency	5	4	3	0	2

The modal class is 1–100 cm of snowfall.

- The **median** is the **middle value** when the data are arranged in order.

Order the snowfall data: 14, 53, 60, 98, 133, 170, 177, 182, 235, 263, 426, 449.
The median is halfway between the two middle values.
The median is 173.5 cm.

- The **mean** is the sum of the values divided by the number of values.

Using a calculator:
Total February snowfall = 2260 cm
Mean = Total ÷ 12 = 2260 ÷ 12 = 188 cm to the nearest cm.

Exercise D3.4

1 Here are the drinks which a group of students have one lunchtime.

Drink	Number of people
Cola	8
Water	9
Juice	5
Lemonade	4

Use the information in the table to find the **mode**.

2 The table below shows midday temperatures.

Day	Mon	Tue	Wed	Thu	Fri
Temperature (°C)	17	20	21	19	11

Find the **range** of these temperatures.

3 A group of five students measure their heights. Here are their results.

Student	Bob	Jo	Asif	Jen	Kay
Height (cm)	152	137	155	144	149

Arrange the heights in order, and find the **median** height.

4 There are five passengers in a lift. The table below shows the weight (in kg) of each passenger.

Passenger	A	B	C	D	E
Weight (kg)	72	48	51	98	67

a Find the **total** weight of the five passengers.
b Now work out the **mean** weight for the passengers.

5 The table below shows the waiting times for 20 patients at a doctor's surgery.

Waiting time (Minutes)	0–9	10–19	20–29	30–39
Number of patients	4	7	5	4

Find the **modal class** of the waiting times.

6 For the data in question 2, find the **median** and **mean**. Which do you think is the better average to use to describe the average temperature? Explain your answer.

D3.5 Communicating results

This spread will show you how to:

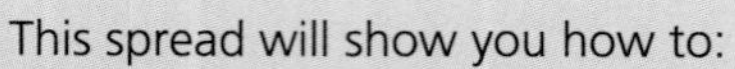

- Solve problems by interpreting data.

KEYWORDS

Average   Statistics
Interpret

Mrs Rogers is preparing a leaflet about the ski trip.
It contains information about the ski resort for students and parents.

Part of the leaflet is about the weather.

Skiing in Canada – It's the snow that counts!

February is a great month for skiing at our chosen resort.

The chart shows the average monthly snowfall over the last 20 years.

As you can see, we can expect lots of snow between November and January. This should mean that there is lots of snow on the ground by February! The lower snowfall in February should mean uninterrupted skiing, but there should be enough fresh snow to keep things crisp and bright!

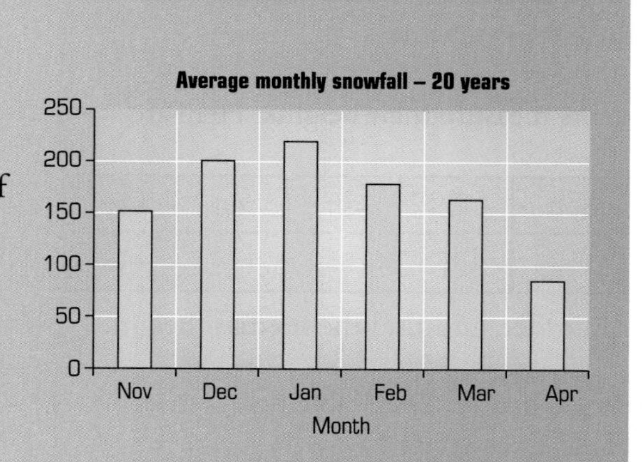

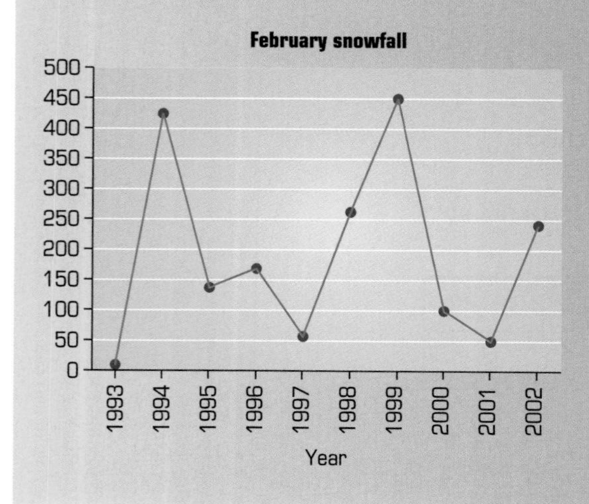

The average February snowfall over this period is 177 cm. However, as this graph shows, it can vary quite a lot.

In a statistical report, you should:

- Illustrate the data with relevant diagrams
- Calculate relevant statistics
- Interpret the findings.

Exercise D3.5

1 A class has a vote to decide where to go on an end-of-year day trip. Here are the results:

Destination	Wonder Park	Seaside	Animal World
Votes	17	6	8

Write a short report explaining where the class will go. Use a bar chart to illustrate the results of the vote.

2 Ten students record the number of times they use the Internet one term. Here are their results.

Student	A	B	C	D	E	F	G	H	I	J
Times used	1	4	0	1	2	1	0	4	1	3

Write a short article about the results for a school newsletter. Your article should include:

- A **bar chart** to show the survey results.
- An **average** for the data.
- The **range** of the data.
- Some **conclusions** about what the data show.

3 George has to carry out a survey of 25 customers each year, and send a report to head office.

'How happy are you with our service?'	Very Unhappy	Unhappy	Satisfied	Happy	Delighted
Last Year	6	6	4	2	7
This Year	7	7	3	0	8

Here is part of the data he has gathered.

It's been another great year here at the Camtown branch! Once again, the most common response from our customers was that they were 'delighted' with our service - and we had even more

Here is the start of his report:

a How is his report unfair?

b Write a fairer report, based on the information in the table.

Summary

You should know how to ...

Solve a problem by representing, extracting and interpreting data in tables, charts and diagrams.

Check out

1 Here are the percentage marks obtained by 10 students who did a maths test and a science test.

Maths Mark	Science Mark
55	48
68	55
47	43
72	67
84	88
61	52
30	41
94	97
49	39
76	63

Which test did the students get the better marks in?

- ▶ You can use averages (mean, median or mode) to compare marks.
- ▶ You can use the range of each set of marks to see how widely spread the marks are.

2 A hotel manager asks guests to fill in a questionnaire about their meals.
Here are the results.

Rating	Poor	Fair	Good	Excellent
Frequency	4	9	18	12

Write a short report about these results.

Remember that you should:

- ▶ Illustrate the data with relevant diagrams
- ▶ Calculate relevant statistics
- ▶ Interpret the findings

14 Probability experiments

This unit will show you how to:

- Use the language associated with probability to discuss events, including those with equally likely outcomes.
- Solve problems by representing, extracting and interpreting data in tables.

You flip a coin to decide fairly who starts a match.

Before you start

You should know how to ...

1 Use the vocabulary and ideas of probability, drawing on your own experience.

2 Understand and use the probability scale from 0 to 1.

Check in

1 In an experiment, tiles marked 1 to 10 are placed in a box, and one tile is picked out at random. Put the following outcomes in order, according to how likely they are:

a The tile chosen has an even number

b The number on the tile is 7

c The number is more than 2.

2 What is the probability of each outcome in question 1? Give your answers as fractions.

D4.1 Theoretical probability

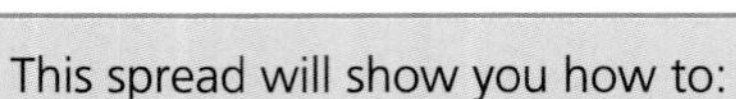

This spread will show you how to:

- Use the language of probability to discuss events with two or more equally likely outcomes.

KEYWORDS
Probability, Random, Outcome, Likelihood, Spinner, Probability scale

Probability is a measure of the likelihood of an outcome. Here are some outcomes:

A score of 7 on an ordinary dice.

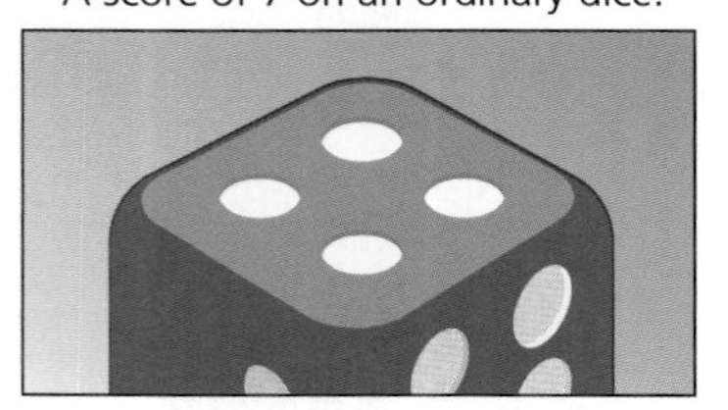

A coin lands on heads.

A single-digit number card shows less than 10.

You can place these possible outcomes on a probability scale.

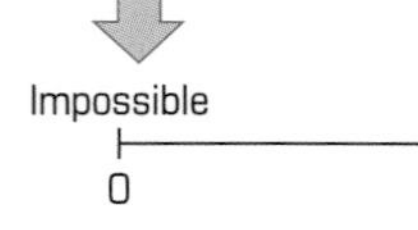

▶ Probability of an outcome = $\dfrac{\text{the number of ways the outcome can happen}}{\text{the total number of possible outcomes}}$

To calculate the probability of an outcome, you need to list all possible outcomes.

example

The numbers 1 to 10 are written on 10 cards.
One card is drawn at random.
What is the probability that it is:

a an even number **b** a multiple of 3 **c** a factor of 12?

There are 10 possible outcomes in total.

a There are 5 even numbers: 2, 4, 6, 8, 10.
Probability of an even number = $\frac{5}{10} = \frac{1}{2}$.

b There are 3 multiples of 3: 3, 6, 9.
Probability of a multiple of 3 = $\frac{3}{10}$.

c There are 5 factors of 12: 1, 2, 3, 4, 6.
Probability of a factor of 12 = $\frac{5}{10} = \frac{1}{2}$.

To remind yourself about factors and multiples look at N5.2.

Exercise D4.1

1 For each of the following situations, say how many possible outcomes there are.

a Kelly rolls an ordinary dice and records the score.

b Pat records whether or not it snows each day.

c Charles puts tickets numbered 1 to 100 in a bag, and then picks one.

d Marcia makes a note of whether or not an elephant walks past her window one morning.

2 For each part of question 1, say whether or not the outcomes are equally likely.
Give reasons for your answers.

3 Mikela writes the letters of her name on a fair 6-sided spinner.
She spins the spinner, and makes a note of the letter touching the table.
What is the probability that the result is:

a M **b** A vowel **c** L or K?

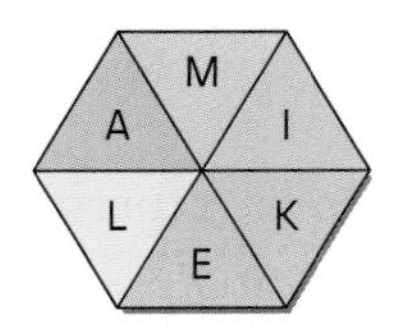

4 Cards numbered 1 to 100 are placed in a box.
What is the probability that a card chosen at random will be:

a 100 **b** An even number

c A number bigger than 20 **d** A number less than 200?

Give your answers as fractions.

5 Mark each of the outcomes from question 4 on a probability scale like this one:

0 ———— $\frac{1}{2}$ ———— 1

6 The letters of the word ENVELOPE are written on cards, and placed in a bag.
What is the probability that a card chosen at random shows:

a E **b** N **c** A vowel **d** J?

Mark each of these probabilities on a probability scale.

Experimental probability

This spread will show you how to:
- Use the language of probability to discuss events.
- Represent data in tables.

KEYWORDS
Experiment, Outcome, Frequency, Spinner, Likely, Tally

James likes cars and he has noticed that there are more of some colours than of others.

James decides to do an experiment. He wants to know:

- What is the most likely colour?
- What is the probability of the next car being purple?

On his way to school one day he completes a tally chart:

Colour	Tally	Frequency
Blue	𝍸 𝍸 𝍸 III	18
Black	𝍸 IIII	9
Purple	𝍸 I	6
Red	𝍸 II	7
Silver	IIII	4
White	III	3
Other	III	3

Total 50 cars

The most likely colour is blue.

There are 6 purple cars out of 50. The probability that the next car is purple is $\frac{6}{50}$, or $\frac{3}{25}$.

A statistical experiment is made up of **trials**.
By observing trials you can estimate probabilities.

Each car James records is a trial.
James observes 50 trials.

In an experiment

- Probability of an outcome = $\dfrac{\text{number of times the outcome occurs}}{\text{number of trials in the experiment}}$

Exercise D4.2

Experiment

- Make a rectangular spinner from card, using this design.

  You can trace the shape onto card.

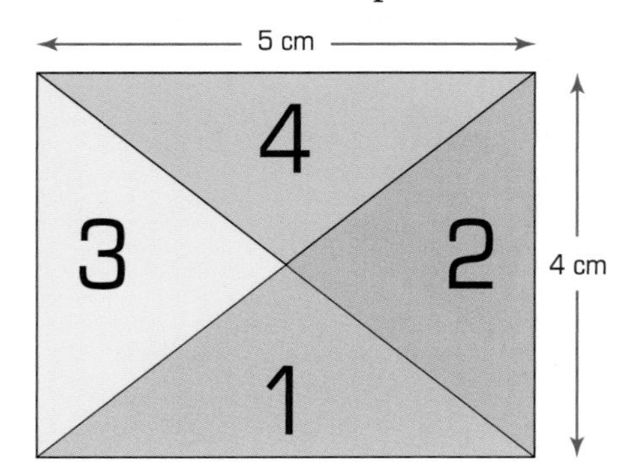

You can use a pencil for the 'spindle';

Use sticky tape to hold it in place, if necessary.

Try to keep everything straight and symmetrical!

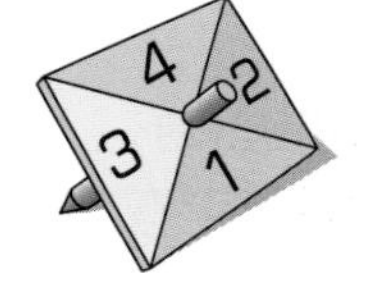

- Test your spinner to estimate the probability for each score.

- Spin your spinner 100 times. Record your results in a table like this:

Score	Tally	Frequency
1		
2		
3		
4		

- When you have finished, **work out an estimate of the probability for each score**.

- Write a report of your findings, explaining what you have found out about your spinner.

Comparing experiment with theory

This spread will show you how to:

- Discuss the difference between the theory of outcomes and the actual, experimental results.

KEYWORDS

Estimate, Experiment, Outcome, Probability, Spinner, Fair

Here is a four-sided square spinner.

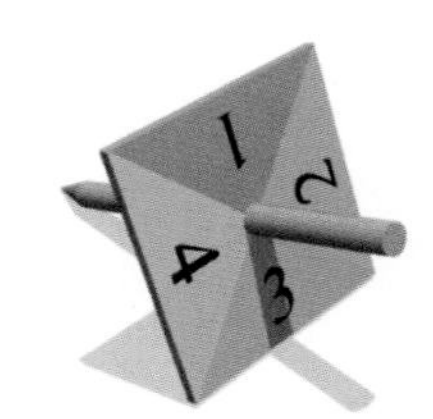

When you spin it, it should be equally likely to give a score of 1, 2, 3 or 4.

In theory, the probability of getting a 3 is $\frac{1}{4}$, or 0.25.

Rob and Maria decide to test the spinner.

- They spin the spinner and record the score.
- They repeat the experiment 50 times and keep a tally.

Here are their results:

Score	Tally	Frequency
1	~~IIII~~ ~~IIII~~ III	13
2	~~IIII~~ ~~IIII~~ ~~IIII~~	15
3	~~IIII~~ ~~IIII~~	10
4	~~IIII~~ ~~IIII~~ II	12

From the experiment, Rob and Maria can estimate the probability of getting a 3.

$$\text{Experimental probability} = \frac{\text{number of times the outcome occurs}}{\text{total number of outcomes}}$$

$$= \frac{10}{50}$$

$$= \frac{1}{5} \text{ or } 0.2$$

Compare with the theoretical probability of 0.25.

The two probabilities are similar.

There is no reason to think that the spinner is unfair.

- **Theoretical and experimental probabilities should be similar, but are unlikely to be exactly the same.**

Exercise D4.3

1 Asif spins a spinner 100 times, and gets these results.

Score	1	2	3	4
Frequency	37	28	12	23

Estimate the probability of each score for the spinner.

2 Sara tosses the same coin 100 times. Here are her results:

Outcome	Frequency
Heads	54
Tails	46

Sara says: 'There were more heads than tails, but I still think the coin is probably fair.'
Explain why Sara is correct.

3 Sam rolls a dice 60 times, and gets these results:

Score	1	2	3	4	5	6
Frequency	8	7	14	11	12	8

a Do you think that Sam's dice is fair?
b What should Sam do if he wants to be more certain?

4 Claire tosses a coin four times, and gets 'heads' every time.
She says:

There must be something funny about this coin.
I should be getting as many tails as heads.

Explain why Claire is wrong.

5 Devra is testing a five-sided spinner to see whether it is fair.
She spins the spinner 50 times. Here are her results.

Outcome	Red	Black	Yellow	Green	Blue
Frequency	10	8	13	13	6

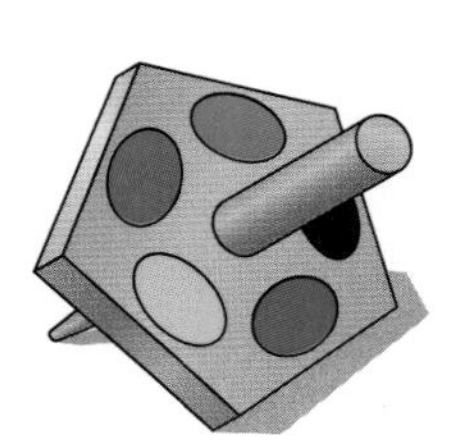

Lionel tests the same spinner, but he spins it 500 times.
Here are his results.

Outcome	Red	Black	Yellow	Green	Blue
Frequency	96	112	90	109	94

Do you think the spinner is fair? Explain your answer.

D4 Summary

You should know how to ...

Solve a problem by representing, extracting and interpreting data in tables.

Check out

1 A computer chooses a number at random from the list below.

34	120	6	20	17
23	5	15	9	24

Find the probability that the number chosen by the computer will be:

a 15

b Less than 100

c 28

d More than 3

2 Chula has three coloured dice.
One of them is biased, but she doesn't know which one.
She decides to roll each dice 240 times to try to find out.
Here are her results.

	Dice		
Score	Red	Blue	Green
1	35	34	44
2	48	63	34
3	33	37	42
4	45	28	42
5	40	40	39
6	39	38	39

a Which one of the three dice do you think is biased?

b Estimate the probability of each score for each dice.

15 Equations and graphs

This unit will show you how to:

- Recognise and extend number sequences.
- Understand and use the relationships between the four operations and the principles of the arithmetic laws.
- Know multiplication facts up to 10 × 10 and derive corresponding division facts.
- Read and plot coordinates in all four quadrants.
- Develop from explaining a generalised relationship in words to expressing it in a formula using letters as symbols.

You can use a formula to change between currencies.

Before you start

You should know how to …	Check in
1 Write algebraic expressions, using algebraic conventions.	1 Write expressions for the following: June works out a number in a board game by: **a** multiplying the score on a dice by 5. **b** adding 6 to the score **c** subtracting 5 from the score **d** dividing the score by 2.
2 Plot coordinate pairs on a grid.	2 Plot the points (1, 6) (2, 7) (3, 8) on a coordinate grid and join them up. What do you notice?

Solving equations

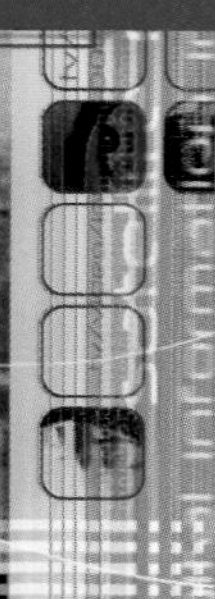

This spread will show you how to:
- Use the relationship between addition and subtraction.
- Understand the operation of multiplication and its relationship to division.

KEYWORDS

Equals (=)	Operation
Equation	Solve
Expression	Value

Jordan swims s lengths in half an hour.
She then swims 4 more lengths.

An expression for the total number of lengths she swims is $s + 4$.

She swims 46 lengths altogether.

This is the value of the expression, so she can write an equation:
$s + 4 = 46$

▶ An **equation** links an expression with its value using an equals sign.

An equation is like a pair of scales.
The equals sign is where the equation balances.

$s + 4 = 46$ so $46 - 4 = s$ and $s = 42$

$s+4$ 46 s $46 - 4$

Jordan swims 42 lengths.

▶ You can solve an equation using a balance.
You perform the same operation on both sides to keep the balance.

example

Solve these equations:

a $x + 4 = 13$ **b** $y - 3 = 10$ **c** $3x = 12$

a

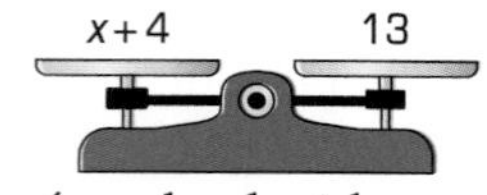

-4 on both sides

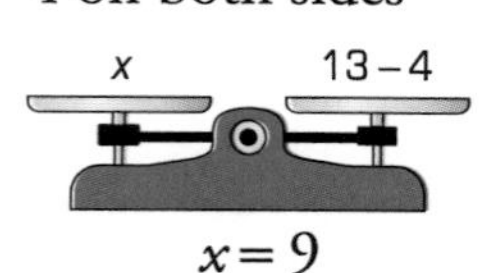

$x = 9$

b

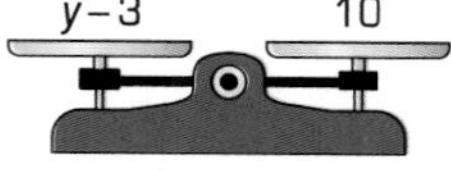

$+3$ on both sides

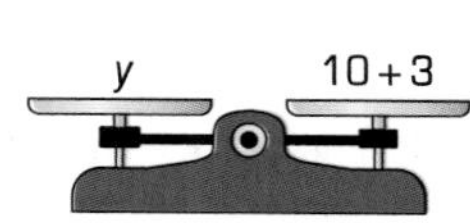

$y = 13$

c

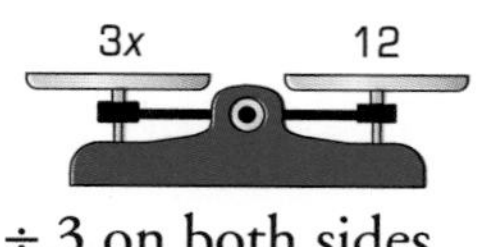

$\div 3$ on both sides

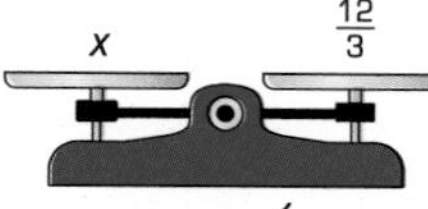

$x = 4$

Exercise A5.1

1 Solve each of these equations.

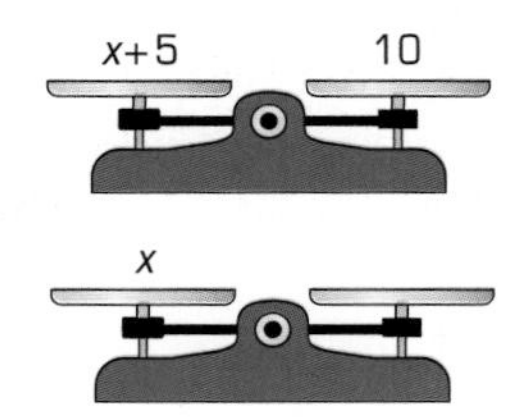

a $x + 5 = 10$ **b** $y + 12 = 15$
c $x + 3 = 9$ **d** $y + 2 = 8$
e $x + 15 = 20$ **f** $y + 5 = 14$
g $y - 4 = 16$ **h** $x - 5 = 9$
i $x - 4 = 11$ **j** $y - 3 = 17$
k $x + 6 = 26$ **l** $y - 5 = 15$

2 Write equations for these sentences then solve them.
The first one is done for you.

a 3 lots of an unknown number make 21.
Let the unknown be n.

$3n = 21$
$3 \times 7 = 21$
So $n = 7$.

b 2 lots of an unknown number make 12
c 4 lots of an unknown number make 32
d 3 lots of an unknown number make 24
e 4 lots of an unknown number make 28
f 3 lots of an unknown number make 33
g 6 lots of an unknown number make 42
h 12 lots of an unknown number make 36
i 10 lots of an unknown number make 90
j 2 lots of an unknown number make 26

3 Solve these equations. Check your solution by multiplying.

a $4x = 12$ **b** $2y = 8$
c $3x = 15$ **d** $6y = 30$
e $4x = 16$ **f** $3y = 18$
g $6r = 24$ **h** $3m = 9$
i $10n = 50$ **j** $6t = 18$
k $5x = 20$ **l** $7y = 28$

4 The diagrams show a square and an equilateral triangle.
The length of all the edges of each shape is x cm.

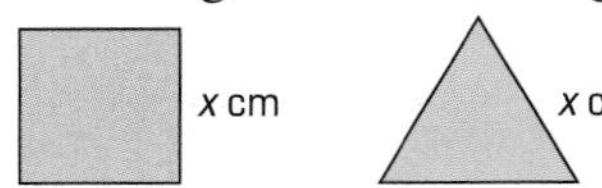

You find the perimeter of the shapes by adding the lengths of the edges together.

a Write an expression for the total perimeter of the square and the triangle.
b What is x if the total perimeter is 28 cm?

A5.2 Using formulae

This spread will show you how to:

- Develop from explaining a generalised relationship in words to expressing it in a formula using letters as symbols.

KEYWORDS

Area, Variable, Substitute, Unknown, Value

Penny uses the 'Anytime 15' tarif on her mobile phone.

She pays £15 line rental per month and 15p per minute for a call.

To work out her bill each month she uses a formula:

Total bill = £10 line rental + number of minutes used × 10p

There are two variables: the total bill and the number of minutes.

- A **formula** is an equation linking variables together.

In January, Penny used her phone for 45 minutes.
Her bill for January was:

Total bill = £10 + 45 × 10p
= £10 + £4.50
= £14.50

- You can substitute known values into a formula to find unknown values.

example

The formula for the area of a rectangle is:

Area = length × width   where all the measures use the same units.

length

Area	width

Use the formula to find the area of rectangles with these dimensions:

a 2 cm by 3 cm   **b** 3 m by 2.1 m   **c** 12 mm by 5 mm

a Area = 2 cm × 3 cm
= 6 cm^2

b Area = 3 m × 2.1 m
= 6.3 m^2

c Area = 12 mm × 5 mm
= 60 mm^2

Exercise A5.2

1 The formula for the cost of buying packets of crisps is:

Total cost = number of packets bought × 30p

What is the cost of:

a 1 packet
b 2 packets
c 5 packets?

2 Mandy has a Pay as You Go mobile phone.
Her calls cost 20p per minute at any time.
The formula for the cost of her calls is:

Total cost = number of minutes used × 20p

Use the formula to work out the cost of using her phone for:

a 5 minutes
b 10 minutes
c 30 minutes
d 1 hour

e How long would a £5 credit last?

3 Joe uses this formula to work out his mobile phone bill:

Total bill = £20 line rental + number of minutes used × 5p

a How much is Joe's line rental?
Work out Joe's bill when he uses:
b 5 minutes
c 20 minutes
d 12 minutes
e 55 minutes

4 Penny uses a formula to work out the cost of using her land line:

Total cost = number of daytime calls × 50p + number of evening calls × 10p + £10 rental

a How much is a daytime call?
b How much is an evening call?
Use the formula to work out the cost of:
c 5 evening calls and 3 daytime calls
d 20 evening calls and 20 daytime calls

Formulae using letters

This spread will show you how to:

- Develop from explaining a generalised relationship in words to expressing it in a formula using letters as symbols.

KEYWORDS
Substitute Value
Unknown

Penny uses her formula every month.

Total bill = £10 line rental + number of minutes use × 10p

She writes the formula simply as:

$B = £10 + m \times 10\text{p}$ where m is the number of minutes and B is the cost.

She makes it even simpler by putting all the money amounts in pence:

$B = 1000 + 10m$ where m is the number of minutes and B is the cost in pence.

If she talks for 56 minutes her bill will be:

$$B = 1000 + 10 \times 56$$
$$= 1000 + 560$$
$$= 1560 \text{ pence}$$
$$= £15.60$$

If her bill is £18 she can work out how many minutes she used.
The line rental is £10 so she spent £8 on calls.

$800 = 10m$ so $m = 800 \div 10$

She talked for 80 minutes.

- **You can substitute known values into a formula to find unknown values.**

example

Use the formula $c = 10m$ to find:

a c when $m = 2$ **b** c when $m = 2.1$ **c** m when $c = 230$

a $m = 2$ so $c = 10 \times 2$
$c = 20$

b $m = 2.1$ so $c = 10 \times 2.1$
$c = 21$

c $c = 230$ so $230 = 10m$
$230 \div 10 = m$
$23 = m$

The formula $c = 10m$
could mean   number of cm = 10 × number of mm
as there are 10 mm in 1 cm.

Exercise A5.3

1 The formula for Mandy's phone bill is:

Total cost = number of minutes used × 20p

Write her formula using letters.

2 The formula for the cost of buying packets of crisps is:

Total cost = number of packets bought × 30p

Write the formula using letters.

3 The formula for the area of a rectangle is:

Area = length × width

Write the formula using letters.

4 The formula for changing pounds (£) to pence (p) is:

Amount in pence = amount in pounds × 100

Write the formula using letters.

5 Joe uses this formula to work out his mobile phone bill:

Total bill = £20 line rental + number of minutes used × 5p

Write Joe's formula using letters.

6 The formula for changing metres to centimetres is:

Length in centimetres = length in metres × 100

a Write the formula using letters.
b Use the formula to change 3.2 metres to centimetres.

7 This formula links feet and inches:

$i = 12f$

a Work out i when $f = 3$.
b How many inches are there in 3 feet?
c Work out the number of inches in 2.5 feet.

A5.4 Generating sequences

This spread will show you how to:

- Recognise and extend number sequences.

KEYWORDS
Generate
Sequence
Rule

Here is a sequence of matchstick patterns:

Pattern 1

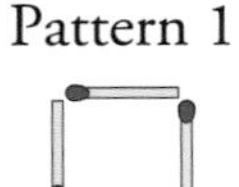

4 matchsticks

Pattern 2

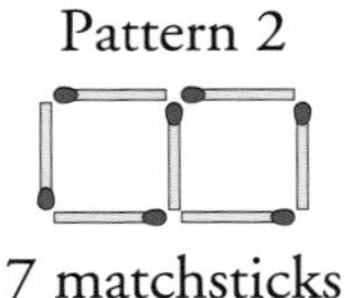

7 matchsticks

Pattern 3

10 matchsticks …

You can describe the way the sequence grows.

> 'You add 3 matchsticks to the number in the current pattern to get the number in the next pattern.'

You can write this using a formula:

> matchsticks in next pattern = matchsticks in current pattern + 3 matchsticks
> or:
> next = current + 3

There are two variables:
the number in the next pattern
the number in the current pattern.

When you know one of the variables you can use the formula to find the other variable.

example

Use the formula to find the number of matches in the:

a 6th pattern **b** 12th pattern if there are 34 in the 11th pattern

a the 4th pattern has $10 + 3 = 13$
the 5th pattern has $13 + 3 = 16$
the 6th pattern has $16 + 3 = 19$ matches

b the 11th pattern has 34, the 12th pattern is the next pattern
next = current + 3
so 12th = 11th + 3
= 34 + 3 = 37 matches

- **You can describe the way a sequence grows using a rule or formula.**

Exercise A5.4

1 Here is a sequence of matchstick patterns:

Pattern: 1 2 3 4

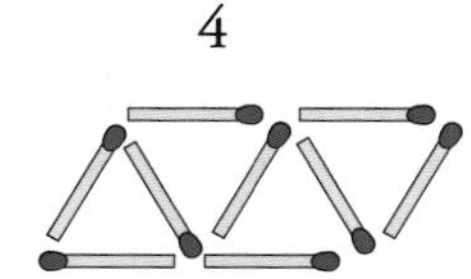

3 matchsticks 5 matchsticks 7 matchsticks 9 matchsticks

a Draw the next two patterns in the sequence.

b Copy and complete this sentence.

You add ____ matchsticks to the current pattern to get the next pattern.

c Write the sentence as a formula:

Next pattern = current pattern +

d Use your formula to write down the number of matchsticks in the 7th pattern.
Check your answer by drawing.

2 For this dot pattern:

Pattern: 1 2 3 4

2 dots 4 dots 6 dots 8 dots

a Draw the next two patterns in the sequence.

b Write a sentence to describe how the sequence grows.

c Write the sentence as a formula:

Next pattern = current pattern +

d Use your formula to write down the number of dots in the 8th pattern.
Check your answer by drawing.

A5.5 Spot the function

This spread will show you how to:

- Develop from explaining a generalised relationship in words to expressing it in a formula using letters as symbols.

KEYWORDS

Function machine
Input
Output

This function machine multiplies everything you put in by 2:

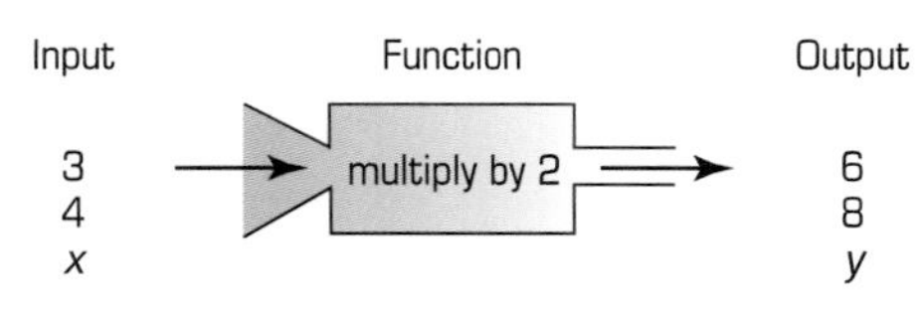

You can write: $3 \times 2 = 6$

$4 \times 2 = 8$

In general: $x \times 2 = y$ or $y = 2x$

This function machine has the function missing:

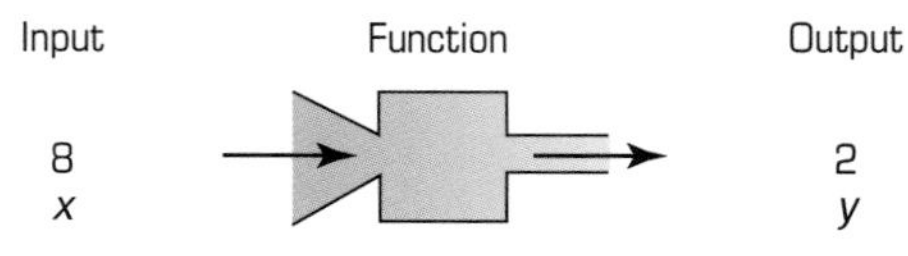

The function could be $-\ 6$: $8 - 6 = 2$ so $x - 6 = y$ or $y = x - 6$

or it could be $\div\ 4$: $8 \div 4 = 2$ so $x \div 4 = y$ or $y = \frac{x}{4}$

If you know more values you can work out the function.

example

Find the function for this machine and write a formula using x and y.

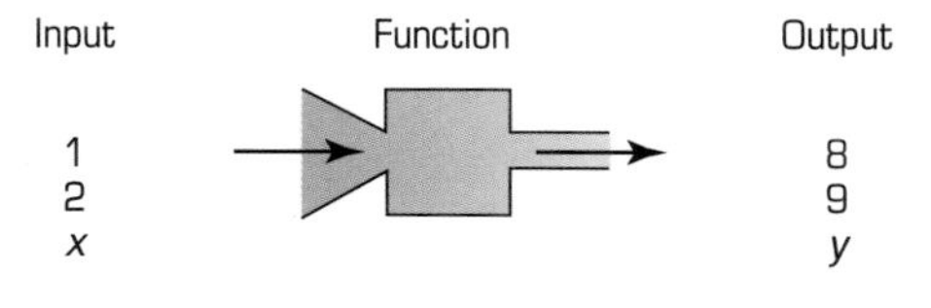

The function could be $+\ 7$: $1 + 7 = 8$

or it could be $\times\ 8$: $1 \times 8 = 8$

Use another pair of values to decide:
$2 + 7 = 9$ so the function is $+\ 7$.

The formula is $x + 7 = y$ or $y = x + 7$

Exercise A5.5

1 For each of these function machines use the rule to calculate the output numbers.

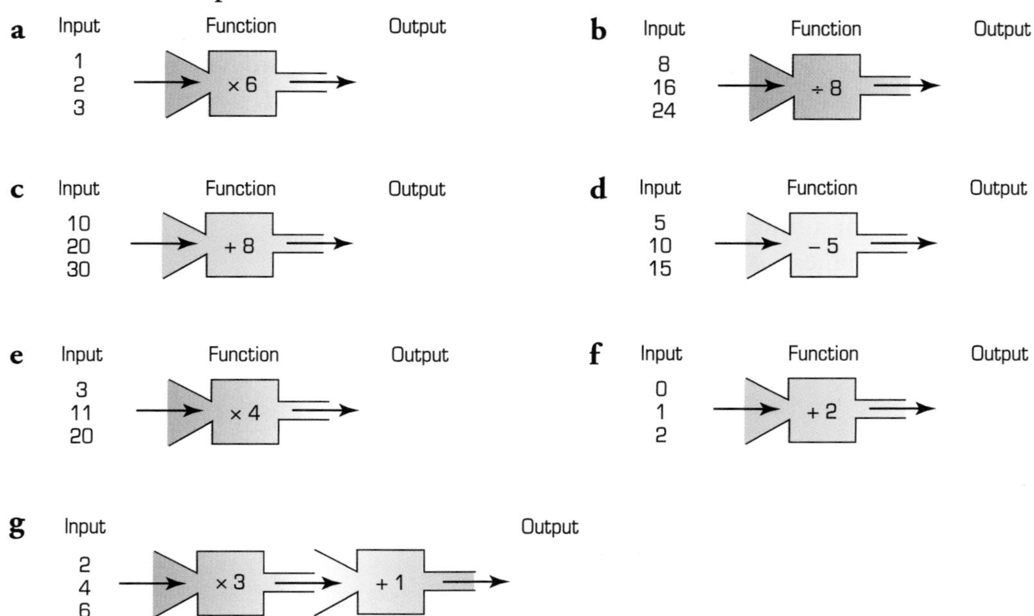

2 For each of these function machines, work out the missing rule.

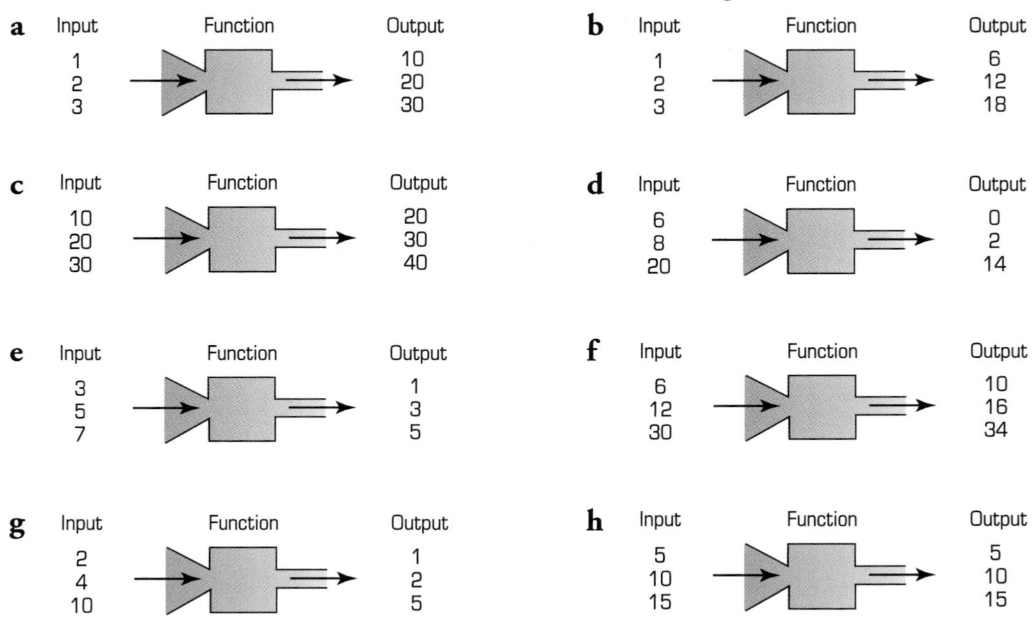

3 Find the formula for each of the function machines in questions 1 and 2.

A5.6 Drawing graphs

This spread will show you how to:

- Develop from explaining a generalised relationship in words to expressing it in a formula using letters as symbols.
- Plot coordinates in the first quadrant.

KEYWORDS
Function machine
Coordinate pair
Grid
Input
Output

You can draw a graph for a function machine.
The input and output values make coordinate pairs.

example

Draw a graph for this function machine:

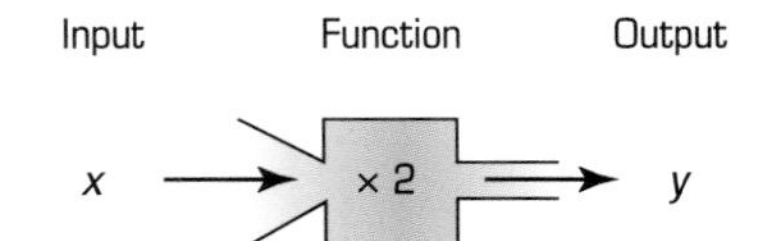

First you need to generate some coordinate pairs.
Choose some values for x.

x	1	2	3	4	5
y	2	4	6	8	10

Find the corresponding values for y: multiply the x values by 2.

Write down the coordinate pairs:

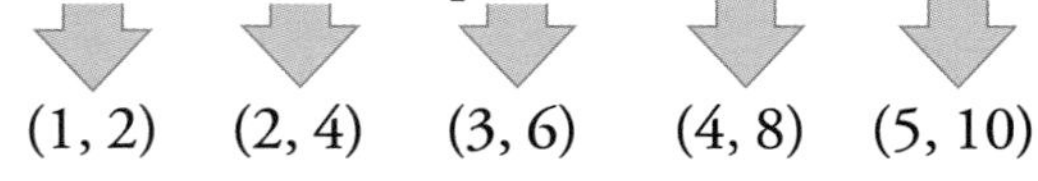

(1, 2) (2, 4) (3, 6) (4, 8) (5, 10)

Plot the points on a coordinate grid:

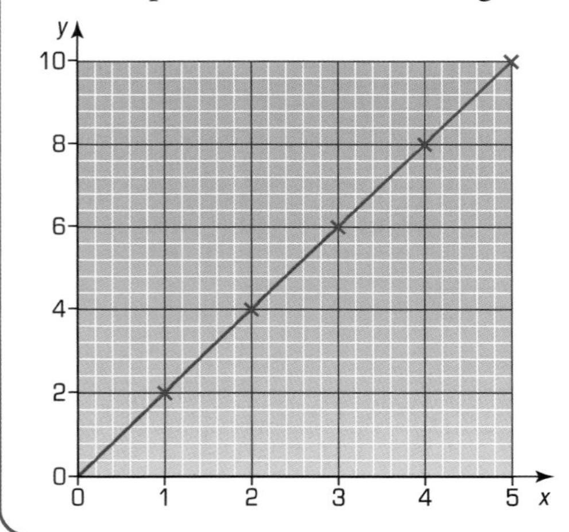

The points make a straight line.

You can name the line using the function machine.

input × 2 = output

$x \times 2 = y$ or $y = 2x$

The line is $y = 2x$

Exercise A5.6

You will need a coordinate grid for each question.

1 The function machine

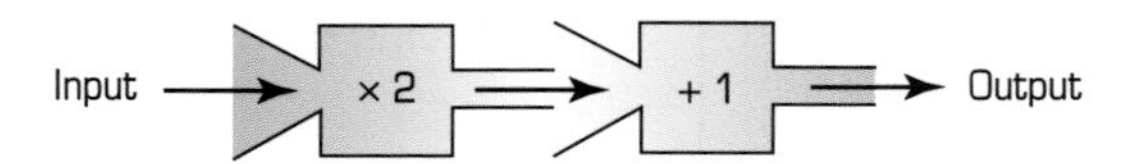

generates this table of values:

Input	1	2	3	4
Output	3	5	7	9

a Copy and complete these coordinate pairs:

(1, 3) (2, 5) (3,) (,)

b Plot the pairs on a graph.
Join them to make a straight line.

2 **a** Copy and complete this table of values for the function:

× 3

Input	1	2	3	4
Output			9	

b Plot the coordinate pairs on a graph.
Join them to make a straight line.

c Name the line. Copy and complete:

Input × ________ = output

x × = y

The line is y = ________

3 **a** Copy and complete this table of values for the function:

– 2.

Input	8	7	6	5
Output			4	

b Plot the coordinate pairs on a graph.
Join them to make a straight line.

c Name the line. Copy and complete:

Input ______ __________ = output

x ______ __________ = y

The line is y = ________

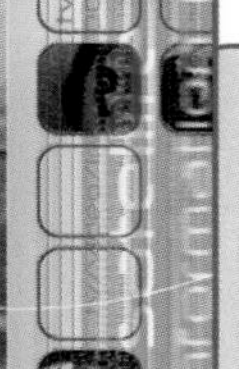

A5.7 Graphs of formulae

This spread will show you how to:

- Develop from explaining a generalised relationship in words to expressing it in a formula using letters as symbols.
- Plot coordinates in the first quadrant.

KEYWORDS
Axes
Graph
Coordinate pair

Penny is saving for her summer holiday in Ibiza.

She wants to limit the amount of money she pays to use her phone.

The formula she uses to work out her phone bill is:

Total bill = number of minutes used × 5p

She can simplify the formula to:

$B = 5m$ where m is the number of minutes and B is the cost in pence.

She draws a graph of the formula, so she can see the total bill at a glance.

This is what she does:

- Makes a table of values for 0 to 50 minutes:

Minutes	0	10	20	30	40	50
Bill (pence)	0	50	100	150	200	250

- Works out the bill – multiplies the minutes by 5.
- Writes down the coordinate pairs:

(0, 0) (10, 50) (20, 100) (30, 150) (40, 200) (50, 250)

- Draws the graph using a scale on both axes:

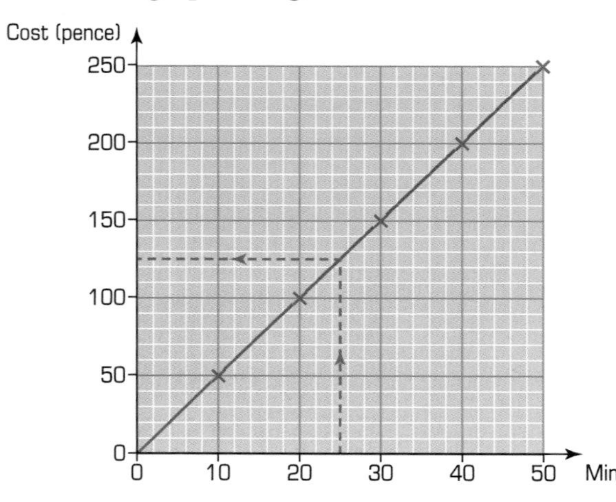

She can read from the graph that 25 minutes would cost 125p or £1.25.

Exercise A5.7

You will need a coordinate grid for each question. You can copy the one on page 230.

1 For the formula

$B = 10m$

a Copy and complete this table of values:

M	0	5	10	15	20
B	0			150	

b Write down the coordinate pairs.
Use them to draw the graph of the function.
Join the points to make a straight line.

2 This formula shows the cost of buying packets of crisps:

$C = 30P$ where C = cost in pence
and P = number of packets

a Copy and complete this table of values:

P	0	2	4	6	8
C	0			180	

b Write down the coordinate pairs.
c Draw the graph of the formula.
d Explain why it does not make sense to join the points.
e Use the graph to find the cost of 3 packets of crisps.

3 The formula for converting feet to inches is:

$i = 12f$

a Copy and complete this table of values:

f	0	5	10	15	20
i	0	60			

b Plot the graph of $i = 12f$.
Join the points to make a straight line.

Use your graph to find:
c the number of inches in 12.5 feet
d the number of feet in 100 inches

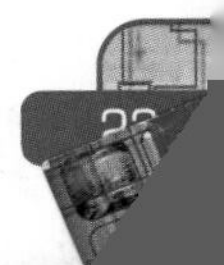

A5.8 All four quadrants

This spread will show you how to:

- Read and plot coordinates in all four quadrants.

KEYWORDS

Axes	*x*-coordinate
Horizontal	*y*-coordinat
Vertical	Quadrant

This vertical line is marked from $^-3$ to $^+3$:

$^+3$, $^+2$, $^+1$, 0, $^-1$, $^-2$, $^-3$

This horizontal line is marked from $^-3$ to $^+3$:

$^-3$, $^-2$, $^-1$, 0, $^+1$, $^+2$, $^+3$

You can put the two lines together to make coordinate axes:
For coordinate pairs you count across first and then up.

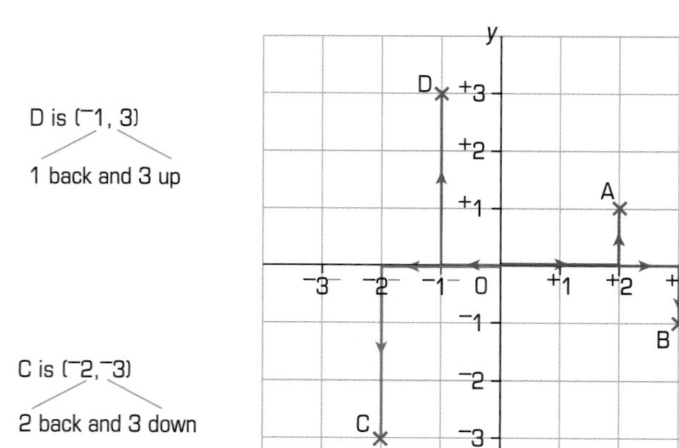

A is (2,1)
2 across and 1 up

B is (3, $^-1$)
3 across and 1 down

To read and plot coordinates you give the:

- horizontal distance first – the *x*-coordinate
- vertical distance second – the *y*-coordinate

Exercise A5.8

1 Write down the coordinates of each letter marked on this coordinate grid:

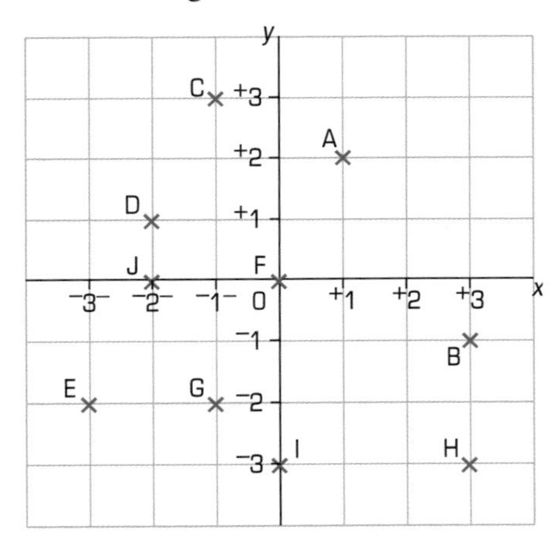

2 Draw x- and y-axes from $^-10$ to $^+10$.
Plot these points on your coordinate grid.
Mark them with the letter.
Remember you go along the x-axis first then the y-axis.

A($^-2$, 3) B(4, $^-6$) C(0, 0)
D(1, $^-7$) E($^-3$, $^-4$) F(5, $^-5$)
G($^-5$, 5) H(5, 5) I($^-5$, $^-5$)
J(0, $^-3$) K($^-6$, 0) L($^-8$, $^-6$)

3 Plot each of these sets of points on the same grid.
Join each set up as you go.

Set A: (3, 4) (6, 4) (6, 7) (3, 7) (3, 4)

Set B: (5, $^-1$) (3, $^-3$) (5, $^-7$) (7, $^-3$) (5, $^-1$)

Set C: ($^-1$, $^-2$) ($^-1$, $^-5$) ($^-4$, $^-5$) ($^-4$, $^-8$) ($^-2$, $^-8$) ($^-2$, $^-9$) ($^-8$, $^-9$) ($^-8$, $^-8$) ($^-6$, $^-8$) ($^-6$, $^-5$) ($^-9$, $^-5$) ($^-9$, $^-2$) ($^-1$, $^-2$)

Set D: ($^-2$, 8) ($^-2$, 7) ($^-5$, 7) ($^-5$, 5) ($^-7$, 5) ($^-7$, 2) ($^-8$, 2) ($^-8$, 8) ($^-2$, 8)

What shapes do your points make?

4 Plot these points on a coordinate grid:

($^-5$, 5) (5, 5) ($^-5$, $^-5$)

The points form three corners of a square.
Write down the coordinates of the 4th corner.

Summary

You should know how to …

Read and plot coordinates in all four quadrants.

Check out

1 The letters of the alphabet correspond to coordinate points on this grid:

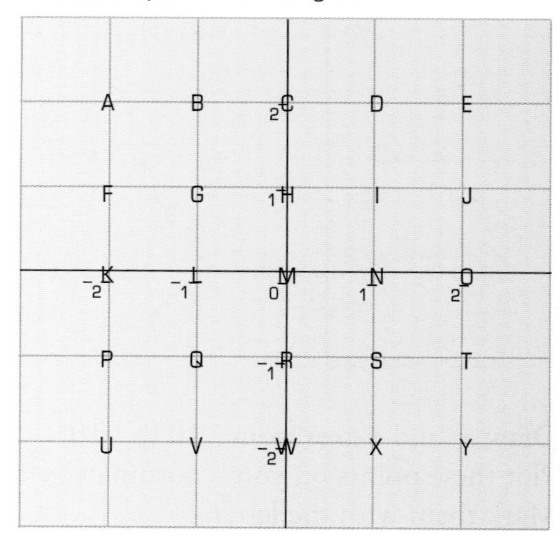

a What do the coordinates spell?
(0, 0) (⁻2, 2) (2, ⁻1) (1, 0) (2, 2) (0, 0) (⁻2, 2) (2, ⁻1) (1, 1) (0, 2) (1, ⁻1)

b Write down the coordinates of the points that will spell your name.
Assume the letter Z is at the point (3, ⁻2).

2 a Use the function machine:

Input → × 2 → − 1 → Output

to copy and complete this table of values:

x	1	2	3	4	5
y			5		9

b Write down the coordinate pairs from your table.

c Plot the points.

d Join them to make a straight line.

e Use your line to find the output when the input is 3.5.

5 Polygons

This unit will show you how to:

- Recognise where a shape will be after a reflection.
- Recognise where a shape will be after a translation.
- Recognise parallel and perpendicular lines.
- Use a protractor to measure and draw angles to the nearest degree.
- Calculate angles in a triangle, around a point and on a straight line.
- Recognise where a shape will be after a rotation.
- Make shapes with increasing accuracy.
- Visualise 3-D shapes from 2-D drawings and identify different nets for a closed cube.
- Classify shapes according to their properties.
- Make and investigate a general statement about familiar numbers or shapes by finding examples that satisfy it.

You find symmetries in nature.

Before you start

You should know how to …

1 Measure and draw acute and obtuse angles to the nearest degree.

2 Identify the nets of an open cube.

Check in

1 Measure these angles:

a *b*

2 What colour is the base in each of these nets of an open cube?

a

b 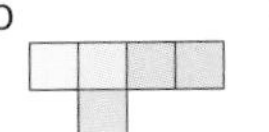

c 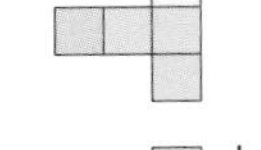

d

e

f

g

h

S5.1 More angle facts

This spread will show you how to:

- Calculate angles on a straight line and around a point.
- Calculate the missing angle in a triangle.

KEYWORDS
Angles on a straight line
Angles around a point
Vertically opposite angles

There are 90° on a corner

90° is a right angle

There are 180° on a straight line

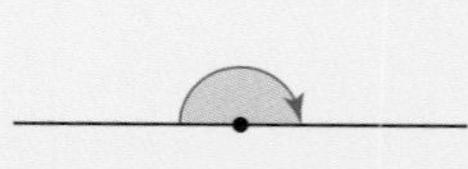

There are 360° around a point

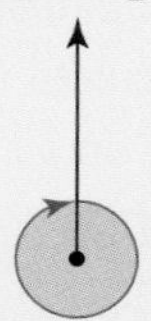

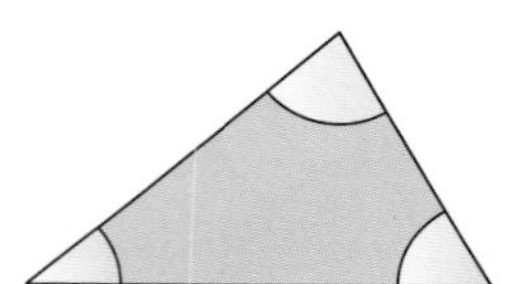

Vertically opposite angles are equal

The three angles in a triangle add to 180°

You can use these facts to solve problems involving angles.

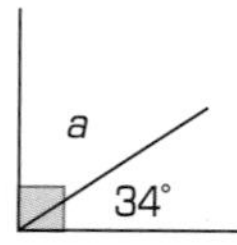

The 2 angles make 90°

So $a + 34° = 90°$
$56° + 34° = 90°$
$a = 56°$

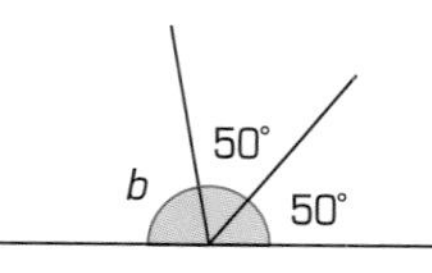

The 3 angles make 180°

So $b + 50° + 50° = 180°$
$80° + 100° = 180°$
$b = 80°$

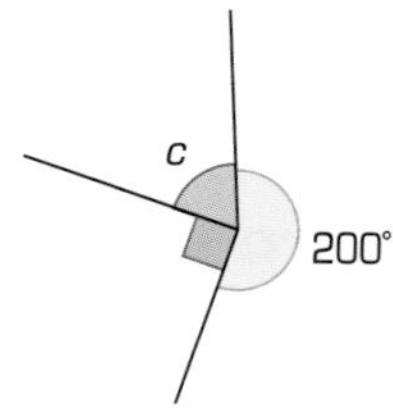

The 3 angles make 360°

So $c + 90° + 200° = 360°$
$70° + 290° = 360°$
$c = 70°$

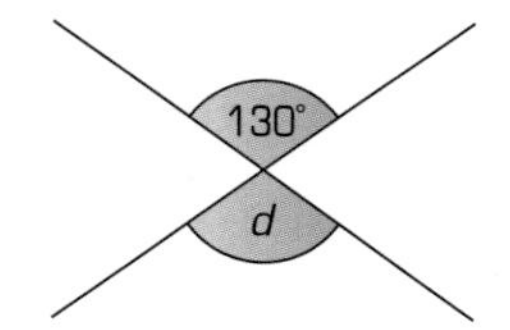

$d = 130°$ as vertically opposite angles are equal

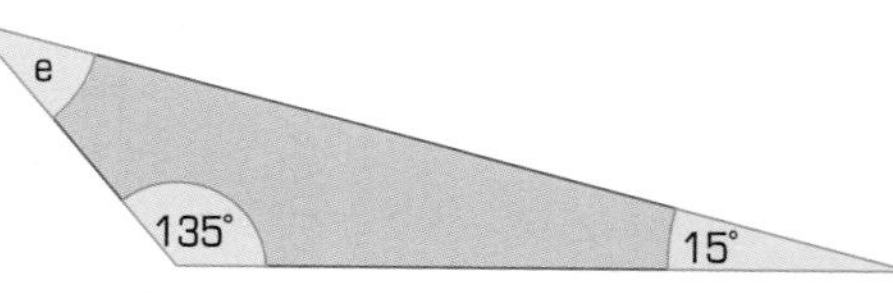

The 3 angles make 180°

So $15 + 135 + e = 180°$
$150 + 30 = 180°$
$e = 30°$

Exercise S5.1

1 Work out the missing angles in these diagrams:
They are not drawn to scale.

a

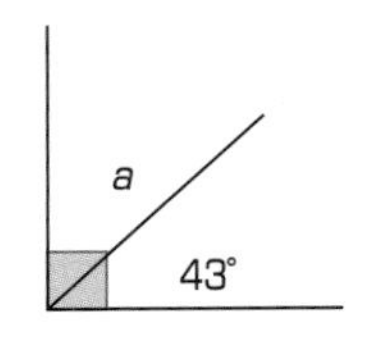

b

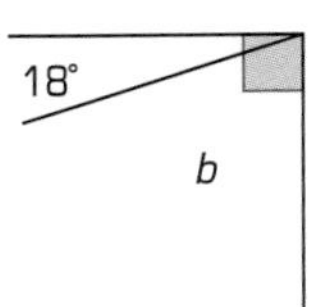

c

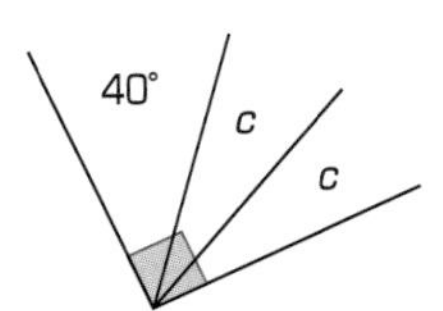

d

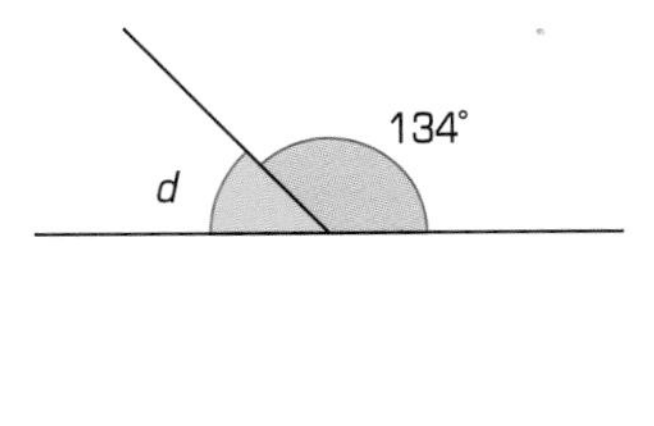

e

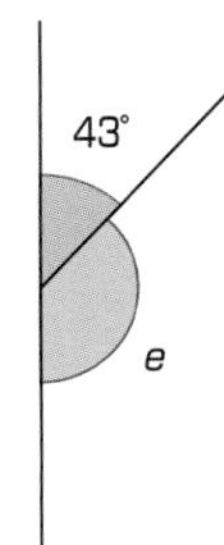

f

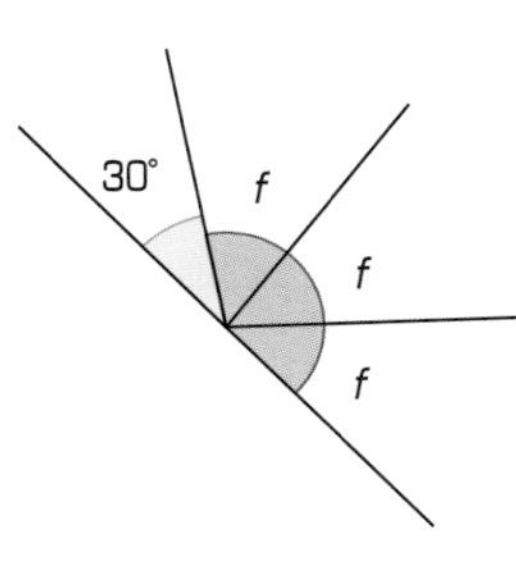

g

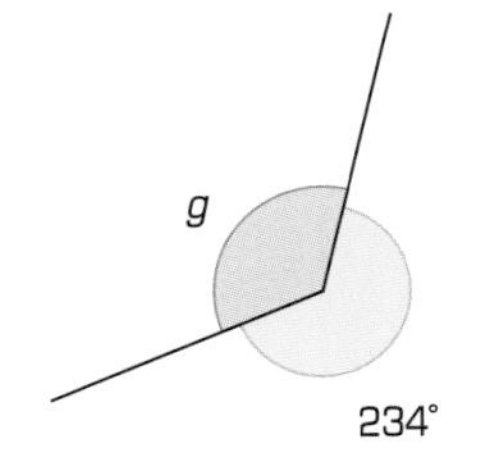

h

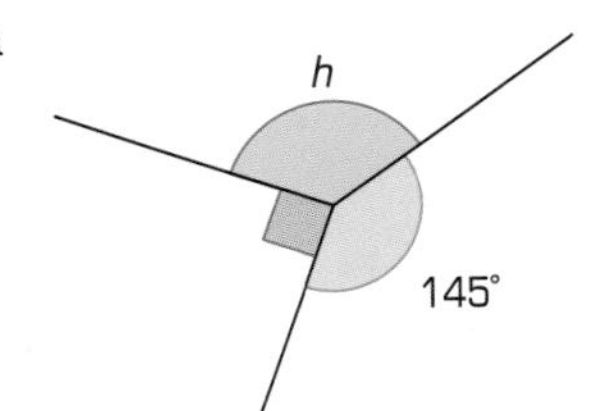

i

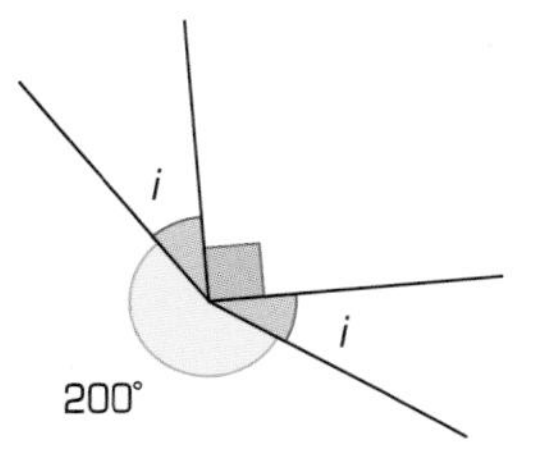

2 Calculate the missing angles in these triangles.

a

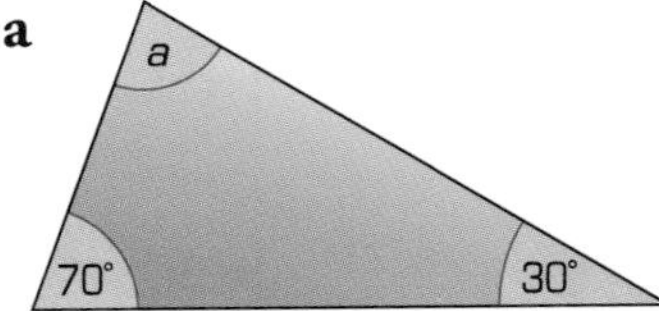

b

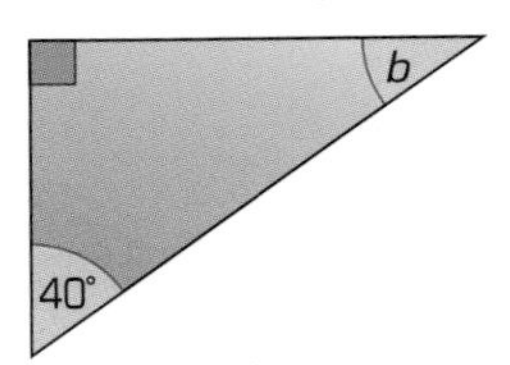

c

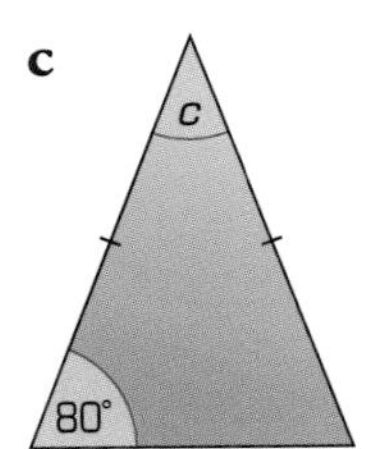

d

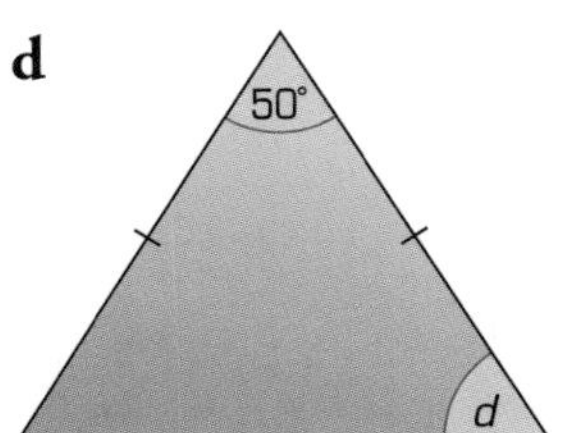

3 Find the missing angles.

a

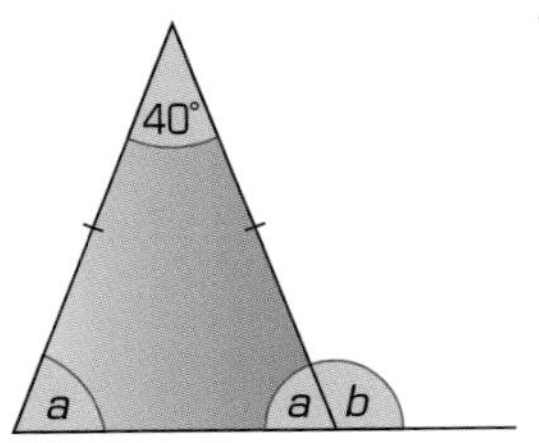

b

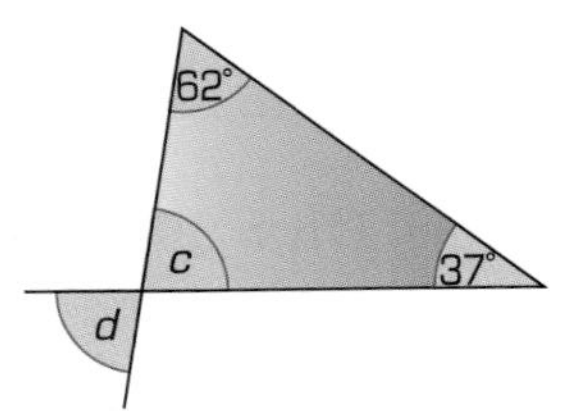

c

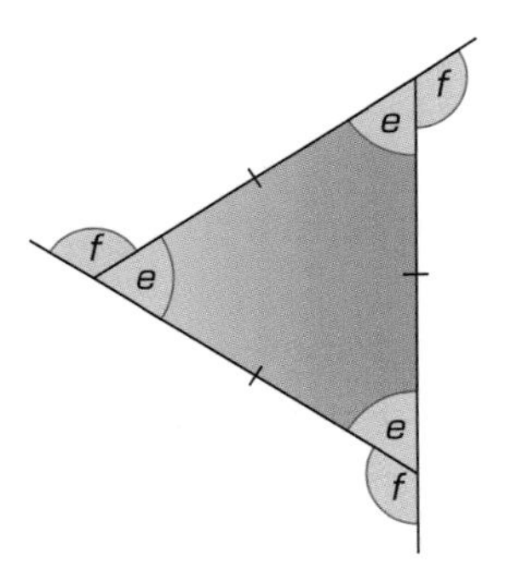

Constructing triangles

This spread will show you how to:

- Use a protractor to measure and draw acute and obtuse angles to the nearest degree.

KEYWORDS
Construct
Triangle
Protractor
Ruler

A right-angled triangle has one angle that is 90°:

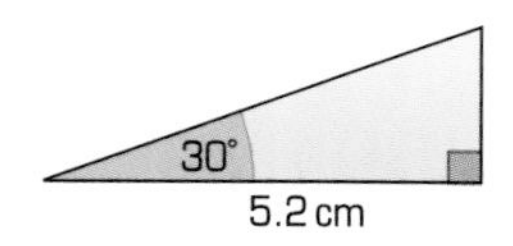

You can construct this triangle using a ruler and protractor.

Draw the base line first – use a ruler.

5.2 cm

Measure 90° using a protractor.

5.2 cm

Draw 30° at the other end of the base.

30°
5.2 cm

The lines join to form a triangle.

30°
5.2 cm

Measure the third angle – it should be 60°.

Exercise S5.2

1 What is the name of each of these triangles?

a

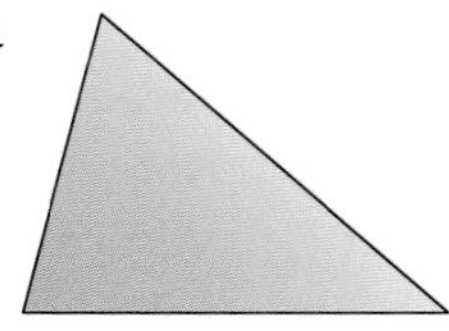

b

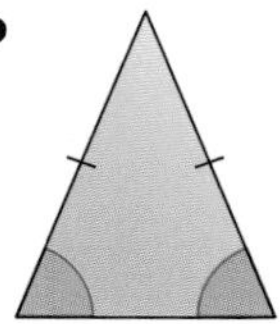

c

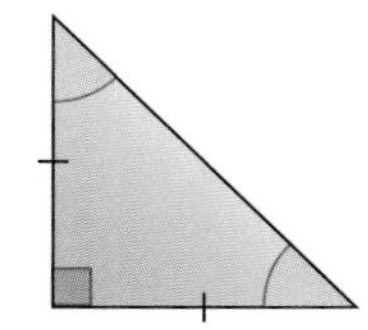

d

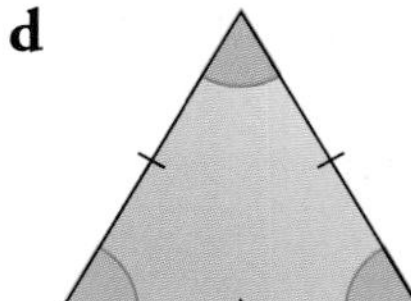

2 Construct each of these triangles accurately.

a

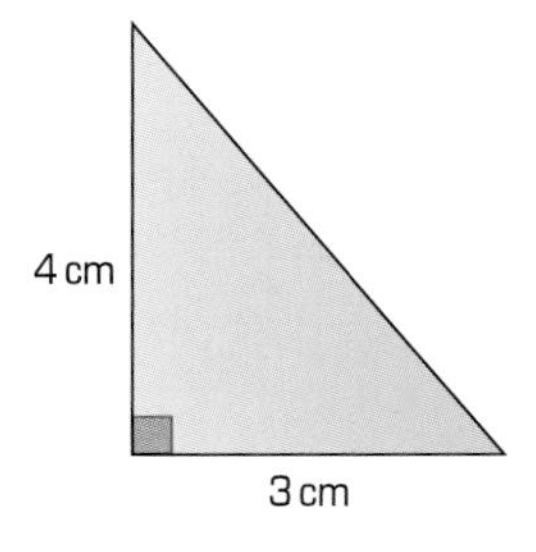

b

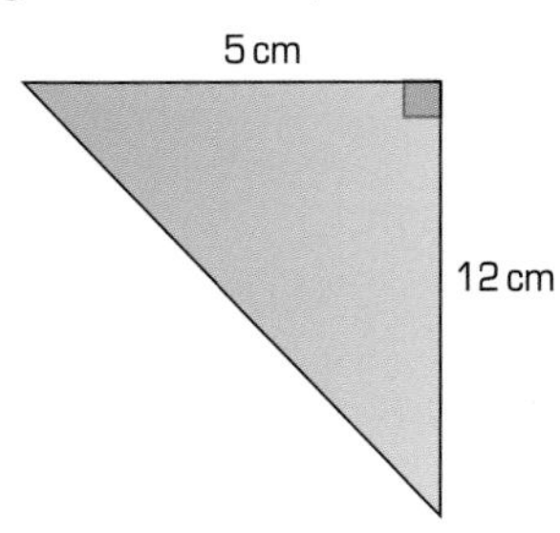

c

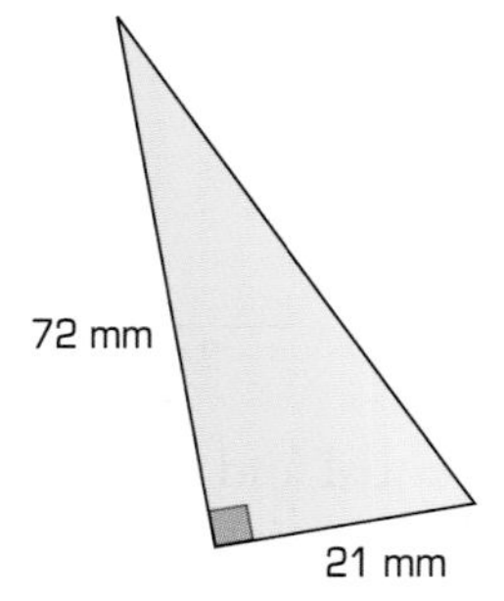

3 For each of the triangles in question 2, measure the long side and the unknown angles.

4 Construct each of these triangles accurately.

a

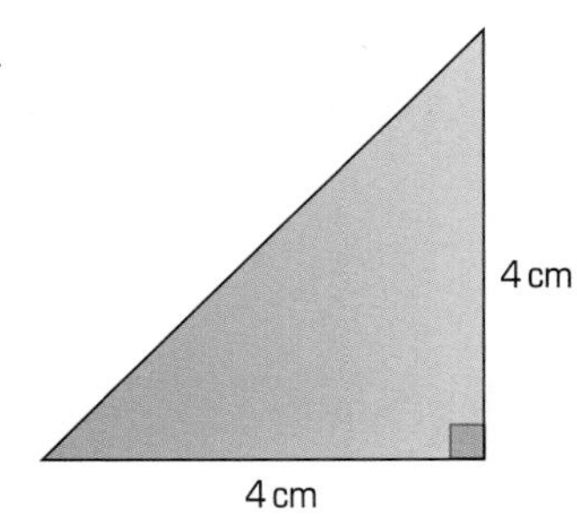

b

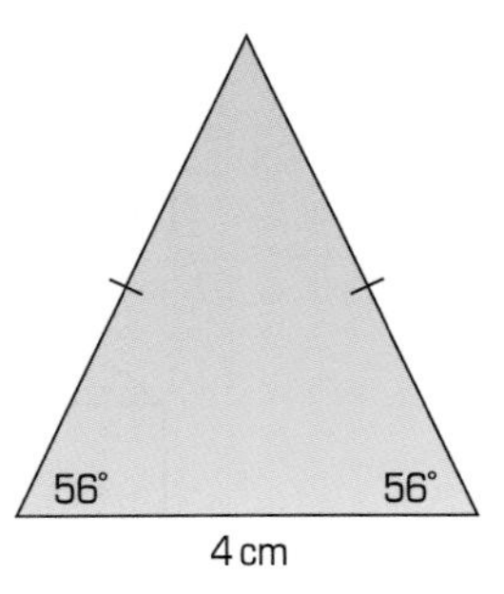

c

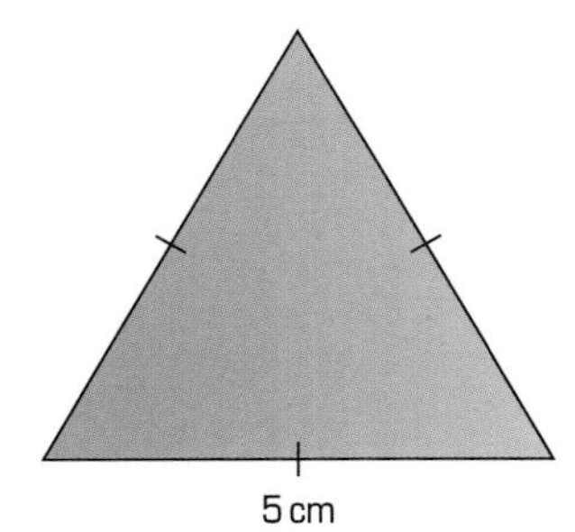

5 Draw a horizontal line 8 cm long.

- At one end, construct an angle of 110°.
- At the other end, construct an angle of 23°.
- Does it matter which scale you use?
  Will you get the same result?
- Change the ends around. Does it make any difference?
  Explain your answer.

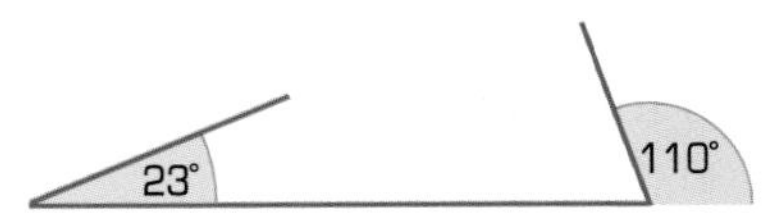

S5.3 Constructing squares

This spread will show you how to:
- Identify different nets for a closed cube.
- Make shapes with increasing accuracy.

KEYWORDS

3-D	Face
Angle	Net
Construct	Vertices
Edge	Solid

You can construct a square using a ruler and protractor:

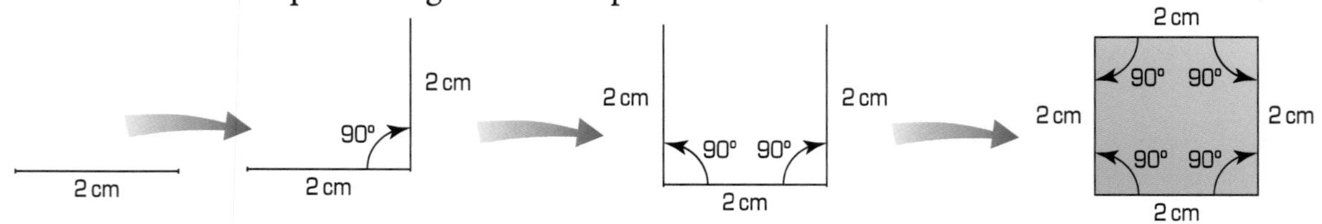

The sides are equal and all angles are 90°.

In a square:
- **All the sides are the same length.**
- **All the angles are the same: 90°.**

A cube is a 3-D shape.
All the faces of a cube are squares.

This box is a cube:

It has:
6 faces

12 edges

8 vertices

This cube has no lid:
It is an open cube.

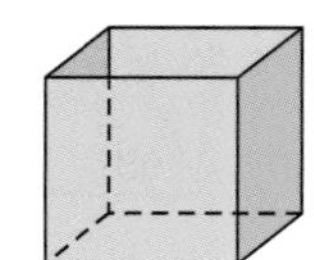

It has 5 faces.

Most boxes start out as flat shapes called **nets**.
You fold up the net to make the box.

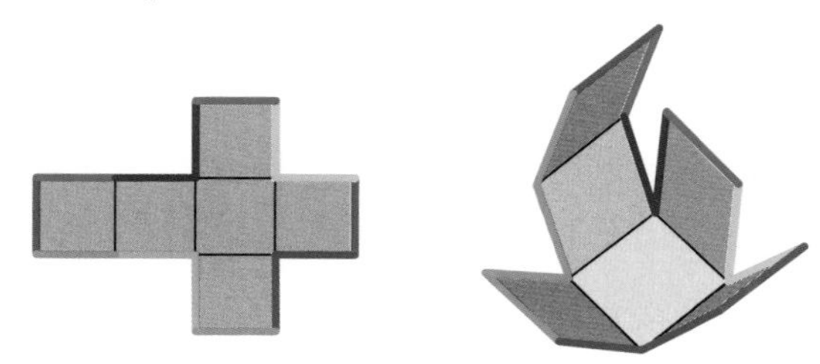

The colours show edges that fit together.

Exercise S5.3

1 Construct a square with a side of 3 cm.

2 Each of these diagrams shows a net of an open cube.
Copy the nets onto squared paper.
Use colour to show which edges fit together to make the open cube.

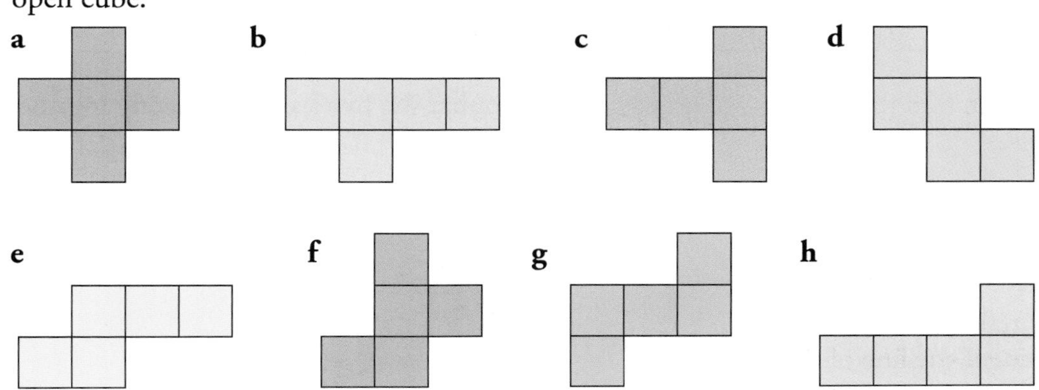

3 Which of these shows the net of a closed cube?

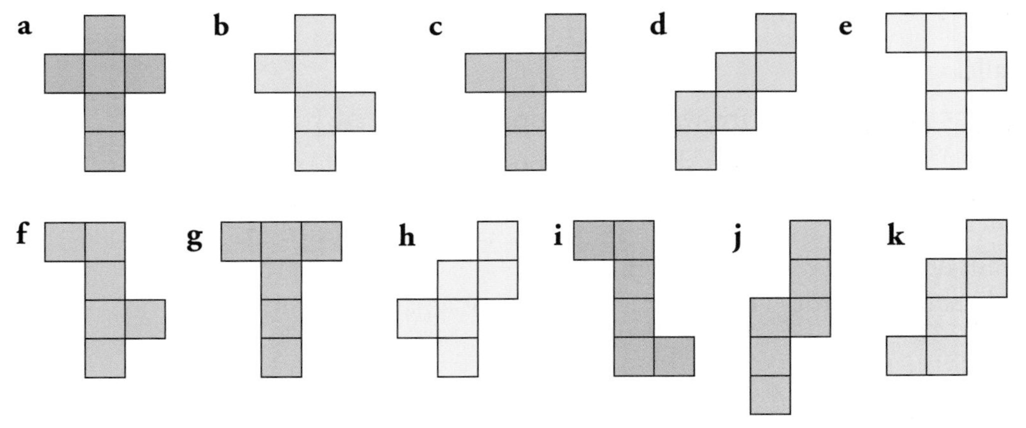

4 Construct the net of a closed cube with an edge of 4 cm.
What is the area of your net?

5 The perimeter of a shape is the distance around the edge.
To find the perimeter you add the lengths of all the edges together.

What is the perimeter of the net you drew in question 4?
Can you find a net for this cube with a smaller perimeter?

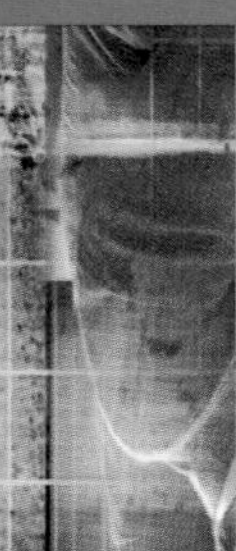

S5.4 Reflection symmetry

This spread will show you how to:

- Recognise reflection symmetry in 2-D shapes.
- Classify 2-D shapes according to their properties.

KEYWORDS

Equal, Opposite, Parallel, Regular, Reflection, Line of symmetry

You can fold this kite ...

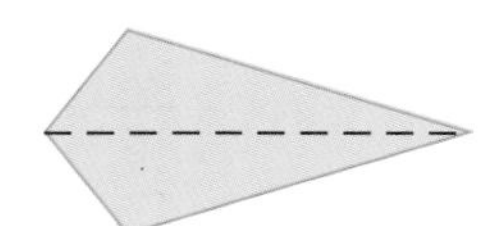

so that the two halves fit exactly together.

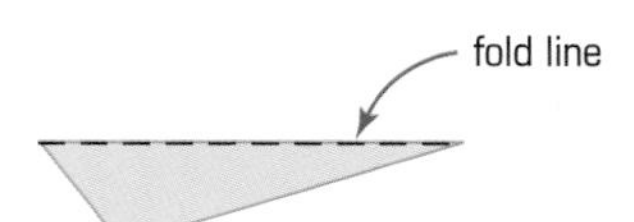

A kite has reflection symmetry.
The fold line is the line of symmetry.

► A shape has **reflection symmetry** if you can fold it so that one half fits exactly on top of the other.

This is a parallelogram.

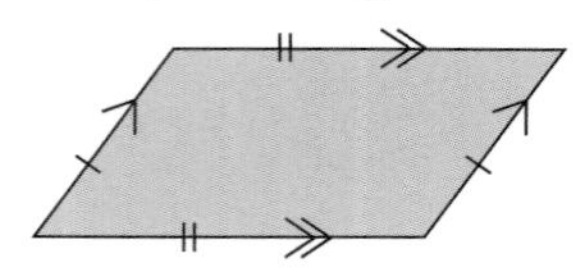

The arrows show the opposite sides are parallel.
The tabs show the opposite sides are equal lengths.
Opposite angles are equal too.

A parallelogram does not have a line of symmetry.
You can check using a mirror. The reflected shape is not a parallelogram!

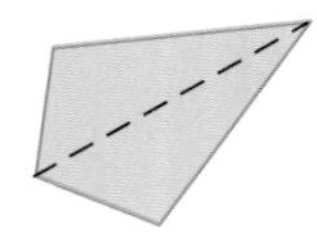

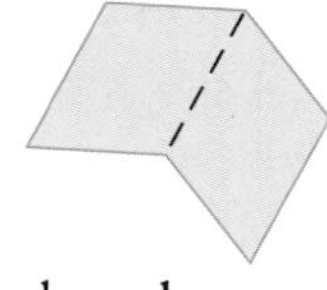

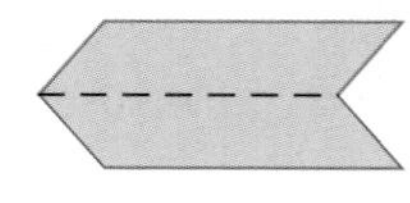

Regular shapes have equal sides and equal angles.

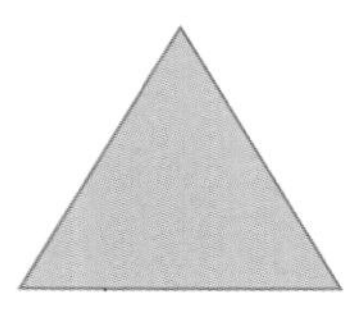

Equilateral triangle

Square

Regular pentagon

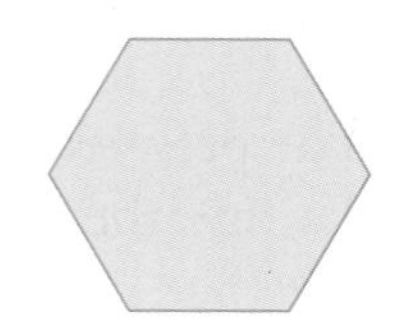

Regular hexagon

A regular hexagon has six lines of symmetry.
Check you can find them all.

Exercise S5.4

1 Copy or trace each of these triangles.
Draw in all the lines of symmetry and name each shape.

a 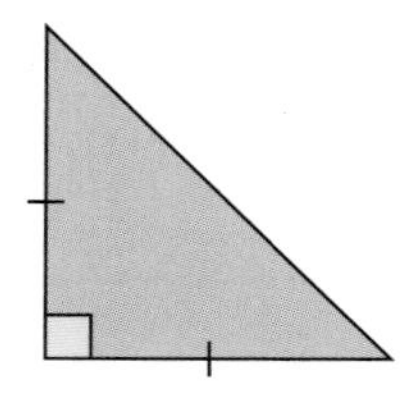

b 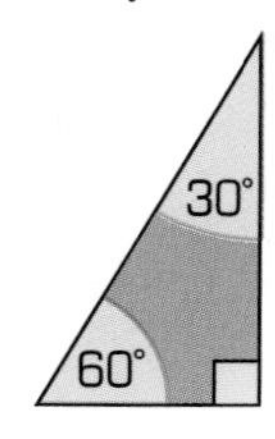

c 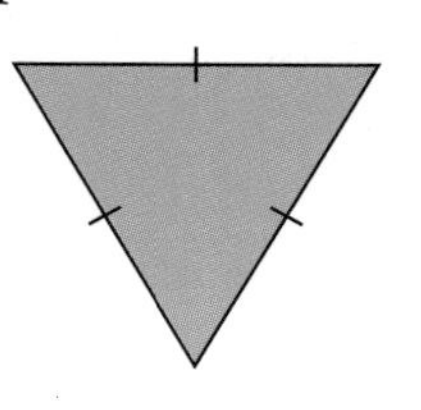

d 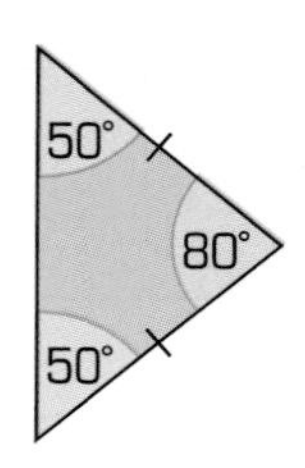

2 Copy or trace each of these quadrilaterals.
Draw in all the lines of symmetry and name each shape.

a 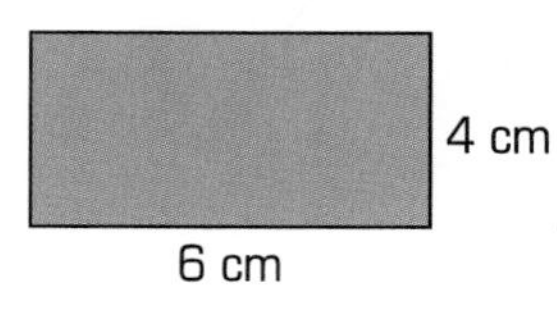

b 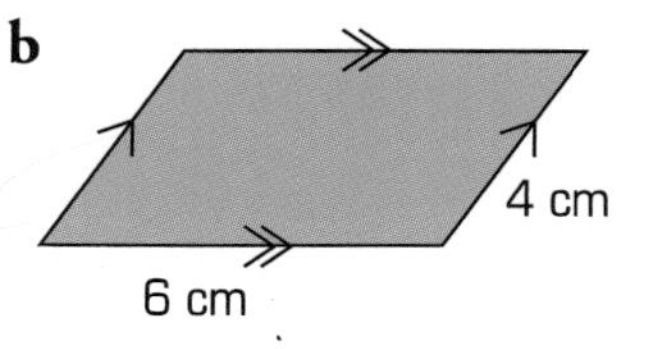

c 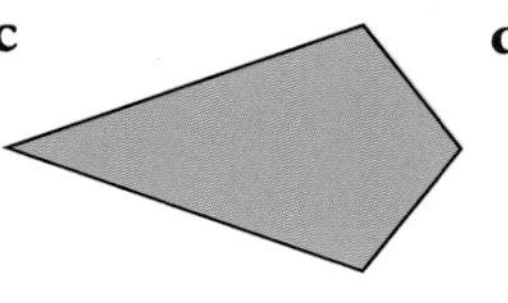

d 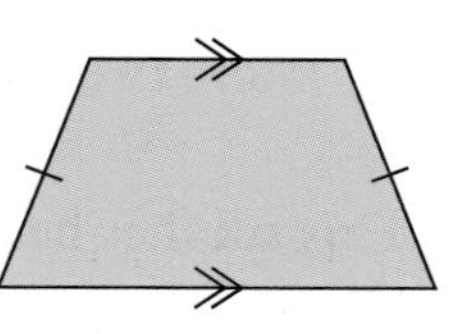

e 

f 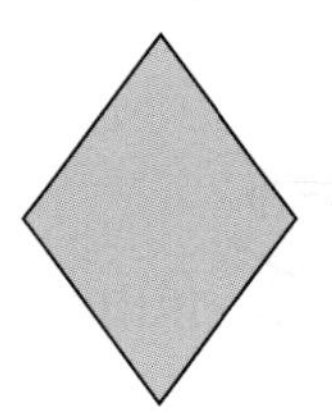

g 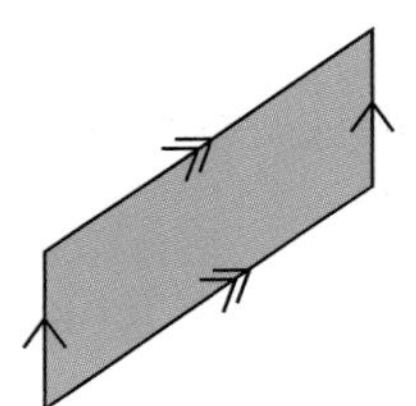

h 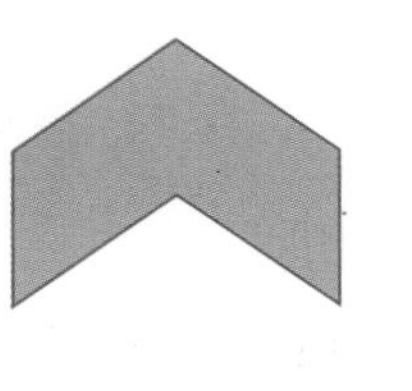

3 Draw a triangle with:
a 1 line of symmetry **b** 3 lines of symmetry
Name each shape that you draw.

4 Draw a quadrilateral with:
a 2 lines of symmetry **b** 4 lines of symmetry
c 1 line of symmetry **d** 0 lines of symmetry
Name each shape that you draw.

5 Copy or trace each of these shapes.
Draw in all the lines of symmetry and name each shape.

a 

b 

c 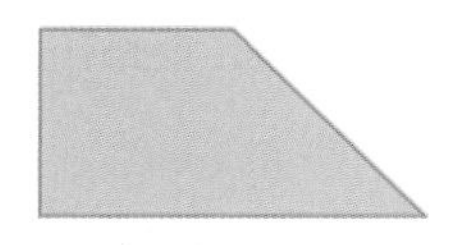

d 

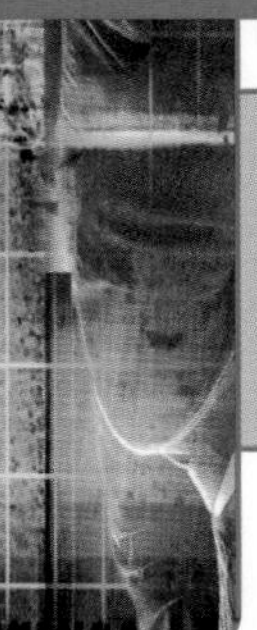

S5.5 Rotational symmetry

This spread will show you how to:

- Recognise where a shape will be after a rotation.

KEYWORDS
Order of rotational symmetry
Rotation symmetry
Spinner

A rotation is a turn.
If you rotate a shape about a point you can make a pattern:

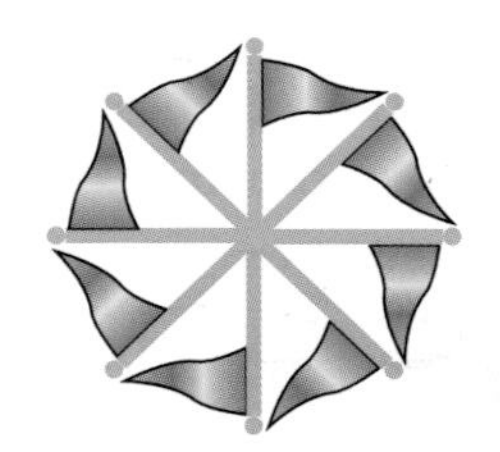

- A shape has **rotational symmetry** if it rotates onto itself more than once in a full turn.

This spinner has rotational symmetry.

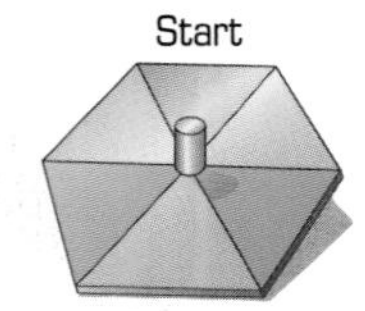

As it rotates through a full turn it looks exactly like itself three times:

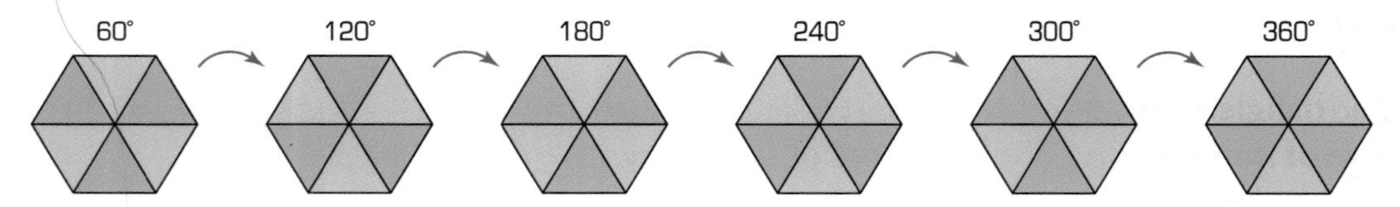

The spinner has rotational symmetry of order 3.

A parallelogram has rotational symmetry:

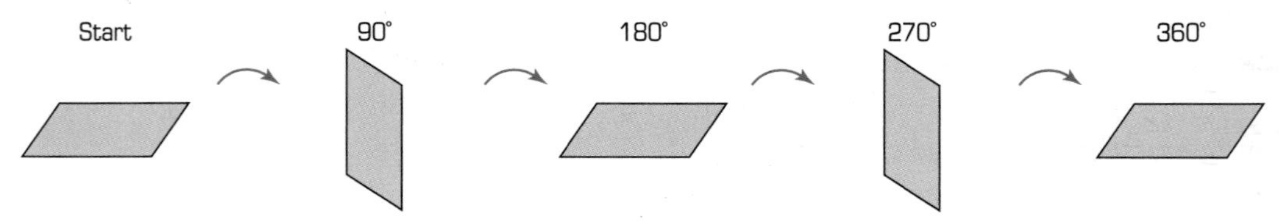

It looks the same after a 180° rotation and a 360° rotation.
The parallelogram has rotational symmetry of order 2.

- The order of rotational symmetry is the number of times the shape looks exactly like itself in a complete turn.

Exercise S5.5

Copy each shape and state the order of rotational symmetry.

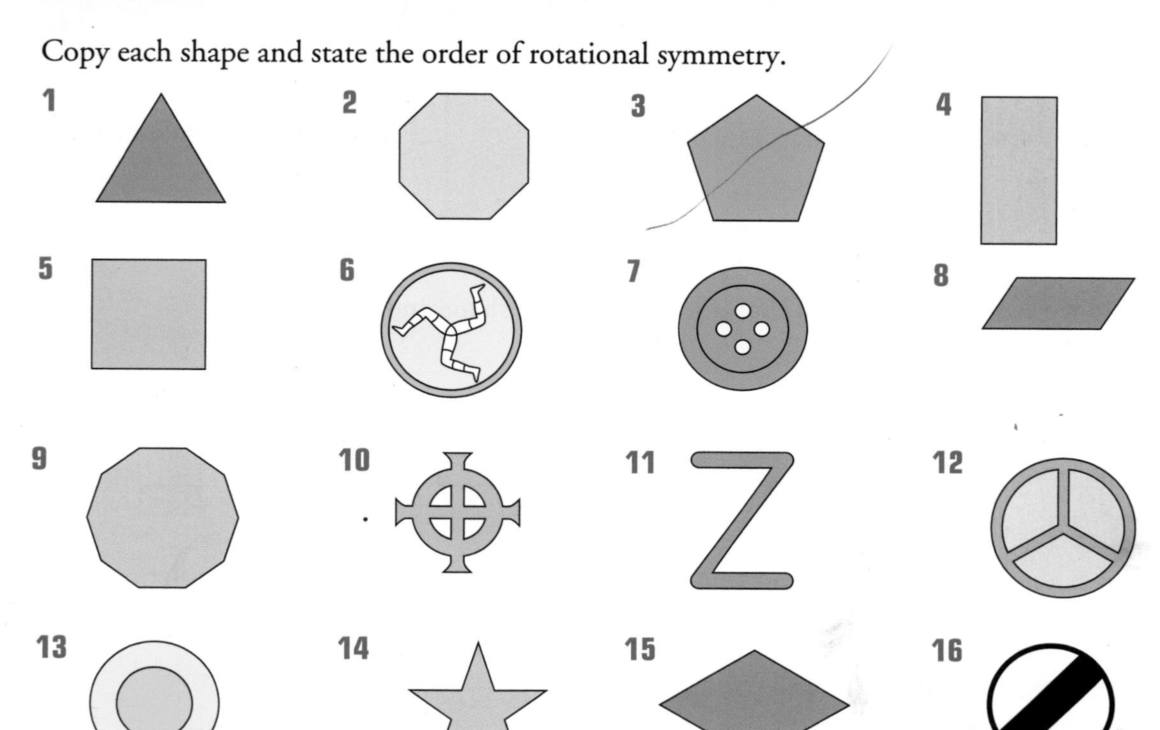

17 Copy these flags.
What is the order of rotational symmetry of each flag?
Draw in any lines of symmetry.

a 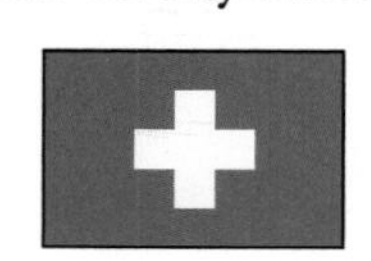

b 

c 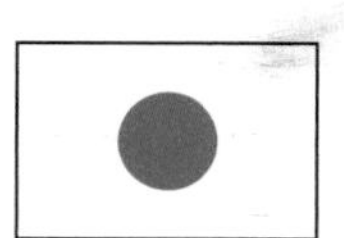

d 

e 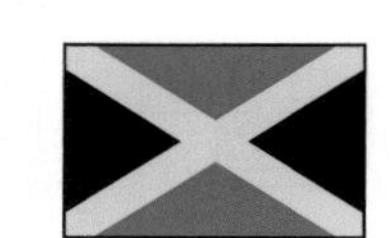

f 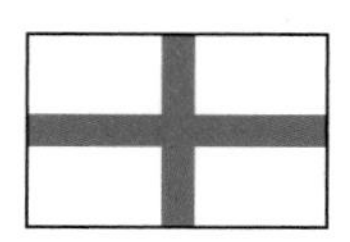

18 The year 1961 has rotational symmetry of order 2.
Find four more years between 1000 and this year that also have rotational symmetry of order 2.

19 List the seven capital letters that have rotational symmetry.

S5.6 Tessellating shapes

This spread will show you how to:

- Recognise reflective symmetry, reflections and translations.

KEYWORDS
Congruent
Translation
Reflection
Rotation

You can move a shape on a grid so that it is still the same shape and size.
You transform the shape.

You can: ...	... flip it;	... turn it;	... slide it.
This is a ...	... reflection	... rotation	... translation.

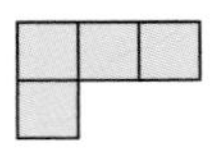

If you cut all the L shapes out they will fit on top of each other exactly.
They are exactly the same shape and size.
They are congruent.

► **Congruent** shapes are exactly the same shape and size.

You can make patterns with the L shape:

The L shape fits together so that there are no gaps or overlaps. It tessellates.

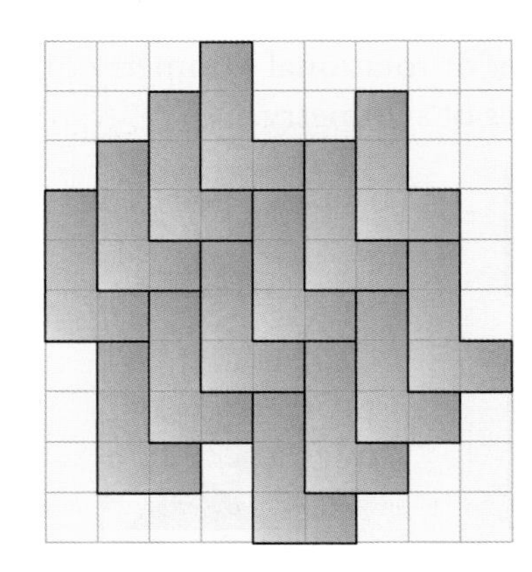

► Congruent shapes **tessellate** if they fit together with no gaps or overlaps.

Only three regular shapes tessellate:

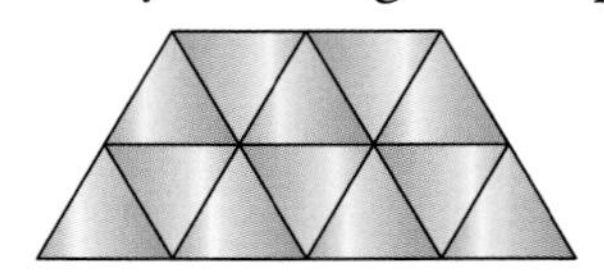

Equilateral triangles

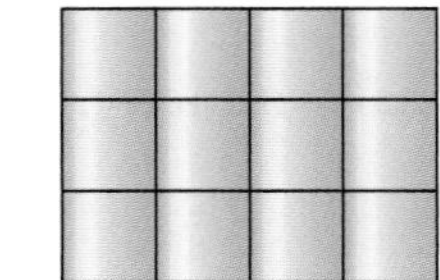

Squares

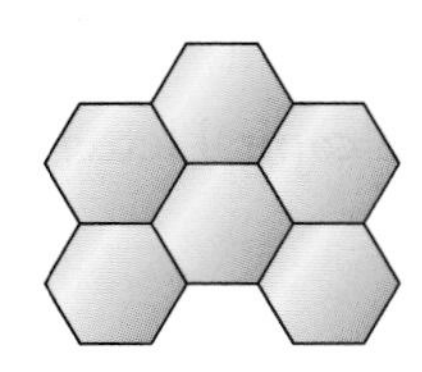

Hexagons

Exercise S5.6

Copy these shapes onto squared paper.
Name each shape
Show how each shape tessellates.

1

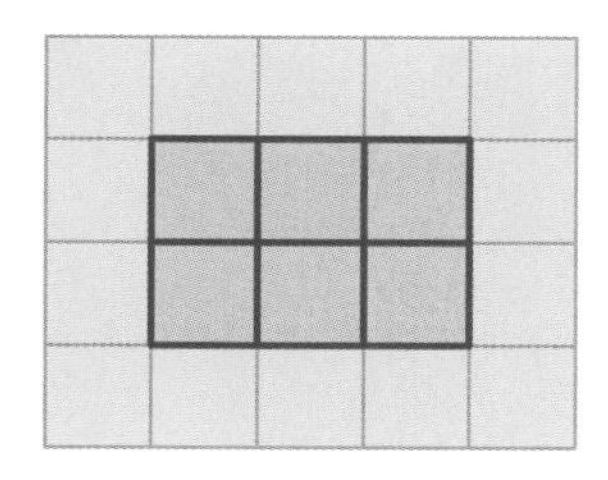

2

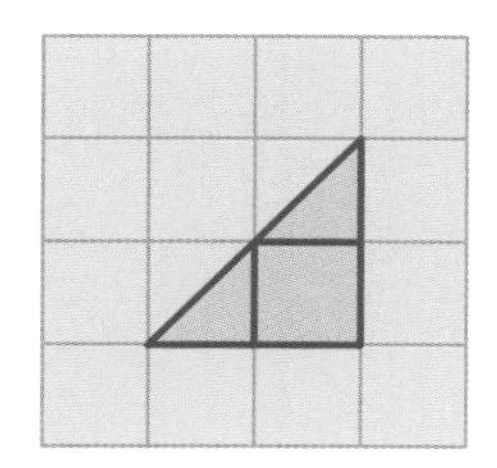

3

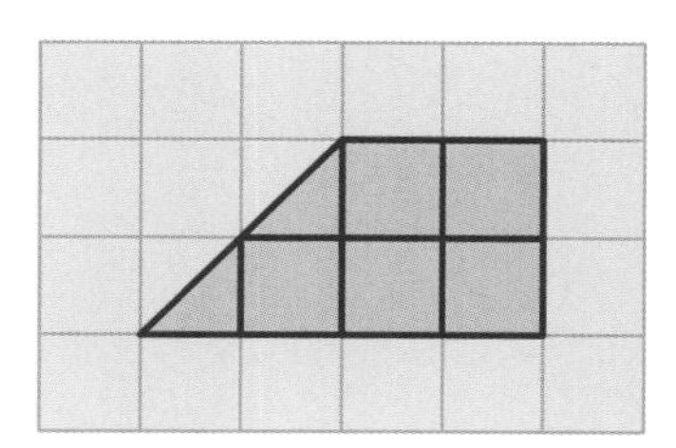

4

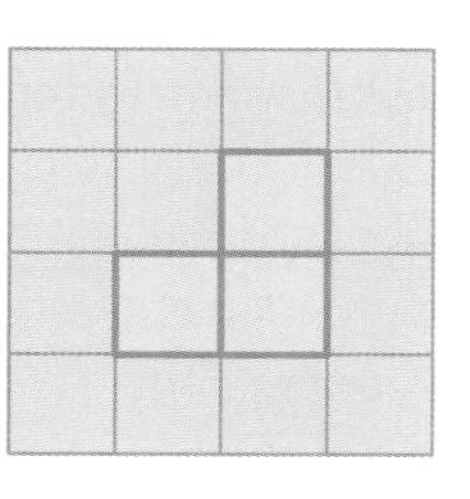

5

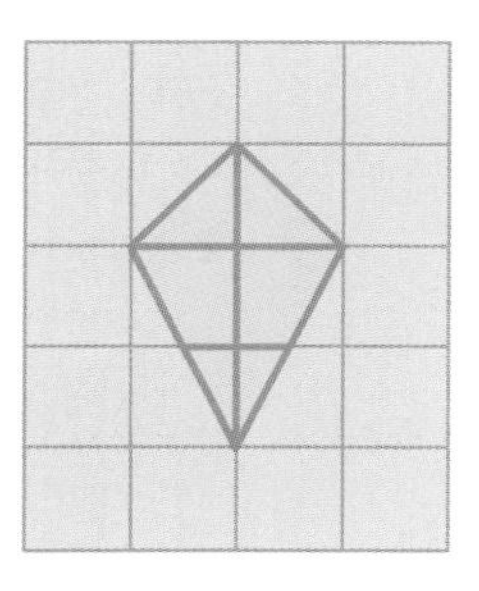

6

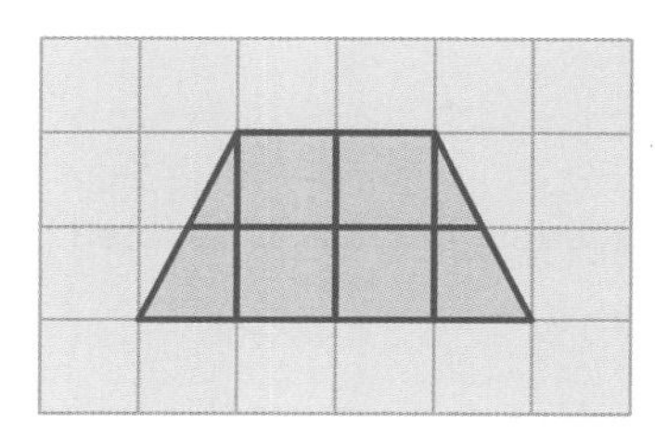

7

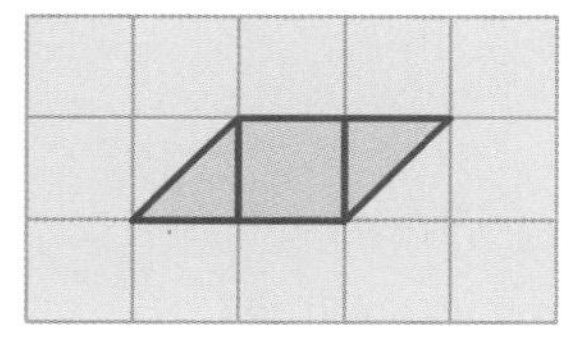

8

9

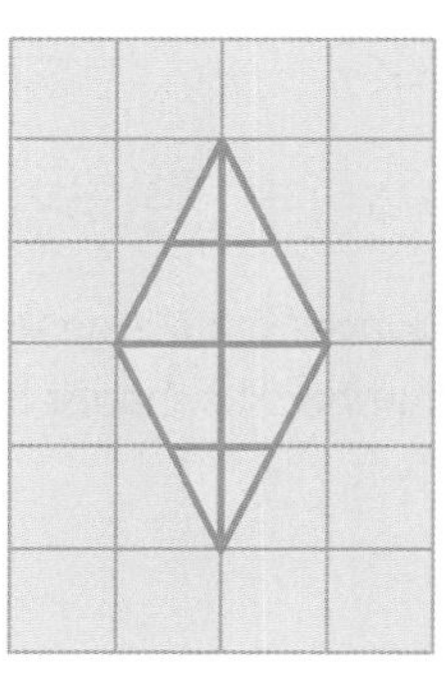

10

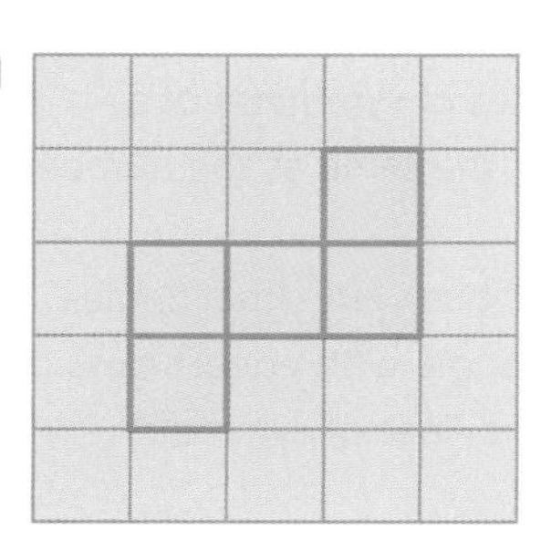

11

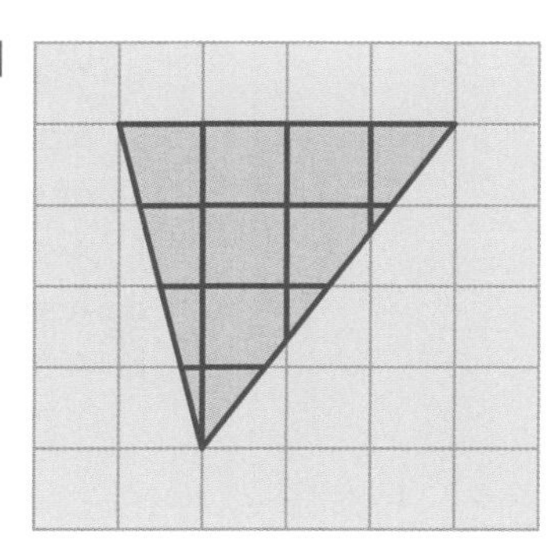

12

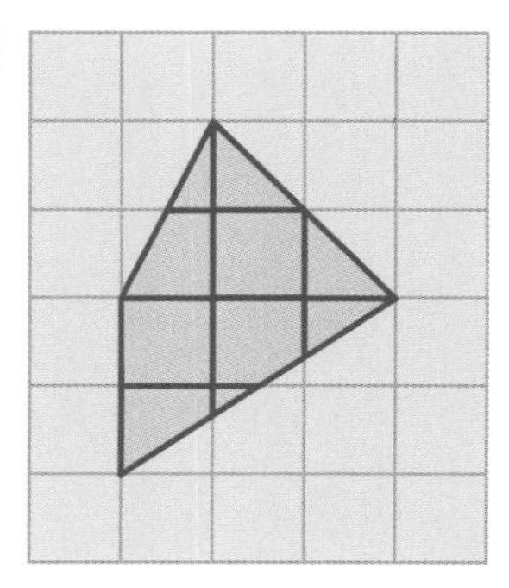

S5 Summary

You should know how to …

1 Recognise the symmetries of 2-D shapes.

2 Identify the nets of a closed cube.

3 Make and investigate a general statement about familiar numbers or shapes by finding examples that satisfy it.

Check out

1 a Copy these shapes:

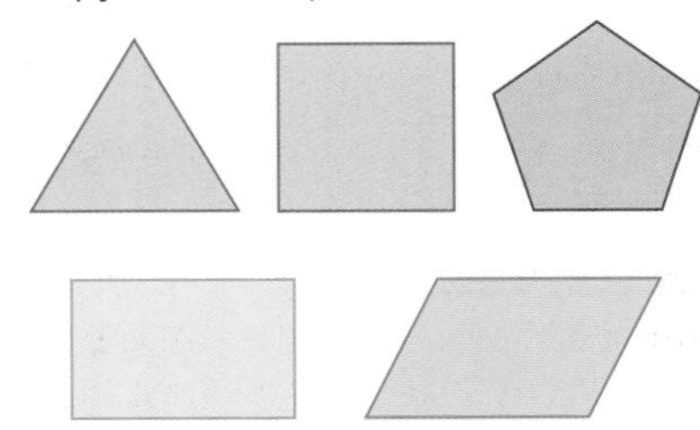

b Draw in all the lines of symmetry.

c State the order of rotational symmetry for each shape.

2 Which of the following nets make a closed cube?

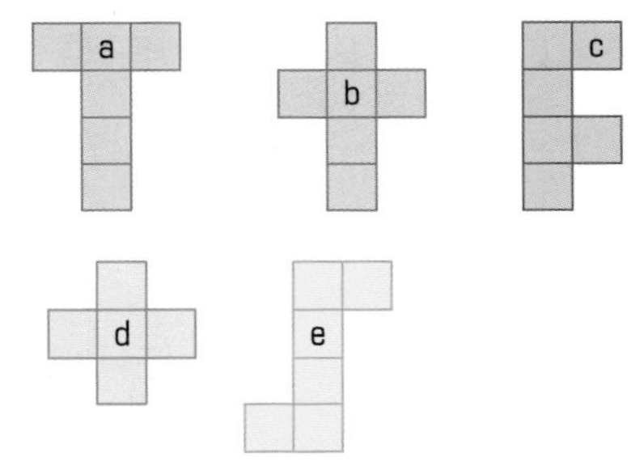

3 Which of these statements are true?
Justify your answers with drawings.

a Any isosceles triangle will tessellate.

b Any parallelogram will tessellate.

c Angles in a triangle add to 180°.

d A trapezium is a quadrilateral with one pair of parallel sides.

e The number of lines of symmetry of a regular polygon is the same as the number of sides.

f A square is a rhombus with equal angles.

g A square is a rectangle with equal sides.

h A trapezium has no lines of symmetry.

add, addition
N1.3, N1.4

Addition is the sum of two numbers or quantities.

algebra
A2.1, A2.2, A4.3

Algebra is the branch of mathematics where symbols or letters are used to represent numbers.

amount
N1.5, N4.3

Amount means total.

angle: acute, obtuse, right, reflex
S2.1, S3.1, S3.3, S5.1, S5.3

An angle is formed when two straight lines cross or meet each other at a point. The size of an angle is measured by the amount one line has been turned in relation to the other.

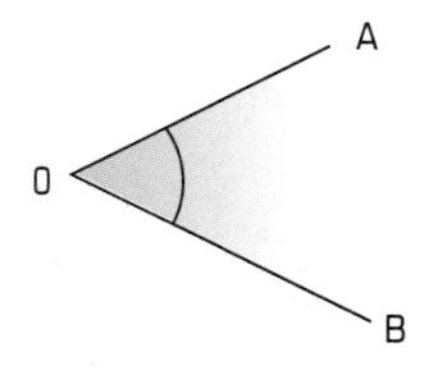

An acute angle is less than 90°.

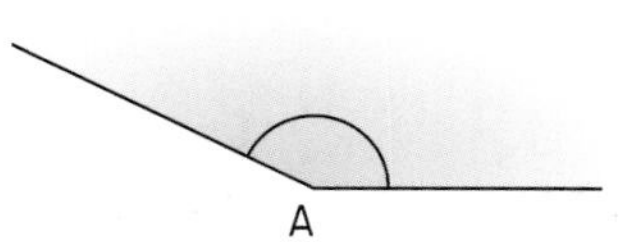

An obtuse angle is more than 90° but less than 180°.

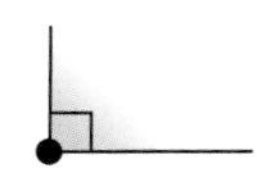
A right angle is a quarter of a turn, or 90°.

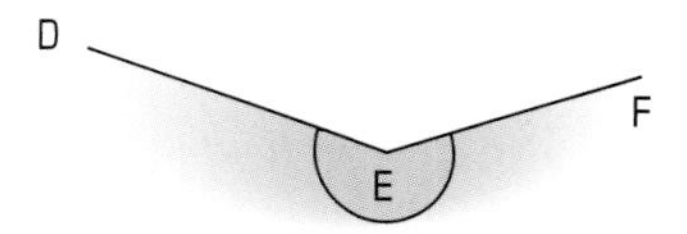

A reflex angle is more than 180° but less than 360°.

angles at a point
S3.2, S5.1

Angles at a point add up to 360°.

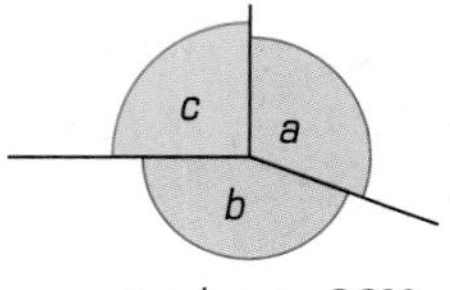

$a + b + c = 360°$

angles on a straight line
S3.2, S3.3, S3.4, S5.1

Angles on a straight line add up to 180°.

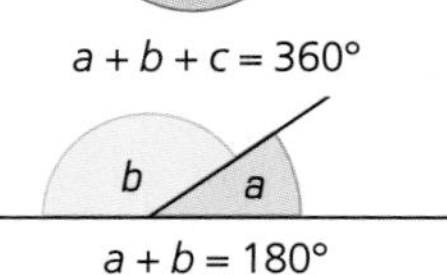

$a + b = 180°$

approximate, approximately
N1.6

An approximate value is a value that is close to the actual value of a number.

area: square millimetre, square centimetre, square metre, square kilometre
S1.1, S1.2, A5.2

The area of a surface is a measure of its size.

average
D1.1, D1.2, D3.4, D3.5

An average is a representative value of a set of data.

axis, axes
A5.7, A5.8

An axis is one of the lines used to locate a point in a coordinate system.

bar chart
D1.3, D2.4, D2.5, D3.2, D3.3

A bar chart is a diagram that uses rectangles of equal width to display data. The frequency is given by the height of the rectangle.

bar-line graph
D2.4
A bar-line graph is a diagram that uses lines to display data. The lengths of the lines are proportional to the frequencies.

between
N1.6
Between means in the space bounded by two limits.

brackets
N3.2, N3.8, A4.3
Operations within brackets should be carried out first.

calculate, calculation
N1.3, N1.5
Calculate means work out using a mathematical procedure.

calculator: clear, display, enter, key, memory
N1.6, N3.8, N4.2
You can use a calculator to perform calculations.

cancel
N4.1
A fraction is cancelled down by dividing the numerator and denominator by a common factor.

For example, $\frac{24}{40} = \frac{3}{5}$ (÷ 8 numerator and denominator)

certain
D1.4
An event that is certain will definitely happen.

chance
D1.4
Chance is the probability of something happening.

compare
N4.5, N5.6
Compare means to assess the similarity of.

congruent
S5.6
Congruent shapes are exactly the same shape and size.

consecutive
A2.3
Consecutive means following on in order.
For example 2, 3 and 4 are consecutive integers.

construct
S3.4, S5.2, S5.3
To construct means to draw a line, angle or shape accurately.

convert
N2.4, N2.6, N3.1, N4.1, N5.7
Convert means to change.

coordinate pair
A3.6, A5.6, A5.7
A coordinate pair is a pair of numbers that give the position of a point on a coordinate grid.
For example, (3, 2) means 3 units across and 2 units up.

coordinates
A3.5, S2.3, S4.3
Coordinates are the numbers that make up a coordinate pair.

data
D1.1, D1.2, D2.4, D3.1
Data are pieces of information.

data collection sheet
D3.1
A data collection sheet is a sheet used to collect data. It is sometimes a list of questions with tick boxes for collecting answers.

decimals, decimal fraction
N1.5, N2.4, N2.6, N4.1, N5.1, N5.4, N5.7
A decimal fraction shows part of a whole represented as tenths, hundredths, thousandths and so on.
For example, 0.65 and 0.3 are decimal fractions.

decimal number
N1.1
A decimal number is a number written using base 10 notation.

decimal place (d.p.)
N3.5, N3.7
Each column after the decimal point is called a decimal place.
For example, 0.65 has two decimal places (2 d.p.)

degree (°)
S2.1, S3.1, S3.2, S4.5, S5.1
A degree is a measure of turn. There are 360° in a full turn.

denominator
N2.1, N2.2, N2.3, N5.6, N5.8
The denominator is the bottom number in a fraction. It shows how many parts there are in total.

diagonal
S2.2
A diagonal of a polygon is a line joining any two vertices but not forming a side.

This is a diagonal.

difference
N1.2, N1.4
You find the difference between two amounts by subtracting one from the other.

digit
N1.1, N5.4
A digit is any of the numbers 0, 1, 2, 3, 4, 5, 6, 7, 8, 9.

direction
S4.5
The direction is the orientation of a line in space.

distance
S1.2, S1.3
The distance between two points is the length of the line that joins them.

divide, division
N2.1, N3.1, N3.2, N3.6, N5.3, N5.5, A3.1
Divide (÷) means share equally.

double, halve
N3.3
Double means multiply by two. Halve means divide by two.

draw
S3.4
Draw means create a picture or diagram

edge (of solid)
S3.5, S5.3
An edge is a line along which two faces of a solid meet.

edge

equal (sides, angles)
S3.3, S5.4
Equal sides are the same length. Equal angles are the same size.

equally likely
D1.5
Events are equally likely if they have the same probability.

equals (=)
A2.5, A3.4, A4.1, A5.1
Equals means having exactly the same value or size.

equation
A4.4, A5.1
An equation is a statement linking two expressions that have the same value.

equivalent, equivalence
N2.2, N2.4, N2.5, N2.6, N4.1, N4.2, N4.3, N5.6, N5.7
Equivalent fractions are fractions with the same value.

estimate
D4.3, S3.1, N1.4, N1.5, N3.5, N3.7, N3.8, N5.4
An estimate is an approximate answer.

experiment
D1.6, D4.2, D4.3
An experiment is a test or investigation to gather evidence for or against a theory.

expression
A2.2, A2.3, A2.4, A4.1, A4.4, A5.1
An expression is a collection of numbers and symbols linked by operations but not including an equals sign.

face
S3.5, S5.3
A face is a flat surface of a solid.

face

factor
N3.6, N5.2, N5.3, A3.1, A3.2
A factor is a number that divides exactly into another number.
For example, 3 and 7 are factors of 21.

fair
D1.6, D4.3
In a fair experiment there is no bias towards any particular outcome.

fraction
N2.1, N2.2, N2.5, N4.1, N4.2, N4.3, N4.4, N4.5, N5.7, N5.8
A fraction is a way of describing a part of a whole.
For example $\frac{2}{5}$ of the shape shown is red.

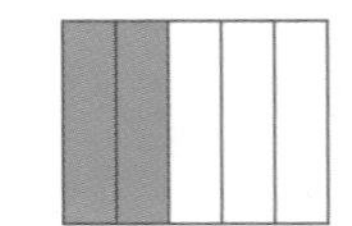

frequency
D1.3, D4.2
Frequency is the number of times something occurs.

function
A3.4
A function is a rule.
For example, + 2, − 3, × 4 and ÷ 5 are all functions.

function machine
A3.4, A3.5, A3.6, A5.5, A5.6
A function machine links an input value to an output value by performing a function.

generate
A3.3, A5.4
Generate means produce.

graph
A5.7
A graph is a diagram that shows a relationship between variables.

greater than (>)
N1.1
Greater than means more than.
For example $4 > 3$.

grid
S2.3, A5.6
A grid is a repeated geometrical pattern used as a background to plot coordinate points. It is usually squared.

horizontal
A5.8, S4.3
Horizontal means flat and level with the ground.

hundredth
N1.1
A hundredth is 1 out of 100.
For example 0.05 has 5 hundredths.

image
S4.2
When a shape is reflected, translated or rotated, the new shape is called the image

object image

impossible
D1.4
An event is impossible if it definitely cannot happen.

improper fraction
N2.3
An improper fraction is a fraction where the numerator is greater than the denominator. For example, $\frac{8}{5}$ is an improper fraction.

increase, decrease
N1.2, N4.3

Increase means make greater. Decrease means make less.

input, output
A3.4, A3.5, A5.5, A5.6

Input is data fed into a machine or process. Output is the data produced by a machine or process.

integer
N2.5

An integer is a positive or negative whole number (including zero). The integers are: ..., $^-3$, $^-2$, $^-1$, 0, 1, 2, 3, ...

interpret
N3.8, D2.5, D3.3, D3.5, D3.2

You interpret data whenever you make sense of it.

intersect, intersection
S2.2

Two lines intersect at the point, or points, that they cross.

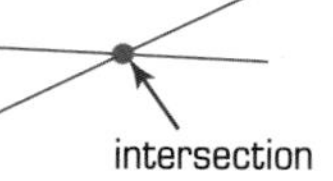

interval
D2.3

An interval is the size of a class or group in a frequency table.

length: millimetre, centimetre, metre, kilometre; mile, foot, inch
S1.1, S1.2, S1.3, A3.2

Length is a measure of distance. It is often used to describe one dimension of a shape.

less than (<)
N1.1

Less than means smaller than.
For example, 3 is less than 4, or $3 < 4$.

likelihood
D4.1

Likelihood is the probability of an event happening.

likely
D1.4, D4.2

An event is likely if it will happen more often than not.

line graph
D3.2, D3.3

A line graph is a set of data points plotted on a graph and joined to show trends.

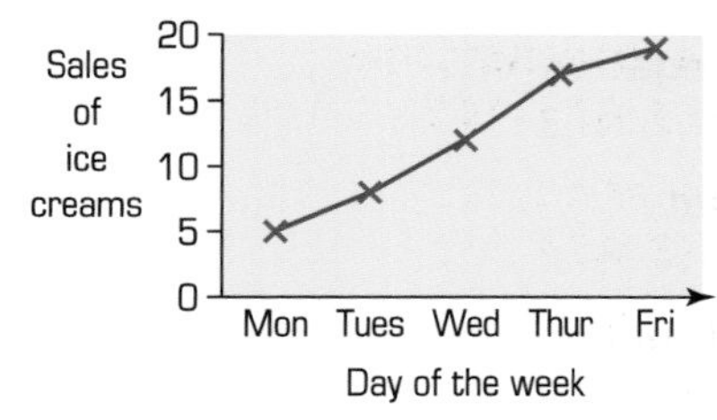

line of symmetry
S4.1, S5.4

A line of symmetry is a line about which a 2-D shape can be folded so that one half of the shape fits exactly on the other half.

line
S3.2, S3.3

A line joins two points and has zero thickness.

mean
D1.2, D3.4

The mean is an average value found by adding all the data values and dividing by the number of pieces of data.

measure
S1.3, S3.1, S3.4

When you measure something you find the size of it.

median
D1.1, D3.4

The median is an average which is the middle value when the data is arranged in size order.

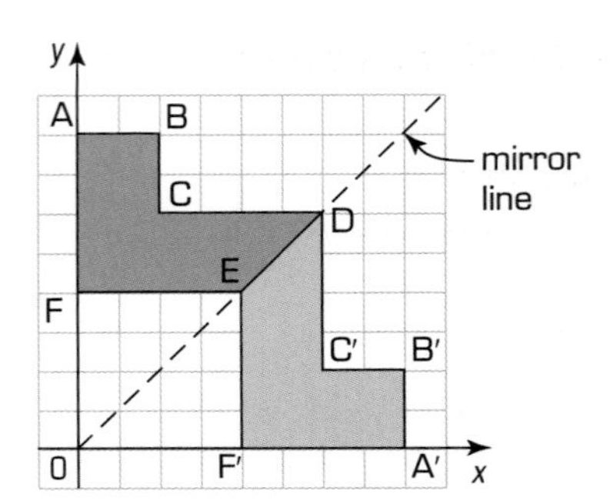

mirror line
S4.1, S4.2, S4.4

A mirror line is a line or axis of symmetry.

mixed number
N2.3

A mixed number has a whole number part and a fraction part. For example, $3\frac{1}{2}$ is a mixed number.

modal class
D3.4

The modal class is the most commonly occurring class when the data is grouped. It is the class with the highest frequency.

mode
D1.1, D3.4

The mode is an average which is the data value that occurs most often.

multiple
A3.4, A3.5, N3.1, N3.2, N3.6, N5.2, N5.6

A multiple of an integer is the product of that integer and any other. For example, these are multiples of 6: $6 \times 4 = 24$ and $6 \times 12 = 72$.

multiply, multiplication
N3.4, N5.4, N5.5

Multiplication is the operation of combining two numbers or quantities to form a product.

nearest
N1.6, N5.1

Nearest means the closest value.

negative
N1.2, N1.3

A negative number is a number less than zero.

net
S1.4, S5.3

A net is a 2-D arrangement that can be folded to form a solid shape.

numerator
N2.1, N2.2, N2.3, N5.6, N5.8

The numerator is the top number in a fraction. It shows how many parts you are dealing with.

object, image
S4.2

The object is the original shape before a transformation. An image is the same shape after a transformation.

operation
A4.2, A4.3, A5.1, N3.2

An operation is a rule for processing numbers or objects. The basic operations are addition, subtraction, multiplication and division.

opposite (sides, angles)
S2.2, S5.4

Opposite means across from.

The red side is opposite the red angle.

order
N1.1, N5.7

To order means to arrange according to size or importance.

order of operations
N3.2

The conventional order of operations is:
brackets first,
then division and multiplication,
then addition and subtraction.

order of rotation symmetry
S5.5

The order of rotation symmetry is the number of times that a shape will fit on to itself during a full turn.

outcome
D1.5, D1.6, D4.1, D4.2, D4.3

An outcome is the result of a trial or experiment.

parallel
S2.2, S5.4

Two lines that always stay the same distance apart are parallel. Parallel lines never cross or meet.

partition; part
N1.4, N3.3, N3.4, N5.3

To partition means to split a number into smaller amounts, or parts. For example, 57 could be split into 50 + 7, or 40 + 17.

percentage (%)
N2.6, N4.1, N4.2, N4.3, N4.4, N5.7, N5.8

A percentage is a fraction expressed as the number of parts per hundred.

perimeter
S1.1, S1.2

The perimeter of a shape is the distance around it. It is the total length of the edges.

perpendicular
S2.2

Two lines are perpendicular to each other if they meet at a right angle.

pie chart
D1.3, D2.4, D2.5, D3.2

A pie chart uses a circle to display data. The angle at the centre of a sector is proportional to the frequency.

place value
N1.1, N3.1

The place value is the value of a digit in a decimal number. For example, in 3.65 the digit 6 has a value of 6 tenths.

point
S3.2, S5.1

A point is a fixed place on a grid or on a shape.

polygon: pentagon, hexagon, octagon

A polygon is a closed shape with three or more straight edges.

A pentagon has five sides.

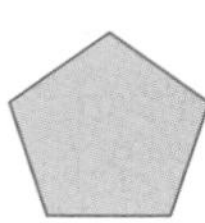

A hexagon has six sides.

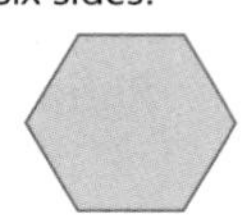

An octagon has eight sides.

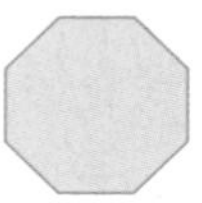

positive
N1.2, N1.3

A positive number is greater than zero.

position
S2.3

A position is a place or location.

prime
N5.2, A3.1

A prime number is a number that has exactly two different factors.

probability
D1.4, D1.5, D4.1, D4.3

Probability is a measure of how likely an event is.

probability scale
D1.5, D4.1

A probability scale is a line numbered 0 to 1 or 0% to 100% on which you place an event based on its probability.

proportion
N4.4, N4.5

Proportion compares the size of a part to the size of a whole. You can express a proportion as a fraction, decimal or percentage.

protractor (angle measurer)
S3.1, S3.2, S3.4, S5.1, S5.2

A protractor is an instrument for measuring angles in degrees.

quadrant
A5.8, S4.3

A coordinate grid is divided into four quadrants by the *x*- and *y*-axes.

quadrilateral: kite, parallelogram, rectangle, rhombus, square, trapezium
S2.3, S3.2, S3.3, S5.2, S5.4

A quadrilateral is a polygon with four sides.

rectangle

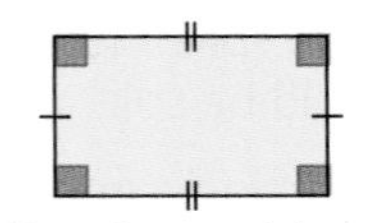

All angles are right angles. Opposite sides equal.

parallelogram

Two pairs of parallel sides.

kite

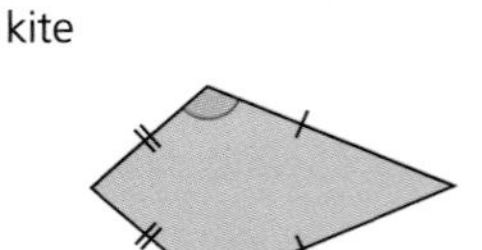

Two pairs of adjacent sides equal. No interior angle greater than 180°.

rhombus

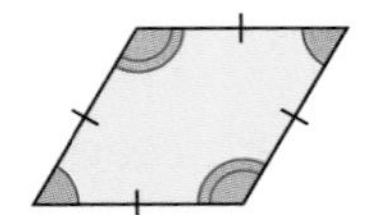

All sides the same length. Opposite angles equal.

square

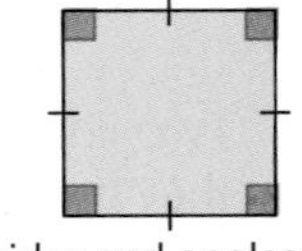

All sides and angles equal and all angles are right angles.

trapezium

One pair of parallel sides.

questionnaire
D2.2

A questionnaire is a list of questions used to gather information in a survey.

quotient
N3.6

A quotient is the result of a division.
For example, the quotient of $12 \div 5$ is $2\frac{2}{5}$, or 2.4.

random
D1.5, D4.1

A selection is random if each object or number is equally likely to be chosen.

range
D2.3, D3.4

The range is the difference between the largest and smallest values in a set of data.

ratio
N4.5

Ratio compares the size of one part with the size of another part.

reflect, reflection
S4.2, S4.6, S4.3, S5.6
S4.4, S5.4

A reflection is a transformation in which corresponding points in the object and the image are the same distance from the mirror line.

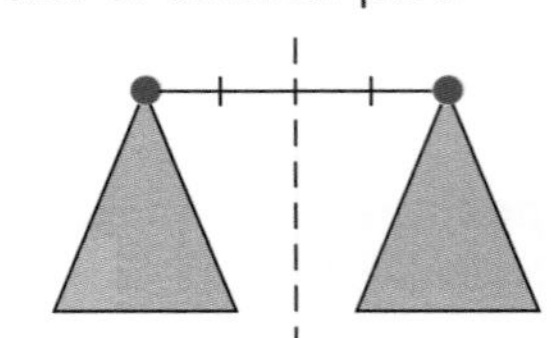

reflection symmetry
S4.1

A shape has reflection symmetry if it has a line of symmetry.

regular
S5.4

A regular polygon has equal sides and equal angles.

relationship
A2.5

A relationship is a link between objects or numbers.

remainder
N3.6, N5.5, A3.1

A remainder is the amount left over when one quantity is exactly divided by another. For example, $9 \div 4 = 2$ remainder 1.

rotate, rotation
S4.5, S4.6, S5.6

A rotation is a transformation in which every point in the object turns through the same angle relative to a fixed point.

rotation symmetry
S5.5

A shape has rotation symmetry if when turned it fits onto itself more than once during a full turn.

roughly
N3.5, N3.7

Roughly means about: 5362 is roughly 5000.

round
N1.6, N3.5, N5.1

You round a number by expressing it to a given degree of accuracy. For example, 639 is 600 to the nearest 100 and 640 to the nearest 10.
To round to one decimal place means to round to the nearest tenth. For example 12.47 is 12.5 to 1 d.p.

rule
A1.2, A2.2, A5.4

A rule describes the link between objects or numbers.
For example, the rule linking 2 and 6 may be +4 or ×3.

ruler
S1.3, S3.4, S5.2

A ruler is an instrument for measuring lengths.

scale
S1.3, N4.2

A scale is a numbered line or dial. The numbers usually increase in sequence.

sequence
A3.3, A5.4

A sequence is a set of numbers or objects that follow a rule.

shape
S4.3, S4.4

A shape is made by a line or lines drawn on a surface, or by putting surfaces together.

side (of 2-D shape)
S3.2

A side is a line segment joining vertices.

simplest form
N2.2, N4.4, N5.6

A fraction (or ratio) is in its simplest form when the numerator and denominator (or parts of the ratio) have no common factors.
For example, $\frac{3}{5}$ is expressed in its simplest form.

simplify
A2.3

To simplify an expression you gather all like terms together into a single term.

solid (3-D) shape: cube, cuboid, prism,
S3.5, S5.3

A solid is a shape formed in three-dimensional space.

cube

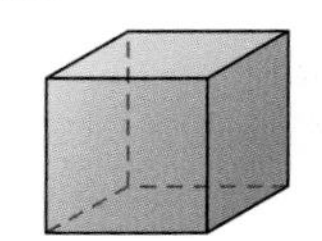

six square faces

cuboid

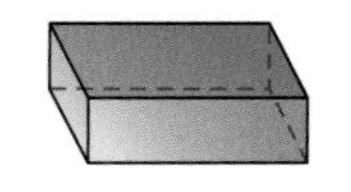

six rectangular faces

prism

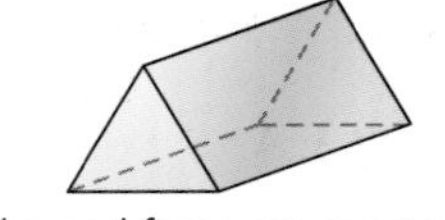

the end faces are constant

solve (an equation)
A5.1

To solve an equation you need to find the value of the variable that will make the equation true.

spin, spinner
D4.1, D4.2, D4.3, S5.5

A spinner is an instrument for creating random outcomes, usually in probability experiments.

square number, squared
A3.2, A3.3

If you multiply a number by itself the result is a square number. For example 25 is a square number because $5^2 = 5 \times 5 = 25$.

statistic, statistics
D1.3, D3.5

Statistics is the collection, display and analysis of information.

straight line
S5.1

A straight line is the shortest distance between two points

straight-line graph
A3.6

When coordinate points lie in a straight line they form a straight-line graph. It is the graph of a linear equation.

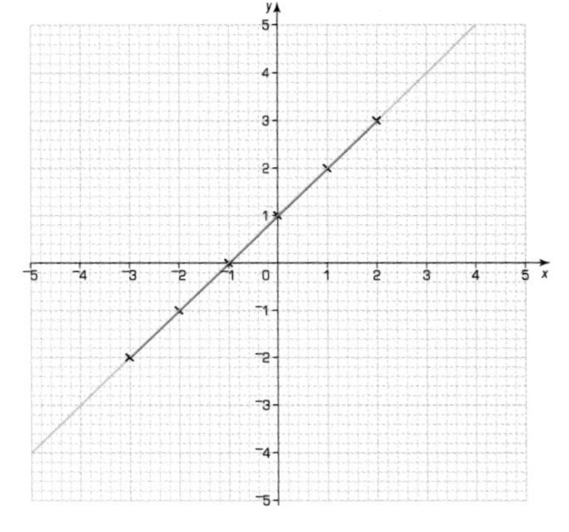

substitute
A2.4, A5.2, A5.3

When you substitute you replace part of an expression with a value.

subtract, subtraction
N1.3, N1.4, N3.6, N5.5

Subtraction is the operation that finds the difference in size between two numbers.

sum
N1.5

The sum is the total and is the result of an addition.

surface, surface area
S1.4

The surface area of a solid is the total area of its faces.

survey
D2.1, D2.2

A survey is an investigation to find information.

table
D1.3, D2.3, A3.6, D3.2, A5.7

A table is an arrangement of information, numbers or letters usually in rows and columns.

tally
D2.3, D1.3, D1.6, D4.2

You use a tally mark to represent an object when you collect data. Tally marks are usually made in groups of five to make it easier to count them.

temperature: degrees Celsius, degrees Fahrenheit

Temperature is a measure of how hot something is.

tenth
N1.1

A tenth is 1 out of 10 or $\frac{1}{10}$.
For example 0.5 has 5 tenths.

term
A2.3

A term is a number or object in a sequence. It is also part of an expression.

three-dimensional (3-D)
S1.4, S3.5, S5.3

Any solid shape is three-dimensional.

time
S2.1

Time is a measure of duration.
There are:
- 60 seconds in a minute
- 60 minutes in an hour
- 7 days in a week
- 28–31 days in a month
- 365 days in most years

total
N1.5, N3.5

The total is the result of an addition.

transformation
S4.6

A transformation moves a shape from one place to another.

translate, translation
S4.4, S4.6, S5.6

A translation is a transformation in which every point in an object moves the same distance and direction. It is a sliding movement.

triangle: equilateral, isosceles, scalene, right-angled
S2.3, S3.3, S3.4, S5.2

A triangle is a polygon with three sides.

equilateral

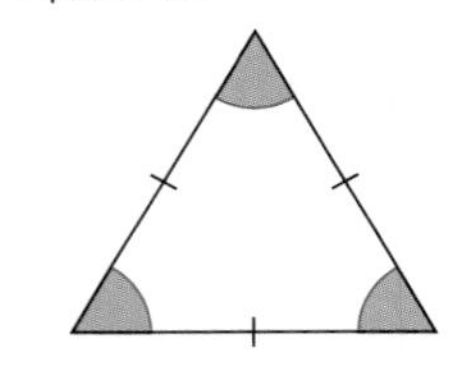

three equal sides

isosceles

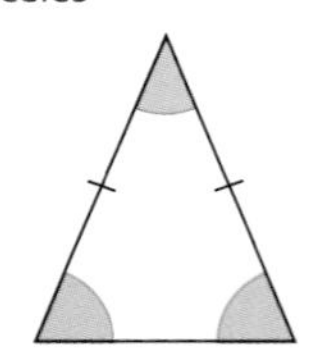

two equal sides

scalene

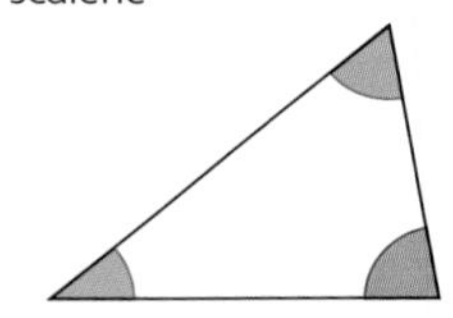

no equal sides

right-angled

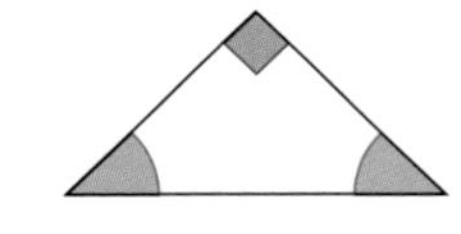

one angle is 90°

triangular number
A3.3

A triangular number is the number of dots in a triangular pattern:
The numbers form the sequence
1, 3, 6, 10, 15, 21, 28 ...

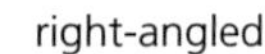

two-dimensional (2-D)
S3.5

A flat shape has two dimensions, length and width or base and height.

unknown
A2.1, A2.2, A2.3, A4.1, A4.2, A4.4, A5.2, A5.3

An unknown is a variable. You can often find its value by solving an equation.

unit
S1.1

A unit is a standard measure or quantity that other quantities can be measured from.

value
A2.2, A2.4, A2.5, A4.1, A4.2, A4.4, A5.1, A5.2, A5.3

The value is the amount an expression or variable is worth.

variable
A2.1, A2.4, A2.5, A4.4, A5.2

A variable is a symbol that can take a range of values.

vertex, vertices
S3.5, S4.3, S5.3

A vertex of a shape is a point at which two or more edges meet.

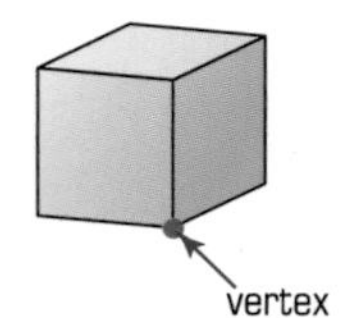

vertical
S4.3, A5.8

Vertical means straight up and down.

vertically opposite angles
S3.3

When two straight lines cross they form two pairs of equal angles called vertically opposite angles.

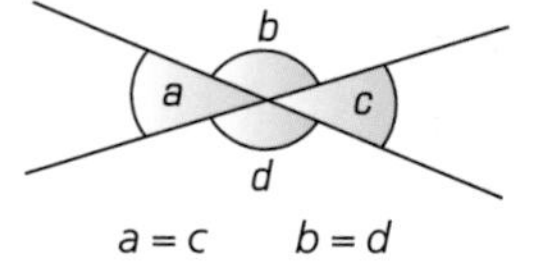

whole number
N2.5

A whole number has no fractional part.

width
S1.1, S1.2, S1.3

Width is a dimension of an object describing how wide it is.

***x*-axis, *y*-axis**
S2.3

On a coordinate grid, the *x*-axis is usually the horizontal axis and the *y*-axis is usually the vertical axis.

x-coordinate, y-coordinate
A5.8

The *x*-coordinate is the distance along the *x*-axis.
The *y*-coordinate is the distance along the *y*-axis.
For example, ($^{-}2$, $^{-}3$) is $^{-}2$ along the *x*-axis and $^{-}3$ along the *y*-axis.

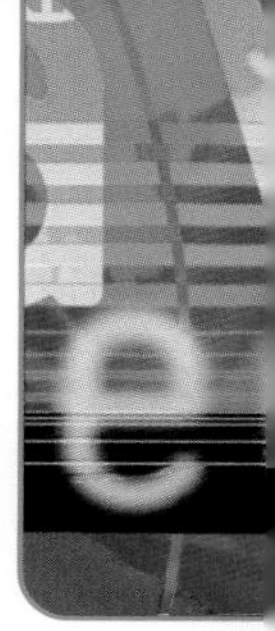

A1 Check in

1 **a** 2, 5, 8, 11, 14
b 20, 16, 12, 8, 4

2 **a** 20 **b** 9 **c** 20 **d** 7
e 10 **f** 6 **g** 30 **h** 8

3 **a** 4 **b** 5 **c** 7 **d** 6
e 9 **f** 8 **g** 6 **h** 8

A1 Check out

1 **a** 0, 4, 8, 12,16 **b** 1, 2, 4, 8, 16
c 10, 7, 4, 1, $^{-}2$ **d** 15, 21, 27, 33, 39
e First term is 3. Rule is +2
f First term is 20. Rule is −3
g First term is 2. Rule is ×2
h First term is 0. Rule is −1

2 **a** There are *n* dogs in a park.
b There is *x* kg of chocolate eaten each week in the UK.
c *n* people own a skateboard.
d There are *x* fleas on a cat.
e There are *n* fish in the sea.

3 **a** 5
b

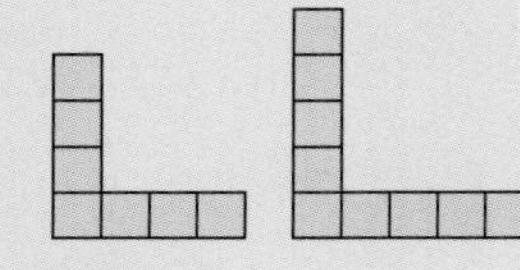

c 11 **d** 39

N1 Check in

1 **a** Forty three
b 87
c 14
d Sixty

2 **a**

0 20
3 5 6 7 11 14 17

3 **a** 40 **b** 25 **c** 19 **d** 63

N1 Check out

1 **a** Three point five six
b Three units
b Six hundredths
b 4

2 **a** 9° C **b** 5° C **c** 9° C **d** $^{-}5$° C

3 **a** 54 **b** 35 **c** £3.50 **d** £5.40
e 339 **f** 2.9

4 **a** 450 **b** 806 **c** 659.7 **d** 438

5 **a** 86 **b** 19 **c** 335 **d** 1350
e 187 **f** 1103 **g** 3679 **h** £3.74
i 283 **j** 439 **k** 3.4 **l** 406

6 1.33 m 1.35 m 1.5 m 1.53 m
1.55 m 2.35 m

S1 Check in

1 9 squares
2 3 cm
3 7

S1 Check out

1 a 75 cm^2
  b 1 cm × 48 cm
  2 cm × 24 cm
  3 cm × 16 cm
  4 cm × 12 cm
  6 cm × 8 cm
2 60 cm^2
3 24 squares

N2 Check in

1 a 12 b 5 c 25 d 18
  e 7 f 42
2 a 35 b 28 c 47 d 53
  e 48

N2 Check out

1 a i $\frac{2}{9}$ ii $\frac{4}{9}$ iii $\frac{3}{9}$
  b i $\frac{7}{20}$ ii $\frac{13}{20}$ iii $\frac{11}{20}$
  iv $\frac{5}{20}$ v $\frac{16}{20}$ vi $\frac{4}{20}$
  c i $\frac{1}{2}$ ii $\frac{50}{200}$ iii $\frac{25}{200}$ iv $\frac{40}{200}$
  d i 43 p ii 15 cm iii 45 cm
  iv 4 p v 18 p vi 36 minutes
2 a i $\frac{1}{2}=\frac{7}{14}$ ii $\frac{1}{4}=\frac{3}{12}$ iii $\frac{2}{3}=\frac{10}{15}$
  b i $\frac{3}{6}=\frac{1}{2}$ ii $\frac{4}{8}=\frac{1}{2}$ iii $\frac{2}{6}=\frac{1}{3}$
  iv $\frac{3}{9}=\frac{1}{3}$ v $\frac{4}{10}=\frac{2}{5}$
3 a i $\frac{3}{7}$ ii $\frac{5}{8}$ iii $\frac{2}{9}$
  iv $\frac{4}{8}$ v $\frac{4}{11}$ vi $\frac{5}{11}$
  b i 0.5 ii 0.25 iii 0.1
  iv 0.75 v 0.7 vi 0.2
4

30%	0.3	$\frac{3}{10}$
40%	0.4	$\frac{2}{5}$
25%	0.25	$\frac{1}{4}$
7%	0.07	$\frac{7}{100}$

D1 Check in

1

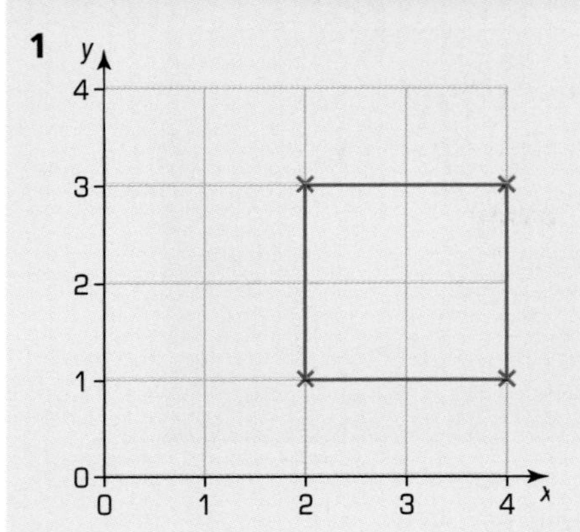

It is a square.

2 A is 5

B is 14

C is 17

D is 22

D1 Check out

1 **a** 17

b 17

c 16.5

d

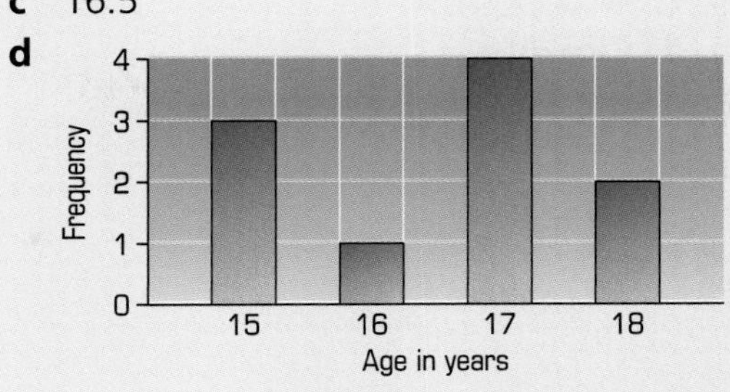

e **i** $\frac{3}{10}$ **ii** $\frac{2}{10}$ or $\frac{1}{5}$ **iii** $\frac{5}{10}$ or $\frac{1}{5}$ **iv** 1

f

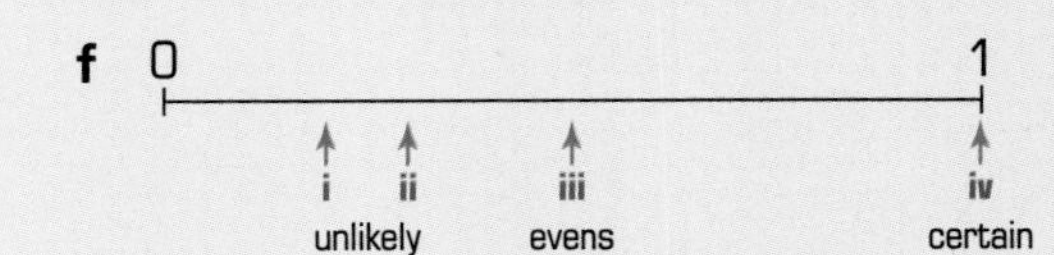

A2 Check in

1 a	5	**b**	18
c	5	**d**	8
e	2	**f**	5
g	8	**h**	8
i	9	**j**	15
k	28	**l**	28
m	90	**n**	90
o	11	**p**	11

A2 Check out

1 **a** $d-5$ **b** $n+2$

c $y+5$ **d** $3x$ **e** $\frac{x}{4}$

2 **a** $x-6$ **b** $y+3$

c $z+4$ **d** $7-f$

e $4x$ **f** $7y$

g $\frac{26}{z}$ **h** $\frac{h}{5}$

i $5y$ **j** $\frac{12}{x}$

k $8-x$ **l** $x-8$

3 **a** $29n$ **b** $35m$ **c** my

S2 Check in

1 a Clockwise
  b Anticlockwise
  c Anticlockwise
  d Clockwise

2 A = (1, 3)
  B = (4, 2)

S2 Check out

1 and **2**

3 a 36 minutes
  b 5 minutes
  c 8 minutes

D2 Check in

1

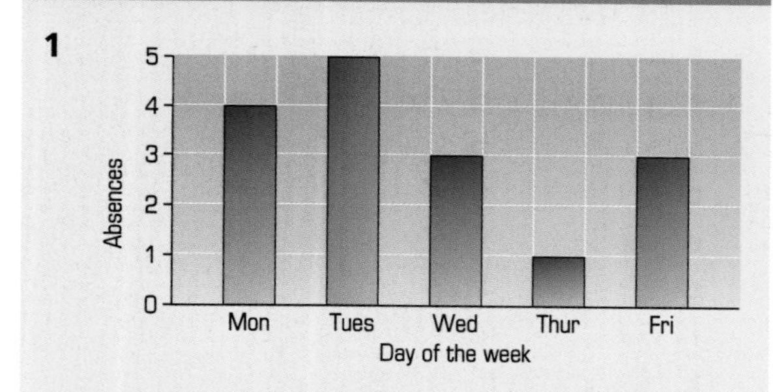

2 It is Tuesday because it has the highest bar.

D2 Check out

1 a 7 students
  b 8 girls
  c The girls were happier overall as more of them rated the meals good or very good.

2 A is true
  B is false – only a quarter of the people were playing either badminton or squash.
  C is true

N3 Check in

1 a i 56 ii 42 iii 72 iv 44
  b i 459 ii 515 iii 485 iv 254
2 a 21 b 8 c 36 d 40 e 5
3 a 1979 b 158
  c 756 d 25 290
  e 523 f 347

N3 Check out

1 a 12 b 4 c 24
  d 35 e 6 f 63
  g 8 h 300 i 6500
  j 32 k 45 l 5.2
2 a 17 b 13
  c 8 d 0
  e 6 f 10
3 a 253 b 405
  c 432 d 517
4 a 228 stickers
  b 900 candles
  c 164 boxes
  d 400 metres, 84 seconds or 1 minute 24 seconds.

A3 Check in

1 a 10 b 70
  c 7 d 5
  e 9 f 10
2 a odd: 15, 2657, 3001, 22 223
  even: 12
  b 12 pens and 4 books
3 An L shape:

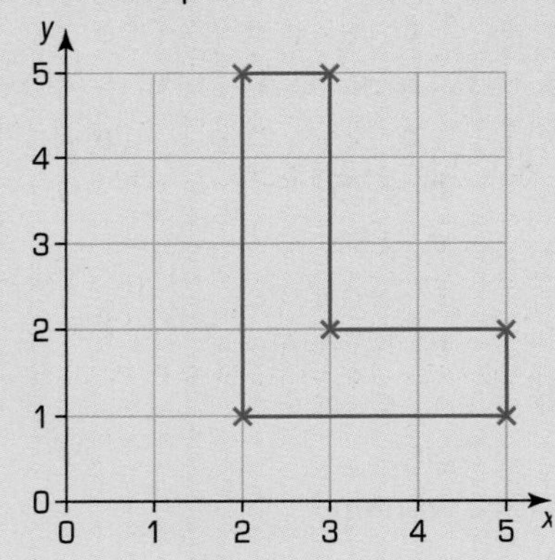

A3 Check out

1 Multiples of 2: 50, 36, 100, 20
  Multiples of 3: 36, 231
  Multiples of 4: 36, 100, 20
  Multiples of 5: 50, 100, 20, 25
  Multiples of 10: 50, 100, 20
2 36: 1×36, 2×18, 3×12, 4×9, 6×6
  49: 1×49, 7×7
  100: 1×100, 2×50, 4×25, 5×20, 10×10
  25: 1×25, 5×5
3

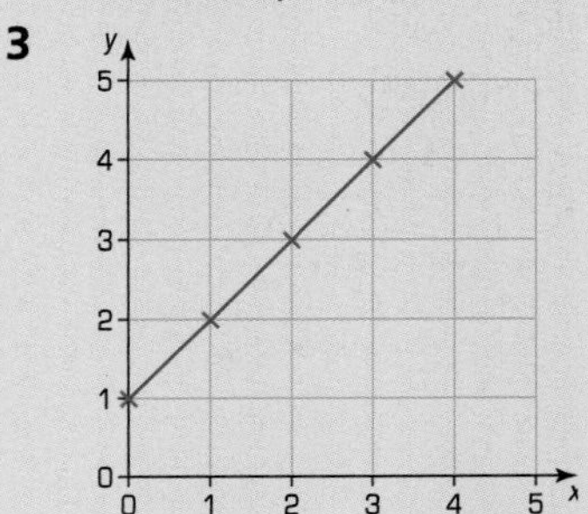

4 a

Pattern Numbers	1	2	3
Number of matchsticks	4	7	10

  b 13 matchsticks

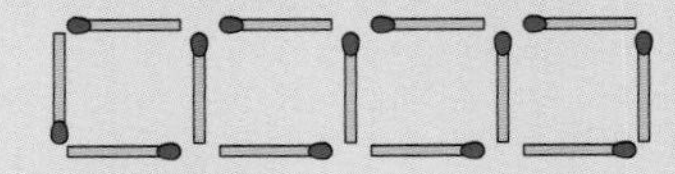

  c 61 matchsticks

S3 Check in

1 **a** Equilateral
b Isosceles
c Right-angled isosceles
d Right-angled
e Scalene

S3 Check out

1 5 cm
2 **a** $a = 34°$
b $b = 34°$
3 **a** $c = 112°$
b $d = 68°$
c 180°
4 10 cubes
5 36 rectangles

N4 Check in

1 **a** 33% **b** 71% **c** 50% **d** 25%
2 $\frac{1}{2} = \frac{4}{8}$, $\frac{2}{20} = \frac{1}{10}$, $\frac{1}{5} = \frac{2}{10}$, $\frac{1}{4} = \frac{3}{12}$, $\frac{3}{9} = \frac{1}{3}$
3 **a** 14 cm **b** 6 minutes
c 45 cm **d** £15
e £20 **f** 45 minutes

N4 Check out

1 **a** $\frac{60}{100}$ **b** $\frac{32}{100}$ **c** $\frac{7}{100}$ **d** $\frac{25}{100}$ **e** $\frac{15}{100}$
f 50% **g** 10% **h** 30% **i** 20% **j** 40%
2 **a** $\frac{3}{5}$ **b** $\frac{8}{25}$ **c** $\frac{7}{100}$ **d** $\frac{1}{4}$ **e** $\frac{3}{20}$
3 **a** £13 **b** £35
c £12 **d** £36
e 6 cm **f** 3 minutes
g 30 cm **h** 60 minutes
4 **a** $\frac{1}{4}$ **b** $\frac{3}{4}$ **c** 1 : 3
5 **a** 18 boys , 12 girls
b 18 men , 42 women
c 15 men

A4 Check in

1. a 21 b 54
   c 36 d 56
   e 6 f 3
   g 9 h 10
   i 5 j 6
   k 36 l 55
   m 18 n 24
   o 7 p 1.5
   q 2 r $\frac{3}{4}$

A4 Check out

1. a $n + 5$ b $n - 3$
   c $n + 2$ d $4 - n$
   e $\frac{n}{2}$ f $\frac{n}{3}$
   g $3n$ h $3n + 4$
   i $5 \times 7 = 35$ so $35 \div 5 = 7$
   j $5 \times x = 35$ so $35 \div 5 = x$
   k $x = 11$ l $y = 17$
   m $x = 6$ n $y = 15$
   0 $x = 3$
2. For example:
   a The total cost of 12 cans of drink costing 37p each.
   b $2 + x$: I have some counters and find two more.
   $n - 5$: I have some counters and lose five of them.
   $3m$: A plant was m metres tall and is now three times as high.
   $\frac{d}{4}$: Four friends share a pizza that costs £d.

S4 Check in

1. TRANSFORMATION
2. a Square
   b Rectangle
   c Kite
   d Isosceles trapezium
   e Quadrilateral
   f Rhombus
   g Parallelogram
   h Trapezium

S4 Check out

1. a (0, 0) (0, 3) (3, 3)
   b (0, 0) ($^-$3, 0) ($^-$3, $^-$3)
   c (1, $^-$1) (1, $^-$4) (4, $^-$4)
2. a translation: 4 right and 4 down
   b reflection in x axis
   c rotation clockwise through 90°, centre (0, 0)
   d translation: 3 right and 3 up.
   e translation: 4 left and 4 up.
3. These complete shapes have line symmetry:

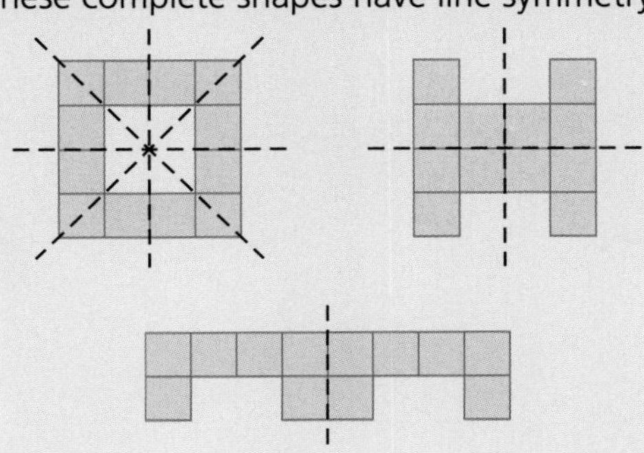

N5 Check in

1 a 25 b 36 c 5 d 4 e 72 f 9

2 a 0.5 b 0.25 c 0.75 d 0.1 e 0.7 f 0.2

3 a 31% b 63% c 7% d 50%
e 25% f 10% g 31% h 63%
i 7% j 50% k 25% l 10%

5 a 930 ($62 \times 10 = 620$
$62 \times 5 = 310$
so $62 \times 15 = 930$)

b £350: 10% = 200
5% = 100
2.5% = 50
so 17.5% = 350

c 703: $37 \times 20 - 37 \times 1$

d £1.35: 45p × 3

6 a $\frac{1}{4}$ b $\frac{5}{7}$ c $\frac{3}{10}$ d $\frac{1}{5}$
e $\frac{3}{5}$ f $\frac{2}{5}$ g $\frac{7}{20}$ h $\frac{3}{20}$

N5 Check out

1 i a 2730 b 2700 c 3000
ii a 5770 b 5800 c 6000
iii a 570 b 600 c 1000
iv a 330 b 300 c 0

2 a 1452 b 5313
c 3105 d 11 891
e 56 f 68
g 52 h 38

3 a £32 b £4.50
c 5 cm d 10 cm
e 2 mins f 150 metres
g 70° h 3 minutes
i 9 mins j 21 cm

4 a 5 b 15
c 7 d 21
e £40 f £280
g 60 litres h 15 kg
i 24 students j 25 students

D3 Check in

1 a A: 40, B: 86, C: 150, D: 180, E: 220, F: 275
b A: 40, B: 90, C: 125, D: 160, E: 173

2

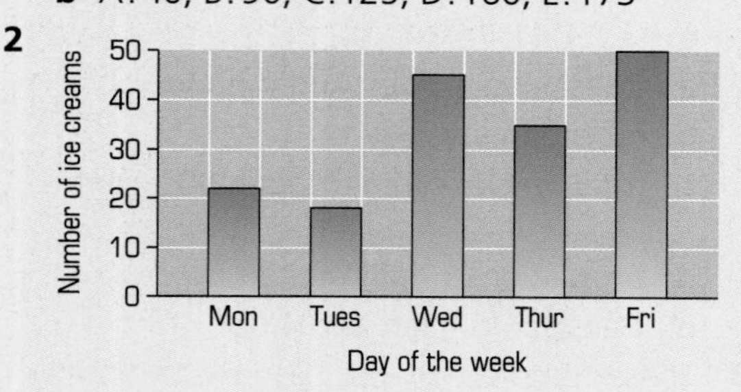

D3 Check out

1 maths: mean = 63.6%
median = 64.5%
range = 64
Science: mean = 59.3%
median = 53.5%
range = 58

In general students did better in maths. The average score was higher and 7 out of 10 students scored 55 or above in maths compared with 5 out of 10 students in science.

2 Most people, 30 out of 43 or 70% rated the food good or excellent but almost 10% of the respondents found the food poor implying the hotel should improve its standards.

A bar chart would be most suitable to display this data.

D4 Check in

1 **b** **a** **c**

2 **a** $= \frac{1}{2}$ **b** $= \frac{1}{10}$ **c** $= \frac{8}{10}$ or $\frac{4}{5}$

D4 Check out

1 **a** $\frac{1}{10}$ **b** $\frac{9}{10}$ **c** 0 **d** 1

2 **a** The blue dice has widely varying results which suggest there may be bias.

b

	Dice		
Score	Red	Blue	Green
1	$\frac{35}{240}$	$\frac{34}{240}$	$\frac{44}{240}$
2	$\frac{48}{240}$	$\frac{63}{240}$	$\frac{34}{240}$
3	$\frac{33}{240}$	$\frac{37}{240}$	$\frac{42}{240}$
4	$\frac{45}{240}$	$\frac{28}{240}$	$\frac{42}{240}$
5	$\frac{40}{240}$	$\frac{40}{240}$	$\frac{39}{240}$
6	$\frac{39}{240}$	$\frac{38}{240}$	$\frac{39}{240}$

A5 Check in

1 **a** $5d$

b $d + 6$

c $d - 5$

d $\frac{d}{2}$

2

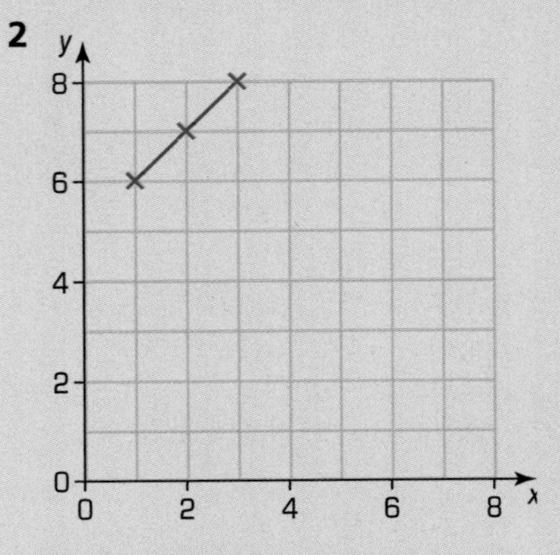

They form a straight line.

A5 Check out

1 **a** MATHEMATICS

b Students own answer.

2 **a**

x	1	2	3	4	5
y	1	3	5	7	9

b (1, 1) (2, 3) (3, 5) (4, 7) (5, 9)

c and **d**

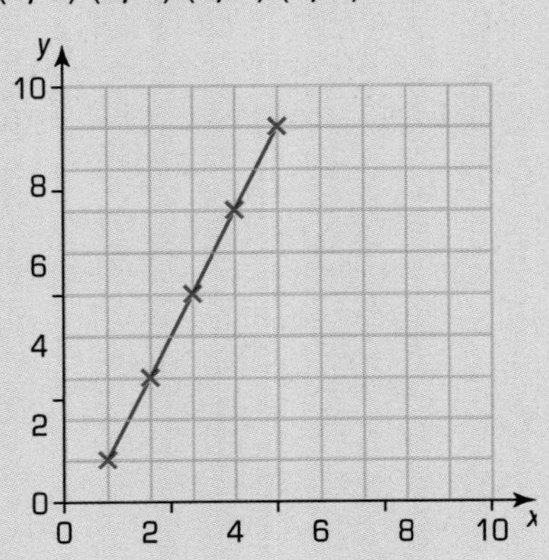

e 6

S5 Check in

1 **a** 136° **b** 44°

2 **a** blue **b** green **c** red **d** purple
e pink **f** yellow **c** blue **d** green

S5 Check out

1 **a**

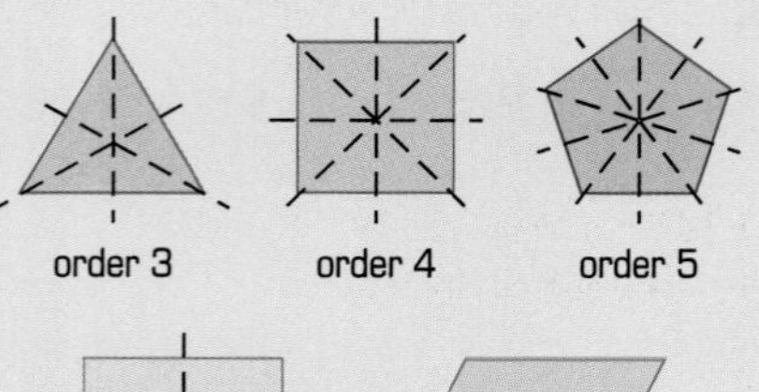

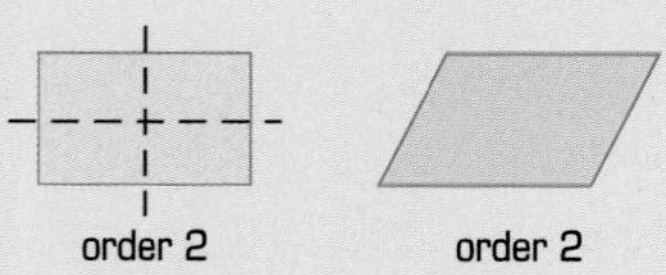

2 **a**, **b** and **e**

3 **a** true
b true
c true
d true
e true
f true
g true
h false

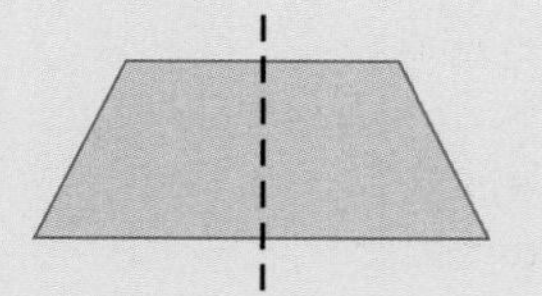

Index

A

acute angles, 132
addition, fractions, 44–5
algebra
  and functions, 12–13
  rules, 68, 70–1
algebraic expressions, 156–7
algebraic operations, 158–9
algebraic terms, 72
angles, 79–86, 235
  drawing, 236–7
  finding, 134–5
  measuring, 132–3
  in triangles, 136–7
approximation, 26
area, 29–38, 235–48
  surface, 36–7
averages, 54–7, 204

B

bar charts, 58–9, 94–8, 197, 200–3
bar-line graphs, 94
brackets, 102, 160–1

C

calculators, 26–7, 114–15
centimetres, 34, 100
chance *see* probability
common factors, 52
common multiples, 190
compensation, 22
congruency, 246
continuous data, 92
conversions, 100, 146
coordinate pairs, 128, 228, 230
coordinates, 84–5, 128, 170, 232
  graphs, 126
cubes, 140, 235, 240
  surface area, 36
cuboids, surface area, 36

D

data, 54
  continuous, 92
  discrete, 92
  handling, 87–98
  non–numerical, 94
  numerical, 94
  primary, 198
  secondary, 198
data collection, 90–1, 198–9
decimals
  and fractions, 16, 20–1, 24–5, 46–7, 50–1, 144–5, 192–3
  and percentages, 20–1, 24–5, 50–1, 144–5, 192–3
degrees, 236
denominators, 39, 42, 190, 194
diagrams
  bar charts, 58–9, 94–8, 197, 200–3
  interpreting, 58–9, 96–7
  pie charts, 58–9, 94–7, 200
  sequences in, 6–7
  statistical, 200–1
  tally charts, 58, 64–5, 92, 212, 214
  see also graphs
discrete data, 92
dividend, 110
division, 99–116, 184–9
  with remainders, 112–13
divisor, 110

E

edges, 140
equations
  solving, 162–3, 218–19
  see also linear equations
equilateral triangles, 84, 138, 242, 246
equivalences, 144
equivalent fractions, 42–3, 50, 190–1
estimation, 64, 108
evaluation, 74
experimental probability, 64–5, 209–16
expressions, 70–5, 162
  algebraic, 156–7
  evaluation, 74
  simplifying, 72–3

F

faces, 140
factor method, 184

factors, 111, 118–19, 182–3
common, 52
formulae, 76–7, 220–3
graphs of, 230–1
fractions, 39–52
addition, 44–5
and decimals, 16, 20–1, 24–5, 46–7, 50–1, 144–5, 192–3
equivalent, 42–3, 50, 190–1
and percentages, 20–1, 24–5, 50–1, 144–5, 192–3
subtraction, 44–5
frequency, 58
frequency diagrams, 64
frequency tables, 92
function machines, 8–11, 124–30, 226–9
functions, 8–14, 124–5
and algebra, 12–13
finding, 10–11, 226–7
graphs of, 126–7

G
graphs
bar-line, 94
drawing, 228–9
of formulae, 230–1
of functions, 126–7
linear, 200–3
straight–line, 128, 228, 230
see also diagrams
grid method, 108

H
hexagons, 242, 246

I
image, 168
input values, 125, 226
intersects, 82
intervals, 92
isosceles trapeziums, 83
isosceles triangles, 84, 122, 138

K
kilometres, 34, 100
kites, 83, 242

L
length, 34
letter symbols, 12, 14, 68–9, 222–3
like terms, 72
likelihood *see* probability
linear equations, 155–64
solving, 162–3
linear graphs, 200–3
lines, 34, 79, 82–3
mirror, 166, 168
number, 15, 18
straight, 128, 134, 236
lines of symmetry, 166

M
mean, 56–7, 204
measures, 34–5
and numbers, 100–1
median, 54, 204
mental arithmetic, 22–3, 28, 104–5, 184–5
metres, 34, 100
metric units, 34, 100
millimetres, 34, 100
mirror lines, 166, 168
modal class, 204
mode, 53, 54, 204
multiples, 111, 124–5, 182–7
common, 190
multiplication, 99–116

N
negative numbers, 18–19
nets, 36–7, 235, 240
number lines, 15, 18
numbers
calculations, 15–28, 179–96
and measures, 100–1
negative, 18–19
patterns, 120–1
positive, 15, 18
prime, 118–19, 182–3
square, 120, 122
triangular, 122
numerators, 39, 42, 190, 194

O

objects, 168
obtuse angles, 132, 235
operations, 102–3
  algebraic, 158–9
operators, 48, 52
ordering, 16–17
outcomes, 62–5, 210, 212
output values, 124, 226

P

parallel lines, 79, 82
parallelograms, 83, 242
partitioning, 22, 104, 106–7, 184
patterns, in numbers, 120–1
pentagons, 242
percentages
  and decimals, 20–1, 24–5, 50–1, 144–5, 192–3
  finding, 146–9
  and fractions, 20–1, 24–5, 50–1, 144–5, 192–3
perimeters, 29–38, 235–48
perpendicular lines, 79, 82
pie charts, 58–9, 94–7, 200
place value, 16–17, 100
positive numbers, 15, 18
primary data, 198
prime numbers, 118–19, 182–3
prisms, surface area, 36
probability, 53–66
  experimental, 64–5, 209–16
  theoretical, 210–11
proportion, 150–1, 152
protractors, 132, 134, 236, 238

Q

quadrants, 79, 232–3
quadrilaterals, 83
quantities, parts of, 194–5
questionnaires, 90
quotient, 110

R

range, 204
ratios, 152–3, 154
rectangles, 83, 86, 122
  area, 30–3, 38
  perimeters, 30–3
reflection symmetry, 166–7, 242–3
reflections, 168–71, 176, 246
reflex angles, 132
remainders, 112–13
repeated subtraction, 110, 188
reports, 206–7
rhombuses, 83
right angles, 132, 236
right-angled triangles, 83, 84, 138
rotational symmetry, 244–5
rotations, 174–5, 176, 246
rounding, 180–1
rules
  algebra, 68, 70–1
  sequences, 2, 4–5

S

scalene triangles, 84, 138
scales, 29, 34
secondary data, 198
sequences, 1–7, 14, 122
  in diagrams, 6–7
  generating, 224–5
  rules, 2, 4–5
  terms, 2–7
shapes, 79–86
  area, 29–38
  perimeters, 29–38
  three-dimensional, 36–7, 140–2
  transformations, 165–78, 246
square numbers, 120, 122
squares, 83, 122–3, 242, 246
  constructing, 240–1
statistical diagrams, constructing, 200–1
statistics, 53–66
  analysing, 197–208
  methods, 88–9
  reports, 206–7
straight lines, 128, 134, 236
straight-line graphs, 128, 228, 230
substitution, 74–5, 222
subtraction
  fractions, 44–5
  repeated, 110, 188
surface area, 36
surveys, 88, 90
symmetry, 166–7, 242–5

T

tables, of values, 128–9
tally charts, 58, 64–5, 92, 212, 214
terms
  algebraic, 72
  like, 72
  sequences, 2–7
tessellations, 246–7
three-dimensional shapes, 36–7, 140–2
time, 80–1
transformations, 165–78, 246
translations, 172–3, 176, 246
trapeziums, 83
trials, 212
triangles, 79, 83, 84, 122–3, 235
  angles in, 136–7
  constructing, 138–9, 238–9
triangular numbers, 122
turns, 80, 132
two-dimensional representations, 140–1

V

values
  input, 125, 226
  output, 124, 226
  tables of, 128–9
variables, 12, 68, 74, 76, 156, 220
vertices, 140, 170

Multiplication Table

×	1	2	3	4	5	6	7	8	9	10
1	1	2	3	4	5	6	7	8	9	10
2	2	4	6	8	10	12	14	16	18	20
3	3	6	9	12	15	18	21	24	27	30
4	4	8	12	16	20	24	28	32	36	40
5	5	10	15	20	25	30	35	40	45	50
6	6	12	18	24	30	36	42	48	54	60
7	7	14	21	28	35	42	49	56	63	70
8	8	16	24	32	40	48	56	64	72	80
9	9	18	27	36	45	54	63	72	81	90
10	10	20	30	40	50	60	70	80	90	100